Mechanics of Micro- and Nano-Size Materials and Structures

Mechanics of Micro- and Nano-Size Materials and Structures

Editors

Mohammad Malikan
Shahriar Dastjerdi

MDPI • Basel • Beijing • Wuhan • Barcelona • Belgrade • Manchester • Tokyo • Cluj • Tianjin

Editors
Mohammad Malikan
Mechanics of Materials and Structures
Gdańsk University of Technology
Gdańsk
Poland

Shahriar Dastjerdi
Civil Engineering
Akdeniz University
Antalya
Turkey

Editorial Office
MDPI
St. Alban-Anlage 66
4052 Basel, Switzerland

This is a reprint of articles from the Special Issue published online in the open access journal *Nanomaterials* (ISSN 2079-4991) (available at: www.mdpi.com/journal/nanomaterials/special_issues/Mechanics_Micro_Nano_Size).

For citation purposes, cite each article independently as indicated on the article page online and as indicated below:

LastName, A.A.; LastName, B.B.; LastName, C.C. Article Title. *Journal Name* **Year**, *Volume Number*, Page Range.

ISBN 978-3-0365-5778-6 (Hbk)
ISBN 978-3-0365-5777-9 (PDF)

Cover image courtesy of Mohammad Malikan

Contents

About the Editors

Mohammad Malikan

Mohammad Malikan obtained his doctoral degree in Civil Engineering from the Department of Mechanics of Materials and Structures at the Gdańsk University of Technology. His research and studies have been focused on investigations of the mechanical properties and responses of advanced and smart materials in a variety of cases. His interests are in FGMs, smart or advanced composites such as piezoelectric, piezomagnetic, etc., particularly those involved with small-scale structures as reinforcers or properties improvers. He is currently working on developing multi-functional composite structures.

Shahriar Dastjerdi

Shahriar Dastjerdi studied Ph.D. in solid mechanics at the Islamic Azad University of Shahrood, Iran. He has been working as a researcher at Akdeniz University since 2018. He is cooperating with several professors and researchers around the world. His main major is seeking novel theoretical methods in geometrical mechanics, composites, FGMs, hyper-elastic and new materials. He has attended many mechanical behaviours such as static, dynamics, chaotic problems, biomechanics, and material science. He is now working on complicated structures for finding a theoretical methodology.

Preface to "Mechanics of Micro- and Nano-Size Materials and Structures"

Nanotechnology knowledge is always looking to expand its boundaries to achieve the most significant benefit to human life and meet the growing needs of today. In this case, we can refer to micro- and nanosensors in micro/nano-electromechanical systems (MEMS/NEMS). These electrical devices can detect minimal physical stimuli up to one nanometer in size. Today, micro/nano-sensor devices are widely used in the environment. For example, sensors made of silicon are suspended in the air for hours and can monitor air pollution. In addition to these, micro-nanosize structures can be employed as a reinforcer in advanced composites, advanced concretes, etc. Except for this use, these small size particles can play the role of a filler and make the possibility of cracking for future working of the materials less and less by filling the pores inside the structure.

On the applications of micro/nanosensors in civil engineering, one can state that small-scale sensors and actuators can be developed and used in construction to monitor and/or control the environmental conditions and the materials/structures' performance. As an example, the sensors can be used to monitor concrete corrosion and micro-cracking. The smart sensor can also be employed for structural health monitoring in bridges and other structures. Therefore, the prediction of the mechanical response of such small size particles in different physical and environmental conditions is momentous. To obtain this, micro- and nano-mechanics enable the scientific basis of the structural response of micro/nanostructures based on different situations.

For design and fabrication purposes, these structures can be investigated computationally through several approaches. First, we mention straightforward experiments that need special and highly precise equipment and result in high costs. Second, computer simulation such as molecular dynamics could be used, which requires a lot of computational efforts, in general, and a powerful computer. Finally, the application of continuum models properly modified for modelling materials and structures at small scales is worth mentioning, e.g., multiscale modeling, nonlocal size-dependent models, etc.

For this reprint, we intend to cover theoretical as well as experimental works performed on small scale to predict the material properties and characteristics of any advanced and metamaterials. New studies on mechanics of small-scale structures such as MEMS/NEMS, carbon and non-carbon nanotubes (e.g., CNTs, Carbon nitride, and Boron nitride nanotubes), micro/nano-sensors, nanocomposites, macrocomposites reinforced by micro-/nano-fillers (e.g., graphene platelets), etc., are included in this reprint.

Mohammad Malikan and Shahriar Dastjerdi
Editors

nanomaterials

MDPI

Article

Cement Composites with Graphene Nanoplatelets and Recycled Milled Carbon Fibers Dispersed in Air Nanobubble Water

Anastasia I. Patrinou [1], Eirini Tziviloglou [2], Athanasios Varoutoglou [1], Evangelos P. Favvas [3], Athanasios C. Mitropoulos [1], George Z. Kyzas [1] and Zoi S. Metaxa [1,]*

[1] Department of Chemistry, International Hellenic University, 65404 Kavala, Greece
[2] Research Unit of Advanced Materials, Department of Financial Engineering, School of Engineering, University of the Aegean, 41 Kountouriotou Str., 82132 Chios, Greece
[3] Institute of Nanoscience and Nanotechnology, NCSR "Demokritos", 15341 Agia Paraskevi, Greece
* Correspondence: zmetaxa@chem.ihu.gr; Tel.: +30-2510-462-227

Abstract: The individual effect of nano- and micro-carbon-based fillers on the mechanical and the electrical properties of cement paste were experimentally examined in this study. The objective of the study was to separately examine the effects of size and morphology (platelets and fibers) of nano- and micro-reinforcement. Three different sizes of Graphene Nanoplatelets (GNPs), at contents of 0.05% and 0.20% and recycled milled carbon fibers (rCFs), at various dosages from 0.1–2.5% by weight of cement, were incorporated into the cementitious matrix. GNPs and rCFs were dispersed in water with air nanobubbles (NBs), an innovative method that, compared to common practice, does not require the use of chemicals or high ultrasonic energy. Compressive and bending tests were performed on GNPs- and rCFs-composites. The four-wire-method was used to evaluate the effect of the conductive fillers on the electrical resistivity of cement paste. The compressive and flexural strength of all the cementitious composites demonstrated a considerable increase compared to the reference specimens. Improvement of 269.5% and of 169% was observed at the compressive and flexural strength, respectively, at the GNPs–cement composites incorporating the largest lateral size GNPs at a concentration of 0.2% by weight of cement. Moreover, the rCFs–cement composites increased their compressive and flexural strength by 186% and 210%, respectively, compared to the reference specimens. The electrical resistivity of GNPs- and rCFs-composite specimens reduced up to 59% and 48%, respectively, compared to the reference specimens, which proves that the incorporation of GNPs and rCFs can create a conductive network within the cementitious matrix.

Keywords: graphene nanoplatelets; recycle carbon fibers; air nanobubbles; cement-based composites and nanocomposites; mechanical properties; electrical properties

Citation: Patrinou, A.I.; Tziviloglou, E.; Varoutoglou, A.; Favvas, E.P.; Mitropoulos, A.C.; Kyzas, G.Z.; Metaxa, Z.S. Cement Composites with Graphene Nanoplatelets and Recycled Milled Carbon Fibers Dispersed in Air Nanobubble Water. *Nanomaterials* **2022**, *12*, 2786. https://doi.org/10.3390/nano12162786

Academic Editors: Mohammad Malikan and Shahriar Dastjerdi

Received: 22 June 2022
Accepted: 10 August 2022
Published: 14 August 2022

1. Introduction

Recognizing cement as one of the dominant materials of the construction industry, researchers from around the world focus on optimizing its structural health. Cement-based materials have been extensively used due to their high availability and low construction and maintenance costs. However, their brittleness and low tensile strength may significantly decrease their functionality [1]. Cement-based materials are prone to cracking, which might affect the integrity of structures over time [2]. The presence of cracks can cause cement degradation by increasing its permeability, and might make cement vulnerable to corrosion, chloride penetration, etc., leading to accelerated deterioration [3]. Therefore, it is important to enhance the cracking resistance of cement-based materials. Reinforcement of cementitious composites in micro- and macro- scale using fibers and other fillers is a common method for the improvement of the intrinsic brittle behavior of cement-based materials [4]. Nano- and micro-reinforcement can increase the tensile strength of the cement matrix by enhancing the load-transfer capacity and the crack bridging mechanism.

Recent advances in materials science and nanotechnology enable the use of various types of carbon-based nanomaterials, such as nanofibers, nanotubes, and nanoparticles, as reinforcing agents in the cement to control the creation of cracks at the nanoscale. Carbon nanomaterials demonstrate outstanding mechanical chemical and physical properties, such as low density, high specific surface area, hardness, and excellent chemical and thermal stability. Carbon nanomaterials incorporated in cement-based materials at relatively low proportions provide important improvement in the compressive, tensile, and flexural strength [5–7], as well as new functions such as stress- and strain-sensing, temperature monitoring, and energy harvesting [8–11].

Lately, graphene-based materials have captured the attention of scientists. Graphene has extraordinary thermal, electrical, and mechanical properties, and it has been described as the most powerful, tough, thin, and dense material in the world [12]. Both its electrical and thermal conductivity are considered superior to the materials used to date. Recently, a number of graphene composites have been developed utilizing the excellent properties of graphene [13]. Graphene Nanoplatelets (GNPs) are of particular interest for combining superior properties, such as two-dimensional planar structure, great aspect ratio, remarkable strength and durability, thermal, electrical, and chemical stability, with low cost and weight, and large-scale production potential. They are also electrical and thermal conductors [14]. Compared to pristine graphene, GNPs are a promising alternative as nanofillers in material science, due to their availability in the market and their affordable prices [15].

By incorporating GNPs in cement to develop innovative and advanced composites, improvements in the stability and longevity of structures are possible [16]. Research has shown that the use of GNPs accelerates the hydration reactions of cement, improves the porosity, and creates a denser microstructure [17,18]. It has also been reported that GNPs are able to increase the compressive and flexural strength of the matrix. Silva et al. [19] found that the incorporation of GNPs in mortar, in a small dosage 0.021% by weight, enhanced the compressive strength by 95.7% after 28 days. Tong et al. [20] used 0.1% by weight GNPs and recorded 19.9% increase in the compressive strength. Other studies refer that cement composites modified with GNPs 0.05% and 0.06% by weight, show flexural strength increase of 15–24% and 27.8%, respectively [21,22]. GNPs have also been used to create conductive cement-based composites, which can monitor their own strain by detecting alterations in the electrical resistivity values [23,24].

Natural and synthetic fibers are widely used in fiber-reinforced cement-based materials [25]. The material, the type, and the amount of the fibers used determine the properties of the composite materials [26]. The implementation of microfibers can greatly improve the mechanical strength of cement and restrain the crack growth into the matrix, however, they cannot prevent crack formation [27]. Due to their unique properties, carbon fibers stand out compared to other types of fibers. Carbon fibers (CFs) have many advantages, including excellent strength-to-weight ratio, high tensile strength and stiffness, low weight, chemical and thermal stability, and low thermal expansion [28,29]. They have high electrical conductivity and can be used to reduce the electrical resistance and provide self-detecting capabilities, as well as to create a shield of electromagnetic protection in building materials [30]. An increase of 85% in flexural strength and 22% in compressive strength has been reported, along with the introduction of a small amount of CFs (0.189% by volume) into concrete [31]. Cholker et al. [32] studied the behavior of "smart" carbon reinforced concrete with microfibers. The specimens were assessed under loading-unloading cycles to examine the stress and damage sensing ability. It was derived that the presence of CFs in 1.5% and 2% by weight of cement provides a sufficient link between stress and strain of concrete and electrical resistivity. In other research, it was found that cement-based composites reinforced with short CFs are capable of sensing their own damage, stress, and temperature [33].

CFs, despite their numerous advantages in modifying the performance of cement composites, have excessive cost, which is a deterrent to their use. Nowadays, the use of recycled CFs (rCFs) as reinforcing material has gained increasing attention owing to their durability and outstanding properties with decreased cost [34]. rCFs reveal superior mechanical performance, attributed to their rugged surface, which creates a stronger bond with the surrounding cement matrix [35]. Faneca et al. [36] used rCFs to fabricate low cost electrically conductive cement composites that can be employed not only in laboratory scale, but also in the civil engineering industry. Akbar and Liew [37] investigated the effect of elevated temperatures on the reinforcement mechanism of cement composites containing rCFs. The amount of fibers and elevated temperatures remarkably affect the mechanical properties and the mass loss of the rCFs–cement composites. Mastali and Dalvand incorporated rCFs 1.25% by volume in plain concrete, which resulted in an increase of 65.10% and of 66.93% on the compressive and the flexural strength of specimens, respectively. Moreover, the impacted resistance increased 6.48 times with rCFs concentration of 1.25% [38]. rCFs represent an alternative to obtain superior performance and low-cost cementitious composites, and they are commercially available by several companies. However, the scientific research on rCFs–cement composites has been limited. The effect of rCFs on the mechanical properties, as well as the interaction between the rCFs and the surrounding cement matrix, has not been investigated yet.

The homogeneous dispersion of carbon materials in cementitious composites is another critical issue that strongly influences the reinforcing efficiency and the ultimate performance of the cement composites. GNPs and rCFs, due to strong Van der Waals forces that are developed among them, have the tendency to agglomerate. Their hydrophobic nature makes their homogeneous dispersion in aqueous suspensions challenging [39]. However, homogeneous dispersion is usually succeeded either mechanically, by using high ultrasonic energy, or chemically, by using various water-reducing agents, such as superplasticizers as surfactants [40–42]. A new promising technology called 'nanobubble technology' is utilized by many important fields for applications, including flotation, nanomaterials dispersion, and crystal growth [43–45]. Nanobubble nucleation is favored at hydrophobic particle surfaces, enhancing the stability of micro-bubble suspensions, and thus, facilitating particle flotation [46]. The use of water enriched with air nanobubbles improves the morphological characteristics of the dispersed nanoparticle clusters without the presence of aggregates [45]. To our knowledge, the application of this technology on dispersing nano- and micro-carbon materials for use in cement pastes has not been previously studied.

Recently, extensive research has been conducted on the improvement of the mechanical and electrical properties of cementitious composites using carbon-based materials. Their size and morphology strongly affect their enhancing action. However, the study on the reinforcing mechanism of different fillers based on their dimensions or size scales is narrow. The influence of the different type and scale of the reinforcing materials on the mechanical and the electrical properties of cement-based composites has not been explicitly clarified yet.

In this research, two commercially available carbon-based materials, GNPs and milled rCFs, were incorporated in the cementitious matrix to serve as reinforcement in nano- and micro-scale. Composite mixtures with three different sized GNPs, at contents of 0.05% and 0.20% and rCFs at proportions from 0.1–2.5% by weight of cement, were fabricated. The prepared specimens were assessed for their compressive strength, flexural strength and electrical resistivity. Overall, recycled milled carbon fibers with improved carbon footprint and lower cost were investigated in the present study, as they differ significantly in size from short carbon fibers reported in the literature (rCFs length is at the μm scale while typical CFs have length at the mm scale) and the research of their effect on the mechanical properties of cement-based materials is limited. Another innovative aspect of this work is the usage of water with air nanobubbles (NBs) for the proper dispersion of GNPs and rCFs. It is noteworthy that compared to frequently used practices, this method does not require the use of chemicals or high ultrasonic energy, and it can be easily applied on the construction site.

2. Materials and Methods

2.1. Materials and NBs Method

A commercial Portland cement CEM II 32,5 N (TITAN Cement Company S.A Athens, Greece) was used for this study. Three types of GNPs were used, namely N008-100-N, N008-100-P10, N008-100-P40 (Angstron Materials Inc. Dayton, OH, USA), exhibiting the same thickness but different lateral size (diameter). The physical properties of the GNPs are shown in Table 1. Moreover, milled rCFs, produced by Haufler Composites GmbH & Co. KG, Blaubeuren, Germany, were utilized in this study. According to the manufacturer CF−milled100 is a mixture of all origins carbon and graphite ex-PAN fibers, obtained from virgin carbon fibers, chosen and milled for compounding. Their mechanical and physical properties are shown in Table 2.

Table 1. Physical properties of GNPs.

GNP Type	Thickness Z (nm)	Diameter X-Y (μm)	Purity (%)	Density (g/cm^3)	Specific Surface Area (m^2/g)
N008-100-N	50–100	<5	98	2.20	13
N008-100-P10	50–100	<10	99	2.20	15
N008-100-P40	50–100	<44	99	2.20	15

Table 2. Properties of the carbon fibers (given by the manufacturer).

Carbon fibers content	100%, (100%)
Carbon content	94% (>92%)
Remaining sizing level	<0.1%
Density (Continuous fiber)	1.7 < d < 2.0 g/cm^3
Mono filament diameter Median length	7 μm +/−2 100 μm +/−20
Tensile strength	3500 MPa
Elongation at break	1.5%
Young's modulus (Tensile)	230 GPa

All specimens were prepared with tap water (low salinity, S < 350 mg/L) enhanced with air nanobubbles (NBs). Air nanobubbles were created using a generator which utilized the counterflow hydrodynamic cavitation. Details about the used NBs generator can be found in our previous work [47]. The mean size and the concentration of the produced NBs are shown in Figure 1. The optimal processing time, regarding the size of the produced NBs, was 30 min [47]. The NBs methodology was chosen for the adequate dispersion of the GNPs and rCFs. Due to the efficacy of the method, no additives, such as superplasticizers, water reducing agents, or surfactants, were employed in this study. Another advantage of this method is that zero energy consumption is required, as the water feed pressure must be at 2.5 bar, equal to the water supply network pressure, and of course it can be easily applied in the cement industry. The difference in the dispersion quality between aqueous solutions with and without NBs is easily noticeable (Figure 2).

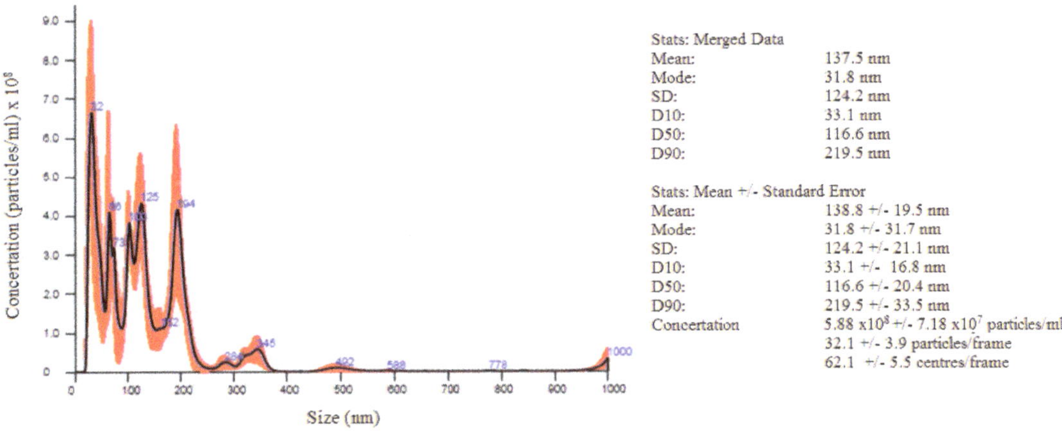

Figure 1. Nanobubbles concentration versus size.

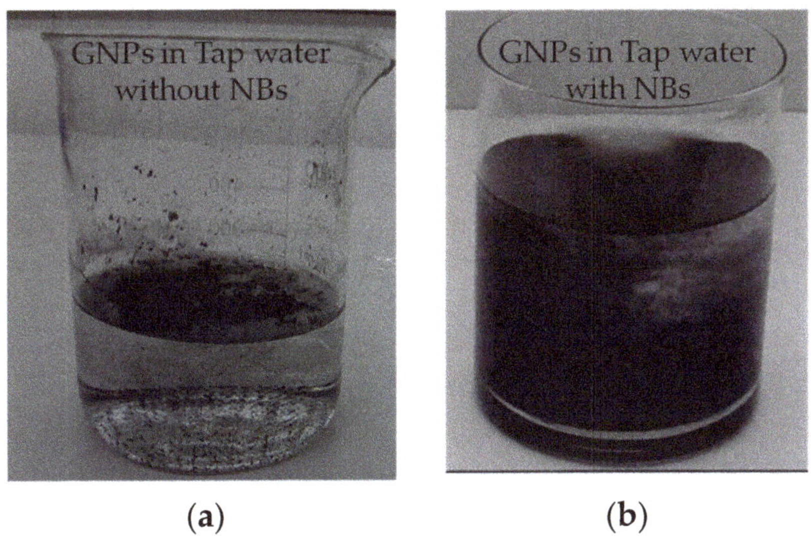

Figure 2. GNP suspensions in (**a**) plain tap water and (**b**) tap water with nanobubbles.

2.2. Preparation of Composites Specimens

Two concentrations of each GNPs-type were used, namely 0.05% and 0.2% by weight (wt.) of cement. Furthermore, six mixtures of milled rCFs were prepared at different concentrations, namely 0.0%, 0.1%, 0.5%, 1.0%, 1.5%, 2.0%, 2.5% by weight of cement. The reinforcing materials were added into water with NBs, and the aqueous suspensions were then poured in the mixer (Figure 3). The pastes were prepared in a stainless-steel standard mixer, according to ASTM C305, at low speed (140 rpm) for 30 s and then at high speed (285 rpm) for 60 s. The water-to-cement ratio (w/c) was constant at 0.50 for all the mixtures. The proportions of the materials used for reference, GNPs- and rCFs-composites are presented in Tables 3 and 4, respectively.

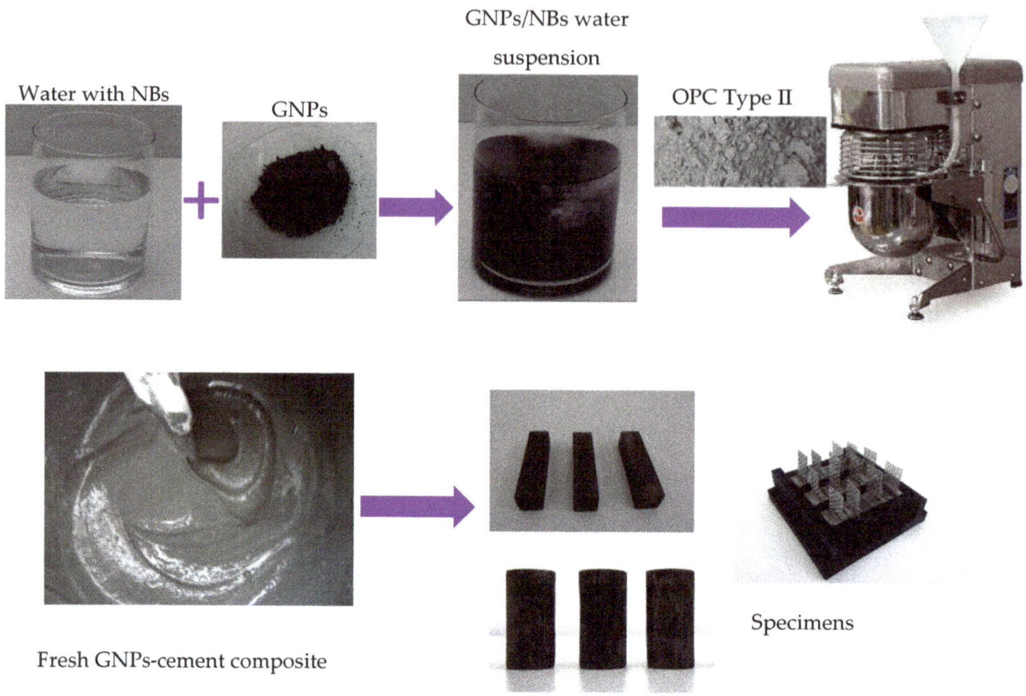

Figure 3. GNP–cement composite preparation.

Table 3. Mix proportions for reference and GNP-composites.

Mixture	GNPs Type	w/c Ratio	Cement g	GNPs wt.%
Ref	-	0.5	470	-
N_0.05	N008-100-N	0.5	470	0.05
P10_0.05	N008-100-P10	0.5	470	0.05
P40_0.05	N008-100-P40	0.5	470	0.05
N_0.2	N008-100-N	0.5	470	0.2
P10_0.2	N008-100-P10	0.5	470	0.2
P40_0.2	N008-100-P40	0.5	470	0.2

Table 4. Mix proportions for reference and rCFs-composites.

Mixture	w/c Ratio	Cement g	rCFs wt.%
Ref	0.5	470	-
rCFs_0.1	0.5	470	0.1
rCFs_0.5	0.5	470	0.5
rCFs_1.0	0.5	470	1
rCFs_1.5	0.5	470	1.5
rCFs_2.0	0.5	470	2.0
rCFs_2.5	0.5	470	2.5

The prepared GNPs- and rCFs–cement composites were cast in 30 mm × 60 mm cylindrical steel molds and 20 mm × 20 mm × 80 mm prismatic plastic molds, to investigate the compressive and flexural strength, respectively. To measure the electrical resistance of the composite specimens, right after molding, four stainless steel electrodes were embedded into the specimens covering their entire cross-section (Figure 4a,b). The stainless steel mesh used had openings of 1.74 mm × 1.74 mm, wire thickness of 0.8 mm, and dimensions of 20 mm × 50 mm (Figure 4c). After 48 h, the specimens were de-molded and placed in tap water tanks with calcium hydroxide, where they were kept for 28 days.

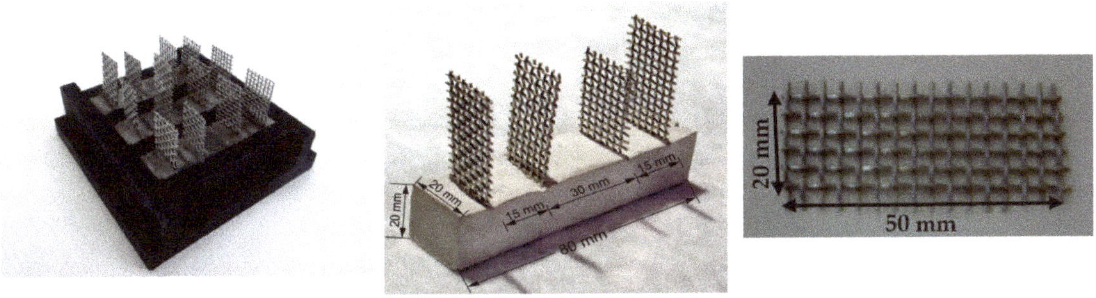

(**a**) (**b**) (**c**)

Figure 4. Specimens for the electrical resistance tests (**a**) inside the mold write after casting, (**b**) after demolding, and (**c**) stainless steel mesh embedded inside the samples.

2.3. *Experimental Measurements*

The effect of GNPs and milled rCFs on the compressive strength pastes was evaluated after 28 days of curing. Three specimens of each testing series were assessed. The compressive strength was measured using the Instron 8801 testing machine (Instron, Norwood, MA, USA) with maximum load capacity of 100 kN at a constant displacement rate of 0.2 mm/min. Loading was applied monotonically until the failure of the specimen (Figure 5).

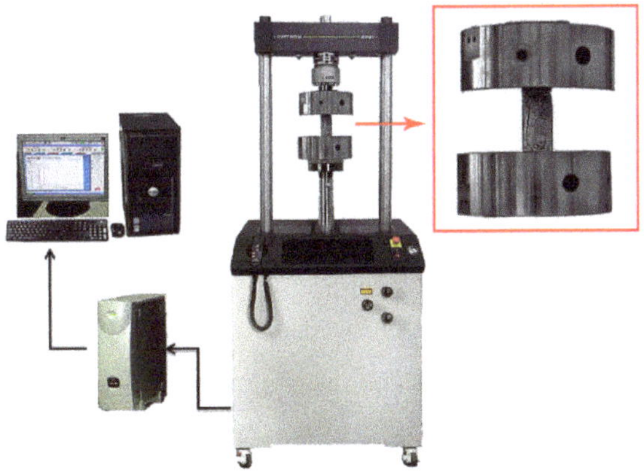

Figure 5. Experimental apparatus for compressive stress test (close up shows a fractured sample).

The effect of GNPs and milled rCFs on the flexural strength of the cement paste was evaluated after 28 days of curing. Flexural strength measurements were carried in a 4-point-bending configuration (Figure 6) using an MTS Insight testing machine (MTS Systems Corporation, Eden Prairie, MN, USA) with maximum load capacity of 10 kN applying a constant displacement rate of 0.002 mm/min (Figure 4b). The average flexural strength of three specimens was recorded and presented as representative flexural strength of the cement composites.

Figure 6. 4-point bending test experimental set up.

The electrical resistance was measured using the four-wire-method (Figure 7). Three specimens of each testing series were tested. After 28 days of curing, the specimens were placed in a laboratory drying oven where they remained for 72 h, at a temperature of 80 °C, to evaporate the water trapped in the pores. A 34450A Keysight Laboratory Digital Bench Multimeter, Keysight Technologies, Santa Rosa, CA, USA)was used for the electrical measurements. The internal electrodes were used to measure the voltage, while the external electrodes were used to supply the current as shown in Figure 7.

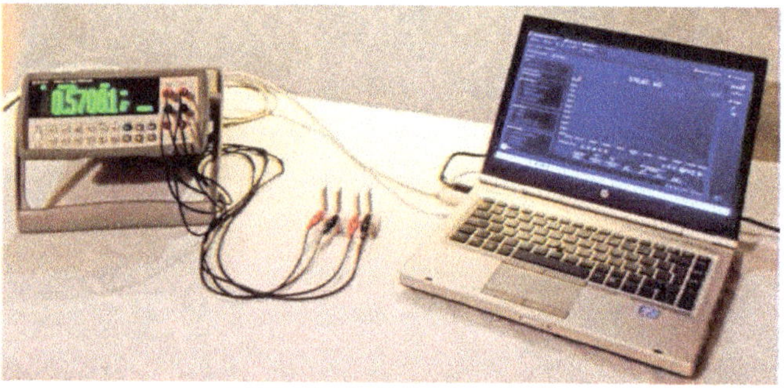

Figure 7. Four-wire method set-up.

During the measurements, a rubber material was placed under the specimens to insulate the specimen. The electrical resistance was recorded every 2 s over a period of 30 min to avoid potential deviations caused by the effects of electric polarization. Data from the last five minutes of the measurements were used to calculate the average resistance values. The resistivity, ρ, of the nanocomposites was calculated using Ohm's law as

$$\rho = R\frac{S}{L}$$

where R is the electrical resistance, S is the cross-section of the sample, and L is the distance between internal electrodes.

3. Results and Discussion

3.1. GNPs Strenthening Mechanism

The average measured values of compressive and flexural strength of cement composites with different sizes and contents of GNPs are illustrated in Figure 8. The abbreviations of N, P-10, and P-40 shown in the graph refer to the N008-100-N, N008-100-P10, and N008-100-P40 GNPs types followed by their concentration 0.05 and 0.2% by weight of cement, i.e., P10_0.05 is the N008-100-P10 GNPs-type at an amount of 0.05%.

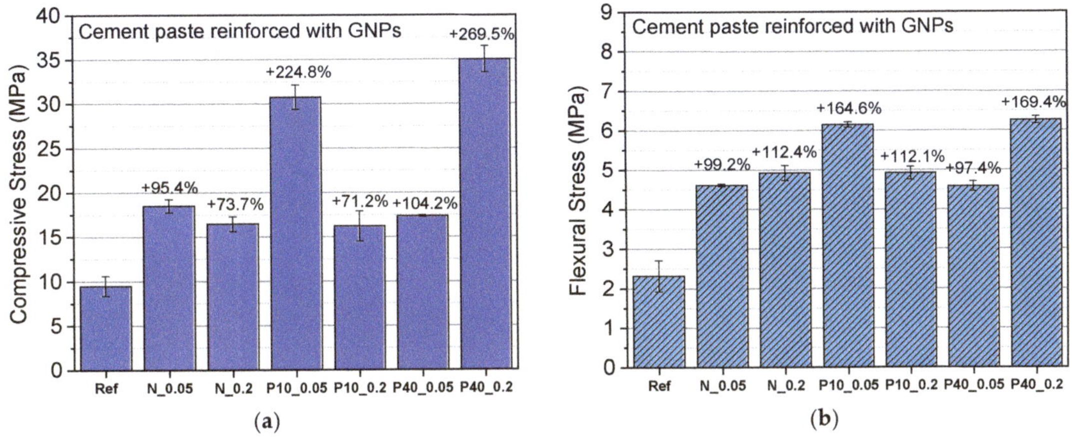

Figure 8. Average (**a**) compressive strength and (**b**) flexural strength of cement composites with GNPs at 28 d.

The incorporation of GNPs into the cement paste is beneficial for its compressive and flexural strength. The compressive strength of the plain cement paste was found to be 9.5 MPa and raised to 30.8 MPa and 35 MPa for the P10_005 and P40_02 specimens, corresponding to an increase of 224.8% and of 269.5%, respectively, which is more than reported in the literature [19,21,48–50]. The GNPs-composites also demonstrated an important increase in the flexural strength compared to the plain cement specimens. The highest flexural strength values varied between 6.14 MPa and 6.25 MPa, corresponding to an increase of 164.6% and 169.4%, for the P10_005 and P40_02 specimens, respectively, compared to the reference specimens (2.32 MPa).

This high improvement in the mechanical properties of GNPs–cement composites can be attributed to the interaction between the GNPs and the paste. GNPs possibly have developed good affinity/bond with the cement matrix, which can enhance the load-transfer capacity by absorbing energy more efficiently and offer a better resistance to crack spreading [17,22,51,52]. Upon load application, GNPs effectively transfer the stress by enhancing the crack bridging mechanism and restraining the crack formation [50,53]. Cracks are blocked and diverted or branched (Figure 9) [19].

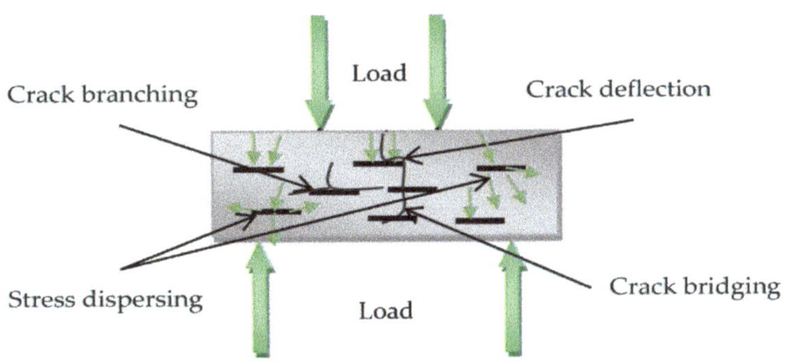

Figure 9. GNPs action under 4-point bending load.

According to the results, there were two reinforcing mechanisms that were developed and associated with the GNPs size. When GNPs with a low average diameter of 5 and 10 μm are incorporated, a greater increase in the compressive strength of cement composites is observed at low concentration (0.05 wt.%). This agrees with other studies claiming that

GNPs demonstrate better performance at lower concentrations, since a more homogeneous GNPs network can be developed at lower concentrations [49,53]. On the contrary, GNPs type N008-100-P40 demonstrated the maximum compressive stress improvement at the higher concentration (0.2 wt.%). The P-40 type demonstrates up to 10 times larger diameter compared to the N and P-10 GNPs-types. This finding is supported by the fact that larger sized nanomaterials are more easy to disperse [54,55]. Therefore, for the P-40 GNPs, the development of a homogeneous network at a larger concentration is feasible.

3.2. rCFs Strenthening Mechanism

The average compressive and flexural strength values of cement composites reinforced with milled rCFs and their deviation from the reference specimens are shown in Figure 10.

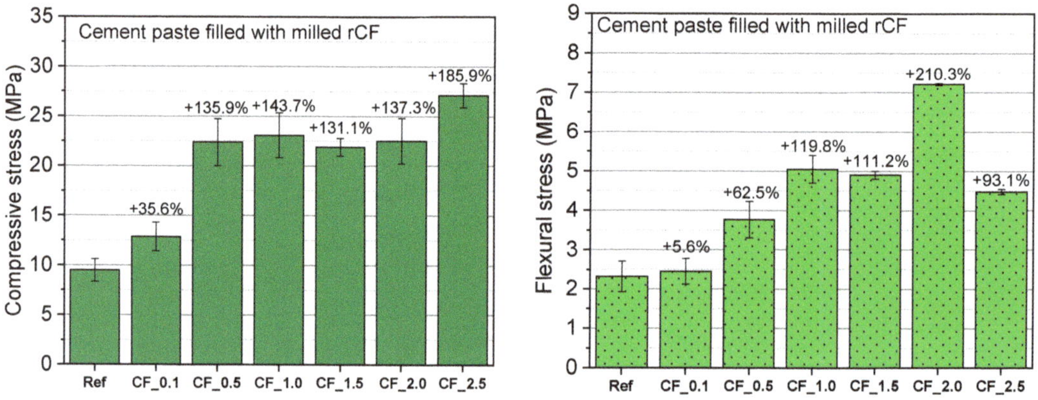

Figure 10. Average (**a**) compressive strength and (**b**) flexural strength of cement composites with milled rCFs at 28 d.

The mechanical strength of cement paste filled with different dosages of rCFs was significantly enhanced. This improvement seems to depend on the rCFs content. The compressive strength of all the specimens was enhanced gradually, with the increase in fiber content ranging from 0.1% to 2.5 wt.%. When the fiber content was 1% and 2.5% by weight of cement, an increase of 143.7% and of 185.9% was recorded, respectively. Similarly, Akbar et al. observed a 47% increase in compressive strength of cement composites by the addition of 1% by volume of milled rCFs [35]. Moreover, Mastali and Dalvand reported 65.10% improvement on the compressive strength of concrete specimens filled with 1.25% by volume rCFs [38].

The rCFs-composites showed considerable improvement in flexural strength compared to the plain cement specimens. A maximum flexural stress of 7.2 MPa was recorded for CF_2.0 mixture (210.3% higher than the reference mixture). This improvement is related to the crack bridging action of CFs. According to the results, the flexural performance of the specimens improves as the fiber content increases. This trend is in line with the findings of Yoo et al. [56], Donnini [30], and Han et al. [57]. As the fiber concentration in the matrix increases, the crack resistance mechanism and the fracture energy absorption improve. However, after a certain concentration, the CFs agglomerate and the mechanical strength of the composite reduces [57].

3.3. Electrical Resistivity of GNPs–Cement Composites

Typical curves of the electrical resistivity over testing time for the plain cement paste for the three different GNPs-types in both concentrations are shown in Figure 11. High electrical resistivity values (1.29 MOhm·cm) were recorded for the reference mixture. At the beginning of the test, increased values of approximately 1,42 MOhm·cm were observed.

This can be explained by the electrical polarization phenomenon. Electric polarization occurs when a dielectric material is exposed to a DC electric field. Next, positive and negative charges are pushed in opposite directions, resulting in separation of positive and negative charges throughout the mass of the material. Consequently, the resistance of the material, especially at the beginning of measurement, is constantly changing. The resistance stabilizes after approximately 3 min; however, marginal scatter of about 5% may be observed in the measurements.

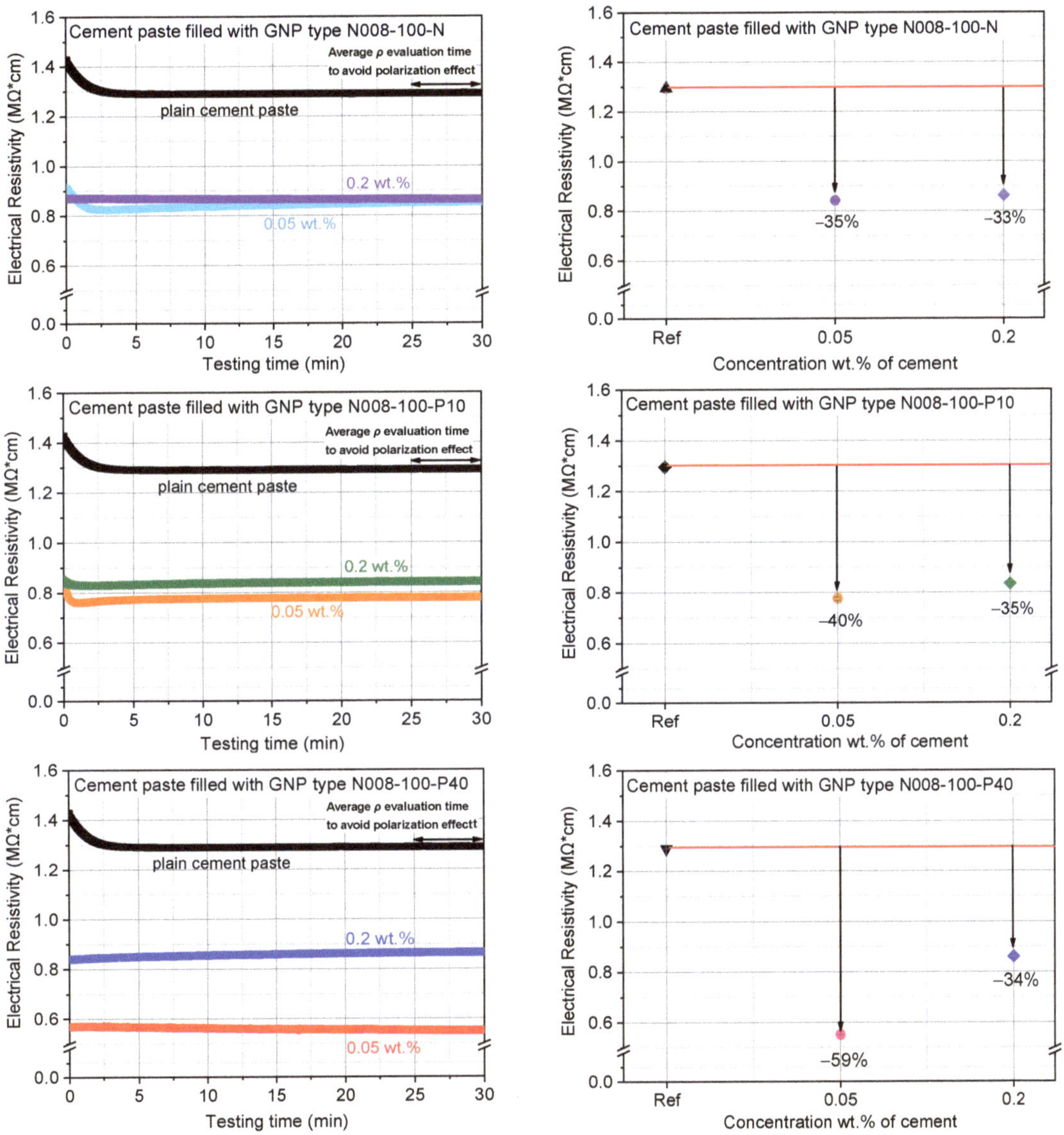

(**a**) (**b**)

Figure 11. (**a**) Typical electrical resistivity over testing time curves of the GNPs–cement composites; (**b**) Average electrical resistivity results of the GNPs–cement composites.

It is evident that GNPs-composites, regardless of the GNPs' type and amount, exhibit lower resistivity values compared to the plain cement paste [24,49]. The values of the resistivity were lower when the concentration of GNPs was 0.05 wt.%, which may be related to the GNPs dispersion state, i.e., the more uniform the dispersion of GNPs, the lower the electrical resistance in the cement matrix. The lowest ρ, 59% reduction compared to the plain cement paste, was measured for the specimens with GNPs-type N008-100-P40. Although the concentration of GNPs remained the same in all specimens, the results indicate that N008-100-P40 created the most efficient conductive network within the cementitious matrix. This is probably associated with its diameter, which was larger (44μm) than the other two types of GNPs. Therefore, the distance between the GNPs is small enough to form conductive paths in the matrix, favoring the passage of the current [17]. A significant decrease in the resistivity of nanocomposites modified with low contents of GNPs has been also reported [51,53].

When the concentration of GNPs increased to 0.2 wt.%, all the composite specimens demonstrated similar reduction in the resistivity close to 35%. This finding can be attributed to the inadequate dispersion of the nanomaterials. According to the literature [9,58,59], the resistivity of a material depends on the quantity and sufficient dispersion of conductive additives (Figure 12). Inadequate dispersion leads to GNPs-agglomeration, which increases the electrical resistance of the composite.

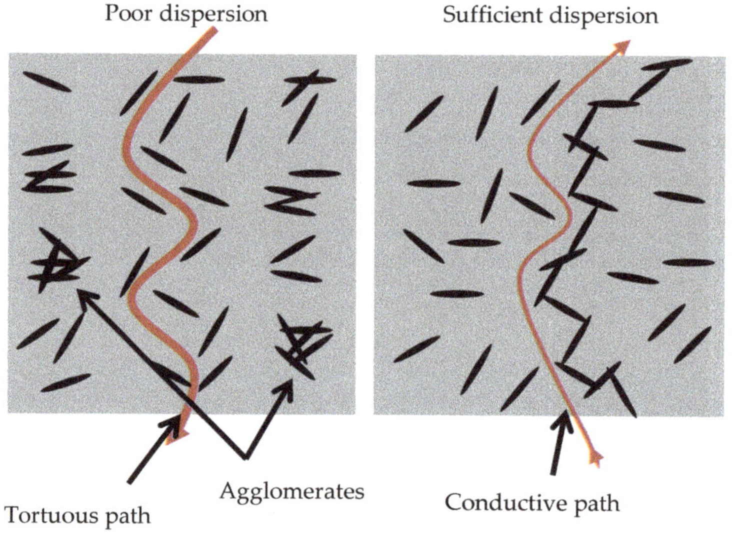

Figure 12. Electrical conductivity of GNPs–cement composites.

3.4. Electrical Resistivity of rCFs–Cement Composites

The typical electrical resistivity curves over testing time for the reference sample, and for rCFs–cement composites, are shown in Figure 13. The rCFs-composites demonstrated reduced resistivity compared to the reference specimens. The rate of resistance reduction is affected by the concentration of rCFs within the matrix. The highest reduction rate of 48% was measured for the specimens with fiber concentration of 1 wt.%. When the fiber content is low, the current flow is difficult, since there might be a large space among the fibers. However, by increasing the concentration of the rCFs, the fibers approach each other, and a stable conductive network is established in the matrix [57]. The value of the resistivity decreases until the percentage of rCFs reaches 1 wt.%, indicating that the percolation threshold, i.e., when the fibers form a continuous electrical network, has been reached. More rCFs cause an increase in the value of the electrical resistance.

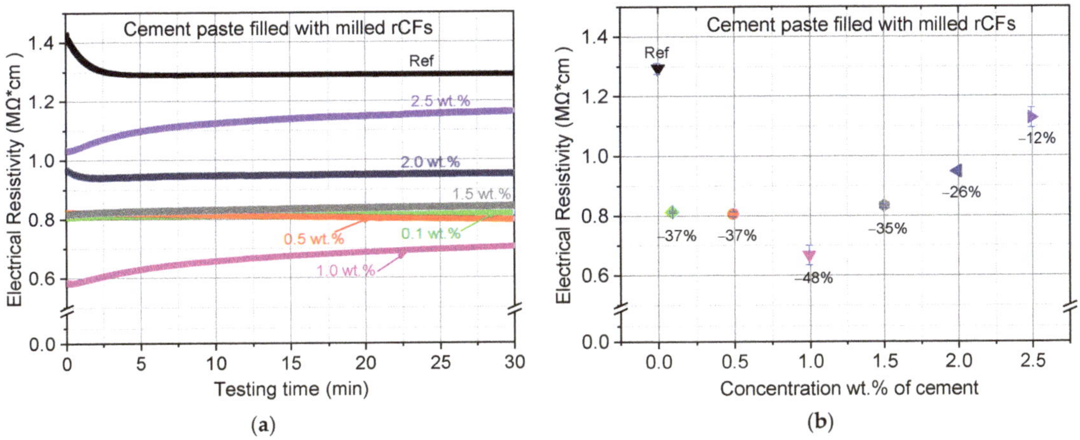

Figure 13. (**a**) Typical electrical resistivity over testing time curves of the investigated rCFs–cement composites; (**b**) Average electrical resistivity results of the GNPs–cement composites.

4. Conclusions

The mechanical and electrical behavior of cement paste reinforced with GNPs and milled rCFs were experimentally examined in this research. The addition of GNPs and rCFs into cement paste had a positive effect on the mechanical and electrical properties.

Among the three types of GNPs used in this study, the most effective proved to be the N008-100-P40 with the larger lateral size. Specimens with 0.2 wt.%. GNPs concentration increased both the compressive and the flexural strength of the cement composites by 269% and 169%, respectively. Furthermore, N008-100-P40 at a content of 0.05 wt.%. created the most effective conductive network within the cementitious matrix, attributed to their larger diameter.

Increasing the content of milled rCFs enhanced the mechanical properties of cement paste. The compressive and flexural strength of the rCFs-composites increased by 186% and 210%, with the incorporation of fibers at contents of 2.5 wt.% and 2 wt.%, respectively. The rate of reduction in electrical resistivity was affected by the concentration of rCFs in the composite. Percolation threshold was reached at a fiber content close to 1 wt.% of cement, as an increase of the fiber dosage was not able to further reduce the electrical resistivity of the cement paste.

Author Contributions: Conceptualization, Z.S.M.; methodology, A.I.P., E.T. and Z.S.M.; investigation, A.I.P., E.T., A.V., E.P.F. and G.Z.K.; resources, G.Z.K. and A.C.M.; writing—original draft preparation, A.I.P., E.T.; writing—review and editing, E.P.F. and Z.S.M.; supervision, Z.S.M.; project administration, A.C.M.; funding acquisition, Z.S.M. All authors have read and agreed to the published version of the manuscript.

Funding: This research was co-financed by the European Regional Development Fund of the European Union and Greek national funds through the Operational Program Competitiveness, Entrepreneurship, and Innovation. Projects: (i) "Nanoreinforced concrete for pavement deicing" (acronym: NEA ODOS; project code: T1EΔK-02692; MIS: 5030194) under the call Research–Create–Innovate.

Acknowledgments: E.P.F., A.M., G.Z.K. and Z.S.M. acknowledges funding from Research–Create–Innovate call under the framework of project NEA ODOS.

Conflicts of Interest: The authors declare no conflict of interest.

References

1. Konsta-Gdoutos, M.S.; Metaxa, Z.S.; Shah, S.P. Highly dispersed carbon nanotube reinforced cement based materials. *Cem. Concr. Res.* **2010**, *40*, 1052–1059. [CrossRef]
2. Alrekabi, S.; Cundy, A.; Lampropoulos, A.; Whitby, R.; Savina, I. Mechanical performance of novel cement-based composites prepared with nano-fibres, and hybrid nano- and micro-fibres. *Compos. Struct.* **2017**, *178*, 145–156. [CrossRef]
3. Meng, W.; Khayat, K.H. Effect of graphite nanoplatelets and carbon nanofibers on rheology, hydration, shrinkage, mechanical properties, and microstructure of UHPC. *Cem. Concr. Res.* **2018**, *105*, 64–71. [CrossRef]
4. Jiang, S.; Zhou, D.; Zhang, L.; Ouyang, J.; Yu, X.; Cui, X.; Han, B. Comparison of compressive strength and electrical resistivity of cementitious composites with different nano- and micro-fillers. *Arch. Civ. Mech. Eng.* **2018**, *18*, 60–68. [CrossRef]
5. Xu, S.; Liu, J.; Li, Q. Mechanical properties and microstructure of multi-walled carbon nanotube-reinforced cement paste. *Constr. Build. Mater.* **2015**, *76*, 16–23. [CrossRef]
6. Metaxa, Z.S.; Seo, J.-W.T.; Konsta-Gdoutos, M.S.; Hersam, M.C.; Shah, S.P. Highly concentrated carbon nanotube admixture for nano-fiber reinforced cementitious materials. *Cem. Concr. Compos.* **2012**, *34*, 612–617. [CrossRef]
7. Li, X.; Wei, W.; Qin, H.; Hu, Y.H. Co-effects of graphene oxide sheets and single wall carbon nanotubes on mechanical properties of cement. *J. Phys. Chem. Solids* **2015**, *85*, 39–43. [CrossRef]
8. Tao, J.; Wang, J.; Zeng, Q. A comparative study on the influences of CNT and GNP on the piezoresistivity of cement composites. *Mater. Lett.* **2020**, *259*, 126858. [CrossRef]
9. Konsta-Gdoutos, M.S.; Aza, C.A. Self sensing carbon nanotube (CNT) and nanofiber (CNF) cementitious composites for real time damage assessment in smart structures. *Cem. Concr. Compos.* **2014**, *53*, 162–169. [CrossRef]
10. Han, B.; Sun, S.; Ding, S.; Zhang, L.; Yu, X.; Ou, J. Review of nanocarbon-engineered multifunctional cementitious composites. *Compos. Part A Appl. Sci. Manuf.* **2015**, *70*, 69–81. [CrossRef]
11. Rehman, S.K.U.; Kumarova, S.; Memon, S.A.; Javed, M.F.; Jameel, M. A Review of Microscale, Rheological, Mechanical, Thermoelectrical and Piezoresistive Properties of Graphene Based Cement Composite. *Nanomaterials* **2020**, *10*, 2076. [CrossRef] [PubMed]
12. Vajtai, R. (Ed.) *Springer Handbook of Nanomaterials*; Springer: Berlin/Heidelberg, Germany, 2013. [CrossRef]
13. Jiménez-Suárez, A.; Prolongo, S.G. Graphene Nanoplatelets. *Appl. Sci.* **2020**, *10*, 1753. [CrossRef]
14. Cataldi, P.; Athanassiou, A.; Bayer, I.S. Graphene Nanoplatelets-Based Advanced Materials and Recent Progress in Sustainable Applications. *Appl. Sci.* **2018**, *8*, 1438. [CrossRef]
15. Döscher, H.; Schmaltz, T.; Neef, C.; Thielmann, A.; Reiss, T. Graphene Roadmap Briefs (No. 2): Industrialization status and prospects 2020. *2D Mater.* **2021**, *8*, 022005. [CrossRef]
16. Du, M.; Jing, H.; Gao, Y.; Su, H.; Fang, H. Carbon nanomaterials enhanced cement-based composites: Advances and challenges. *Nanotechnol. Rev.* **2020**, *9*, 115–135. [CrossRef]
17. Tao, J.; Wang, X.; Wang, Z.; Zeng, Q. Graphene nanoplatelets as an effective additive to tune the microstructures and piezoresistive properties of cement-based composites. *Constr. Build. Mater.* **2019**, *209*, 665–678. [CrossRef]
18. Chougan, M.; Marotta, E.; Lamastra, F.R.; Vivio, F.; Montesperelli, G.; Ianniruberto, U.; Bianco, A. A systematic study on EN-998-2 premixed mortars modified with graphene-based materials. *Constr. Build. Mater.* **2019**, *227*, 116701. [CrossRef]
19. Silva, R.E.; Guetti, P.D.C.; da Luz, M.S.; Rouxinol, F.; Gelamo, R.V. Enhanced properties of cement mortars with multilayer graphene nanoparticles. *Constr. Build. Mater.* **2017**, *149*, 378–385. [CrossRef]
20. Tong, T.; Fan, Z.; Liu, Q.; Wang, S.; Tan, S.; Yu, Q. Investigation of the effects of graphene and graphene oxide nanoplatelets on the micro- and macro-properties of cementitious materials. *Constr. Build. Mater.* **2016**, *106*, 102–114. [CrossRef]
21. Wang, B.; Jiang, R.; Wu, Z. Investigation of the Mechanical Properties and Microstructure of Graphene Nanoplatelet-Cement Composite. *Nanomaterials* **2016**, *6*, 200. [CrossRef] [PubMed]
22. Wang, B.; Shuang, D. Effect of graphene nanoplatelets on the properties, pore structure and microstructure of cement composites. *Mater. Express* **2018**, *8*, 407–416. [CrossRef]
23. Dong, W.; Li, W.; Zhu, X.; Sheng, D.; Shah, S.P. Multifunctional cementitious composites with integrated self-sensing and hydrophobic capacities toward smart structural health monitoring. *Cem. Concr. Compos.* **2021**, *118*, 103962. [CrossRef]
24. Ozbulut, O.E.; Jiang, Z.; Harris, D.K. Exploring scalable fabrication of self-sensing cementitious composites with graphene nanoplatelets. *Smart Mater. Struct.* **2018**, *27*, 115029. [CrossRef]
25. Ahmad, W.; Khan, M.; Smarzewski, P. Effect of Short Fiber Reinforcements on Fracture Performance of Cement-Based Materials: A Systematic Review Approach. *Materials* **2021**, *14*, 1745. [CrossRef] [PubMed]
26. Ali, B.; Raza, S.S.; Hussain, I.; Iqbal, M. Influence of different fibers on mechanical and durability performance of concrete with silica fume. *Struct. Concr.* **2021**, *22*, 318–333. [CrossRef]
27. Khan, M.; Cao, M.; Ali, M. Cracking behaviour and constitutive modelling of hybrid fibre reinforced concrete. *J. Build. Eng.* **2020**, *30*, 101272. [CrossRef]
28. Bhatt, P.; Goe, A. Carbon Fibres: Production, Properties and Potential Use. *Mater. Sci. Res. India* **2017**, *14*, 52–57. [CrossRef]
29. Divate, D.D. A Review of Carbon Fiber. *Int. Adv. Res. J. Sci. Eng. Technol.* **2017**, *4*, 170–177. [CrossRef]
30. Donnini, J.; Bellezze, T.; Corinaldesi, V. Mechanical, electrical and self-sensing properties of cementitious mortars containing short carbon fibers. *J. Build. Eng.* **2018**, *20*, 8–14. [CrossRef]
31. Chen, P.-W.; Chung, D.D.L. Concrete reinforced with up to 0.2 vol% of short carbon fibres. *Composites* **1993**, *24*, 33–52. [CrossRef]

32. Cholker, A.K.; Kulkarni, P.; Ravekar, V. Experimental behaviour of smart concrete embedded with micro carbon fibres as a sensing material. *J. Phys. Conf. Ser.* **2021**, *1921*, 012095. [CrossRef]
33. Chung, D.D.L. Cement reinforced with short carbon fibers: A multifunctional material. *Compos. Part B Eng.* **2000**, *31*, 511–526. [CrossRef]
34. Pimenta, S.; Pinho, S.T. Recycling carbon fibre reinforced polymers for structural applications: Technology review and market outlook. *Waste Manag.* **2011**, *31*, 378–392. [CrossRef] [PubMed]
35. Akbar, A.; Kodur, V.; Liew, K. Microstructural changes and mechanical performance of cement composites reinforced with recycled carbon fibers. *Cem. Concr. Compos.* **2021**, *121*, 104069. [CrossRef]
36. Faneca, G.; Segura, I.; Torrents, J.; Aguado, A. Development of conductive cementitious materials using recycled carbon fibres. *Cem. Concr. Compos.* **2018**, *92*, 135–144. [CrossRef]
37. Akbar, A.; Liew, K. Influence of elevated temperature on the microstructure and mechanical performance of cement composites reinforced with recycled carbon fibers. *Compos. Part B Eng.* **2020**, *198*, 108245. [CrossRef]
38. Mastali, M.; Dalvand, A. The impact resistance and mechanical properties of self-compacting concrete reinforced with recycled CFRP pieces. *Compos. Part B Eng.* **2016**, *92*, 360–376. [CrossRef]
39. Metaxa, Z.S.; Kourkoulis, S.K. Dispersion of graphene nanoplatelets reinforcing type II cement paste. *Procedia Struct. Integr.* **2018**, *13*, 2011–2016. [CrossRef]
40. Gao, J.; Wang, Z.; Zhang, T.; Zhou, L. Dispersion of carbon fibers in cement-based composites with different mixing methods. *Constr. Build. Mater.* **2017**, *134*, 220–227. [CrossRef]
41. Nazari, B.; Ranjbar, Z.; Hashjin, R.R.; Moghaddam, A.R.; Momen, G.; Ranjbar, B. Dispersing graphene in aqueous media: Investigating the effect of different surfactants. *Colloids Surf. A Physicochem. Eng. Asp.* **2019**, *582*, 123870. [CrossRef]
42. Metaxa, Z.S.; Boutsioukou, S.; Amenta, M.; Favvas, E.P.; Kourkoulis, S.K.; Alexopoulos, N.D. Dispersion of Multi-Walled Carbon Nanotubes into White Cement Mortars: The Effect of Concentration and Surfactants. *Nanomaterials* **2022**, *12*, 1031. [CrossRef] [PubMed]
43. Kyzas, G.Z.; Bomis, G.; Kosheleva, R.I.; Efthimiadou, E.K.; Favvas, E.P.; Kostoglou, M.; Mitropoulos, A.C. Nanobubbles effect on heavy metal ions adsorption by activated carbon. *Chem. Eng. J.* **2019**, *356*, 91–97. [CrossRef]
44. Michalopoulou, A.; Favvas, E.; Mitropoulos, A.; Maravelaki, P.; Kilikoglou, V.; Karatasios, I. A comparative evaluation of bottom-up and break-down methodologies for the synthesis of calcium hydroxide nanoparticles for the consolidation of architectural monuments. *Mater. Today Proc.* **2018**, *5*, 27425–27433. [CrossRef]
45. Michalopoulou, A.; Michailidi, E.; Favvas, E.P.; Maravelaki, P.; Kilikoglou, V.; Karatasios, I. Comparative evaluation of the morphological characteristics of nanolime dispersions for the consolidation of architectural monuments. *Int. J. Arcitectural Herit.* **2020**, *14*, 994–1007. [CrossRef]
46. Alheshibri, M.; Al Baroot, A.; Shui, L.; Zhang, M. Nanobubbles and nanoparticles. *Curr. Opin. Colloid Interface Sci.* **2021**, *55*, 101470. [CrossRef]
47. Michailidiab, E.D.; Bomisc, G.; Varoutoglouc, A.; Kyzas, G.; Mitrikas, G.; Mitropoulos, A.C.; Efthimiadou, E.K.; Favvas, E.P. Bulk nanobubbles: Production and investigation of their formation/stability mechanism. *J. Colloid Interface Sci.* **2019**, *564*, 371–380. [CrossRef] [PubMed]
48. Qureshi, T.S.; Panesar, D.K. Nano reinforced cement paste composite with functionalized graphene and pristine graphene nanoplatelets. *Compos. Part B Eng.* **2020**, *197*, 108063. [CrossRef]
49. Dong, W.; Li, W.; Wang, K.; Shah, S.P. Physicochemical and Piezoresistive properties of smart cementitious composites with graphene nanoplates and graphite plates. *Constr. Build. Mater.* **2021**, *286*, 122943. [CrossRef]
50. Khalid, A.; Khushnood, R.A.; Saleem, S.; Farooq, S.Z.; Shaheen, N. Improving the mechanical properties of cementitious composites with graphite nano/micro platelets addition. *IOP Conf. Series Mater. Sci. Eng.* **2018**, *431*, 062005. [CrossRef]
51. Chougan, M.; Marotta, E.; Lamastra, F.R.; Vivio, F.; Montesperelli, G.; Ianniruberto, U.; Ghaffar, S.H.; Al-Kheetan, M.J.; Bianco, A. High performance cementitious nanocomposites: The effectiveness of nano-Graphite (nG). *Constr. Build. Mater.* **2020**, *259*, 119687. [CrossRef]
52. Ranjbar, N.; Mehrali, M.; Mehrali, M.; Alengaram, U.J.; Jumaat, M.Z. Graphene nanoplatelet-fly ash based geopolymer composites. *Cem. Concr. Res.* **2015**, *76*, 222–231. [CrossRef]
53. Sun, S.; Ding, S.; Han, B.; Dong, S.; Yu, X.; Zhou, D.; Ou, J. Multi-layer graphene-engineered cementitious composites with multifunctionality/intelligence. *Compos. Part B Eng.* **2017**, *129*, 221–232. [CrossRef]
54. Evgin, T.; Turgut, A.; Hamaoui, G.; Spitalsky, Z.; Horny, N.; Micusik, M.; Chirtoc, M.; Sarikanat, M.; Omastova, M. Size effects of graphene nanoplatelets on the properties of high-density polyethylene nanocomposites: Morphological, thermal, electrical, and mechanical characterization. *Beilstein J. Nanotechnol.* **2020**, *11*, 167–179. [CrossRef] [PubMed]
55. Horszczaruk, E.; Łukowski, P.; Seul, C. Influence of Dispersing Method on the Quality of Nano-Admixtures Homogenization in Cement Matrix. *Materials* **2020**, *13*, 4865. [CrossRef]
56. Yoo, D.-Y.; Lee, J.-H.; Yoon, Y.-S. Effect of fiber content on mechanical and fracture properties of ultra high performance fiber reinforced cementitious composites. *Compos. Struct.* **2013**, *106*, 742–753. [CrossRef]
57. Han, B.; Zhang, L.; Zhang, C.; Wang, Y.; Yu, X.; Ou, J. Reinforcement effect and mechanism of carbon fibers to mechanical and electrically conductive properties of cement-based materials. *Constr. Build. Mater.* **2016**, *125*, 479–489. [CrossRef]

58. Metaxa, Z.S. Polycarboxylate Based Superplasticizers as Dispersant Agents for Exfoliated Graphene Nanoplatelets Reinforcing Cement Based Materials. *J. Eng. Sci. Technol. Rev.* **2015**, *8*, 1–5. [CrossRef]
59. Lamastra, F.R.; Chougan, M.; Marotta, E.; Ciattini, S.; Ghaffar, S.H.; Caporali, S.; Vivio, F.; Montesperelli, G.; Ianniruberto, U.; Al-Kheetan, M.J.; et al. Toward a better understanding of multifunctional cement-based materials: The impact of graphite nanoplatelets (GNPs). *Ceram. Int.* **2021**, *47*, 20019–20031. [CrossRef]

nanomaterials

Article

Study on Nanoporous Graphene-Based Hybrid Architecture for Surface Bonding

Xiaohui Song [1,*], Mingxiang Chen [2], Jingshuang Zhang [3], Rui Zhang [3] and Wei Zhang [1]

[1] Institute of Applied Physics, Henan Academy of Science, Zhengzhou 450008, China; 13513808093@163.com
[2] School of Mechanical Science and Engineering, Huazhong University of Science and Technology, Wuhan 430074, China; chimish@163.com
[3] School of Mechanical Engineering, Zhengzhou University, Zhengzhou 450001, China; zhangjingshuang@hniot.com (J.Z.); lyzr@zzu.edu.cn (R.Z.)
* Correspondence: xhsong@foxmail.com

Abstract: Graphene-copper nanolayered composites have received research interest as promising packaging materials in developing next-generation electronic and optoelectronic devices. The weak van der Waal (vdW) contact between graphene and metal matrix significantly reduces the mechanical performance of such composites. The current study describes a new Cu-nanoporous graphene-Cu based bonding method with a low bonding temperature and good dependability. The deposition of copper atoms onto nanoporous graphene can help to generate nanoislands on the graphene surface, facilitating atomic diffusion bonding to bulk copper bonding surfaces at low temperatures, according to our extensive molecular dynamics (MD) simulations on the bonding process and pull-out verification using the canonical ensemble (NVT). Furthermore, the interfacial mechanical characteristics of graphene/Cu nanocomposites can be greatly improved by the resistance of nanostructure in nanoporous graphene. These findings are useful in designing advanced metallic surface bonding processes and graphene-based composites with tenable performance.

Keywords: surface bonding; nanoporous graphene

Citation: Song, X.; Chen, M.; Zhang, J.; Zhang, R.; Zhang, W. Study on Nanoporous Graphene-Based Hybrid Architecture for Surface Bonding. *Nanomaterials* **2022**, *12*, 2483. https://doi.org/10.3390/nano12142483

Academic Editors: Mohammad Malikan and Shahriar Dastjerdi

Received: 28 June 2022
Accepted: 10 July 2022
Published: 20 July 2022

1. Introduction

Surface bonding is an essential step in three-dimensional electronic packaging, as it provides mechanical support, heat transfer, and electrical integration [1]. A low-temperature bonding technology and reliable connection interface greatly influence electronic systems' performance and service life, especially when packaging density dramatically increases with device scaling [2,3]. Traditional surface bonding techniques in electronic assembly rely on high-temperature processes such as reflow soldering [4,5] and thermo-compression bonding [6], which can lead to undesirable thermal damage, toxic solder materials pollution, and a thermal mismatch at the bonding interface. Recently various nanometal materials such as metal nanowires, nanoparticles, and nanocones-based surface bonding are being studied to lower the bonding temperature and pressure [7–10]. However, these bonding technologies introducing low deformation resistance joints have been entangled in thermo-mechanical stresses and aging degradation issues, limiting their reliability. Therefore, it is necessary to develop high stretch and shear deformation resistance of interconnection materials architecture and a low-temperature bonding process for a complex interface of three-dimensional packaging structures.

Since its isolation as a two-dimensional (2D) system, graphene-copper nanolayered composites have been widely acclaimed and proved to be the effective alternatives to pure metal materials owing to their outstanding mechanical and heat transfer properties [11–15], showing great promise for next-generation electronic and optoelectronic device packaging. However, the reinforcing effect of graphene is limited by vdW non-bonded interfacial interaction between graphene and Cu, resulting in a low load transfer rate from Cu matrix to graphene during mechanical deformation [16]. To pave the way toward practical applications, a variety of approaches have been developed, such as chemical functionalization and metal coating [17,18]. The controllable and precise solution that uses graphene-metal composites as bonding layers remains a major challenge, despite considerable progress.

Here, we report a simple new Cu-nanoporous graphene-Cu based bonding technology that is of low bonding temperature and high reliability. We transfer nanoporous graphene to the copper surface and form copper nanoparticles on nanopores by depositing copper atoms before thermocompression bonding. Since the copper nanoparticles deposited in the nanopores form a metal bond connection with the copper bonding surface, they effectively enhance the interfacial connection and, consequently, lead to higher interfacial shear strength between graphene and Cu surface. At the same time, the metal nanoparticles contribute to solid-state diffusion and intermixing of surface atoms between the interfaces at low temperatures. The present study aims to quantitatively investigate the above-mentioned process. Comprehensive molecular dynamic simulations are carried out, and the results reveal that the copper-nanoporous graphene-copper sandwich structure can be bonded at low temperature by the thermocompression method, and the interfacial shear strength of the interface can be significantly increased by nanoporous-nanometal particles hybrid structure.

2. Computational Methods

For molecular dynamics simulation of Cu-nanoporous graphene-Cu bonding process and the reinforcing effect evaluation, the setup of the simulations performed in the present work is shown in Figure 1. Initially, the composite model is constructed by covering single-layer graphene with nine nanopores onto the single crystal copper surface that contains 34,924 atoms with the size of 161.78 Å × 99.65 Å × 23.35 Å, as shown in Figure 1a. The boundary condition for the simulation box is periodic in the x and y directions while the nonperiodic boundary condition is applied along the z-direction. The equilibrium structures are achieved using the canonical ensemble (NVT) [19] of the constant volume and temperature for 20 ps with a time step of 0.5 fs.

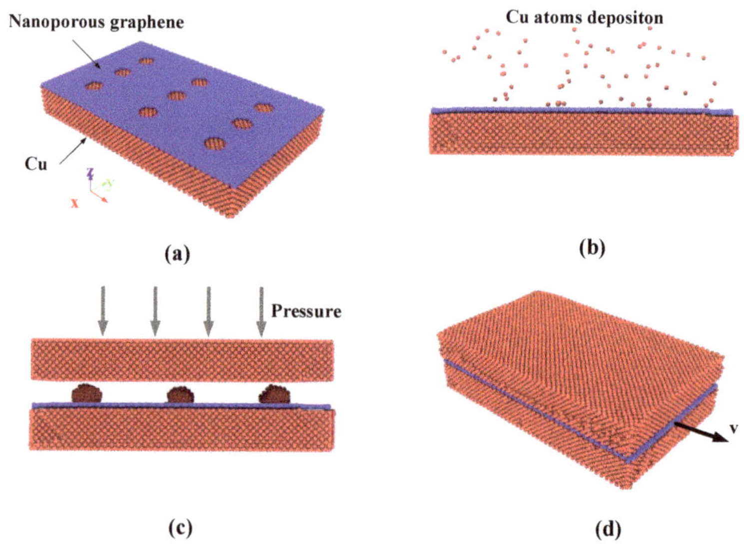

Figure 1. Atomic configurations. (**a**) Cu-nanoporous graphene composite model; (**b**) Cu atoms deposition onto the nanoporous graphene surface; (**c**) Cu-nanoporous graphene-Cu thermocompression bonding; (**d**) Pull-out simulation model.

Next, the bottommost two layers are fixed. The other atoms above the fixed layers are defined as temperature control layers that control the system temperature using the canonical ensemble (NVT), and the time step is set to 0.5 fs. In the condition of copper sputtering deposition simulation, the deposition rate of incident Cu atoms is 0.25 atoms/ps. The coordinates of incident Cu atoms were distributed randomly within the defined insertion volume, which is 50 nm above the graphene surface, as shown in Figure 1b. In each simulation, after all Cu atoms are released, a relaxation process of 2500 ps is conducted to enable the deposited atoms to reach full thermal equilibrium.

Then the bonding model is constructed by introducing a single crystal copper cell of the same size as the previous composite model as shown in Figure 1c. The distance between the two surfaces is set to nm. The uppermost two atomic layers are set as a fixed layer to produce the pressure of the system, and five atomic layers connected to the fixed layers are set up as temperature-controlled layers using the canonical ensemble (NVT). The other free copper atoms and nanoporous graphene at the bonding interface use the micro-canonical ensemble (NVE) of the constant volume and energy with a time step of 0.5 fs. During the first stage of simulation, no pressure is loaded on the fixed layer in 2 ns to reduce internal stress and make the model be in a steady state at a certain temperature. The pressure is loaded on the upper fix layer for 4 ns. At last, the pressure is unloaded and the model holds for 5 ns.

Finally, the pull-out simulation is used to investigate the interfacial shear strength of Cu nanocomposite. During the pull-out simulations, the upmost two layers and the bottommost two layers of Cu atoms are fully fixed. The pull-out simulations using the canonical ensemble (NVT) are performed by applying a constant velocity to the graphene along the x-direction until it is completely pulled out from the bonding interface as shown in Figure 1d. To ensure that the model is fully relaxed in each step and eliminate the effect of velocity on the pull-out force, a relatively low velocity of 0.8×10^{-5} Å/fs is adopted to 245 atoms treated as a rigid body on the rightmost side of the graphene. A temperature of 300 K is employed for all simulations, and a time step of 0.5 fs is adopted throughout the whole simulation process. The boundary condition for the simulation box is periodic in the y directions while the nonperiodic boundary condition is applied along the x and z directions.

The embedded atomic method (EAM) [20] models successfully used in modeling various bulk metals are used to describe the Cu-Cu interactions. For the vdW interactions between the graphene surface and Cu atoms, the Lennard–Jones (LJ) pair potential involving nonbonded long-range interactions is used with parameters $\sigma = 3.0825$ Å, $\varepsilon = 0.02578$ eV [21], and a cutoff radius = 4 σ. Our simulations are based on the massively parallel LAMMPS code [22]. The visualization is based on OVITO [23].

3. Results and Discussion

First, the Cu atoms deposition onto the nanoporous graphene surface is investigated. As shown in Figure 2a, in the initial state, the Cu atoms prefer to deposit in the nanoporous graphene and form metallic connections with the attached copper surface until the porous are fully filled. The reason for this is that Cu-Cu interactions are stronger than Cu-graphene. More copper atoms are then deposited onto the copper structure in the nanopores and gradually grow to form nanoislands. The growth pattern observed in this simulation approximates the insular growth shown in Figure 2b. When the nanoislands reach a certain size, they join with neighboring nanoislands to form larger ones as shown in Figure 2c,d.

Then thermocompression bonding simulations are performed to exploit the bonding mechanisms of nanoporous graphene-based hybrid architecture. A constant temperature of 300 K and pressure of 0.5 MPa are applied to the model. In the initial state, two contact interfaces form between nanoislands and the bulk copper surface as shown in Figure 3a. Compression deformation of the nanoislands increases until the bulk copper surface and the graphene are joined together as shown in Figure 3b,c. In this process, the nanoislands are gradually dispersed on the graphene surface and fill the interface while atomic diffusion occurs with the bulk copper surface. During this state, nanoislands undergo a typical process from elastic deformation to plastic deformation, and the crystal structure of nanoislands has been mostly destroyed. As a result, more and more active copper atoms on the nanoislands surface make contact with the bonding surface to form metallic bonding connections with time increasing.

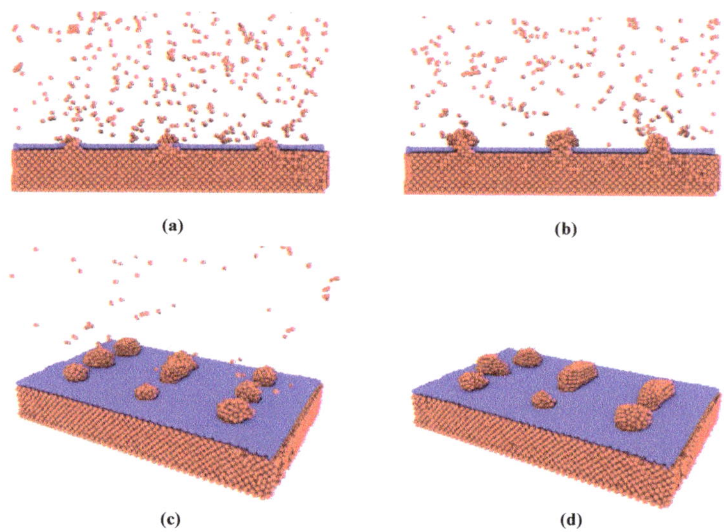

(a) (b) (c) (d)

Figure 2. Atomic configurations of Cu atoms deposition onto the graphene at various stages. (**a**) Filling of nanopores at the initial stage; (**b**) Insular growth to nanoislands; (**c**) Further growing of nanoislands; (**d**) Joining with neighboring nanoislands.

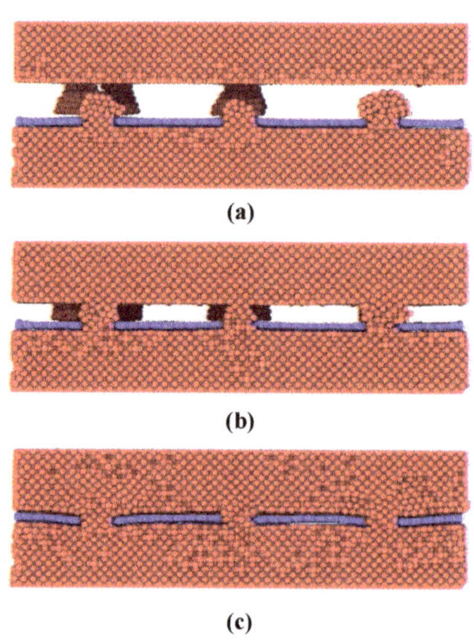

(a) (b) (c)

Figure 3. Cross-section configurations of Cu-nanoporous graphene-Cu during the bonding process at various stages. (**a**) Contacting of bonding surface; (**b**) Compression deformation of the nanoislands; (**c**) Final bonding structure.

Pull-out simulations of the nanoporous graphene from the metal matrix are used to investigate the reinforcing mechanisms of Cu-nanoporous graphene-Cu nanocomposites. Along the x-axis, a constant velocity is delivered to the graphene. Figure 4 shows the pull-out force for graphene/Cu composites with and without nanoporous surfaces in terms of sliding distance. The atomic configurations during the nanoporous pull-out process are depicted in the insets of Figure 4. The pull-out force without nanoporous grows fast until the sliding distance reaches roughly 8 Å as illustrated in Figure 4. After this, it swings at around 13 nN for the next 140 Å. The pull-out force then steadily declines until graphene is totally pulled out in about the last 10 Å of sliding distance. In this condition, the van der Waals interaction between the graphene surface and the copper atoms at the interface dominates the pull-out force. The pull-out force with nanoporous, on the other hand, rises quickly to about 60 nN at the start of the pull-out process, nearly six times higher than without nanoporous. Copper atoms in nanopores and their van der Waals contact with the graphene surface are credited with resisting the pull-out motion of nanoporous graphene. In particular, the graphene pull-out process requires breaking through the resistance of the nanostructures in the pores, which greatly increases the pull-out force. Therefore, much larger pull-out force is generated in graphene. As a result, a larger pull-out force is needed for the nanoporous graphene, giving rise to better interfacial properties than that of pure graphene.

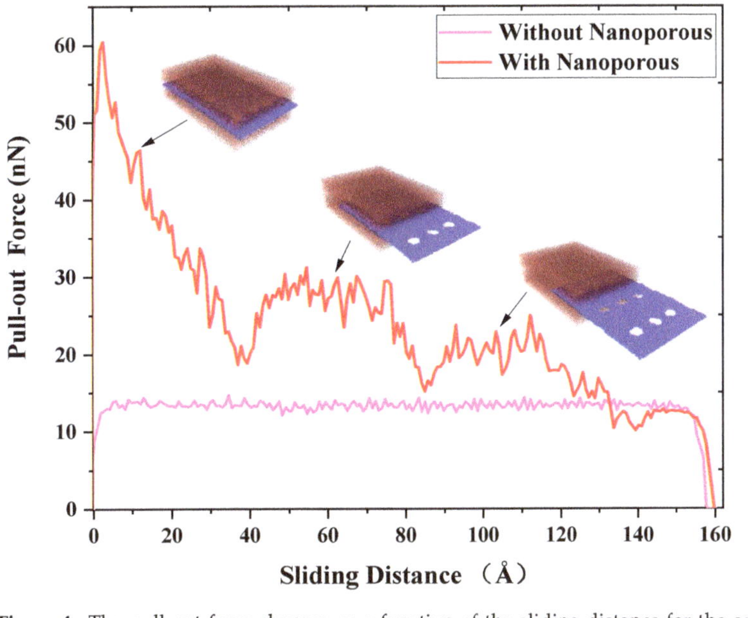

Figure 4. The pull-out force changes as a function of the sliding distance for the graphene/Cu composite with and without nanoporous.

When the sliding distance is larger than 4 Å, the force decreases to about 18 nN when the first column of nanoporous is fully pulled out. Figure 5 presents the displacement of copper atoms at the interface of graphene/Cu nanocomposites for this process. As the sliding distance increases, the atoms in the nanoporous are largely deformed until they move along with the graphene. The fluctuation of the pull-out force corresponds to the deformation-movement process of copper structures in nanoporous. Consequently, van der Waals interactions play an increasingly important role, and the copper atoms in nanopores are becoming less resistant to graphene. Next, as the sliding distance increases, there are two processes of increasing and decreasing pull-out force until it becomes 0. It is noted that these two processes occur after the first and second columns of nanoporous are pulled out, respectively, as shown in Figure 3. Since the graphene is regarded as a flexible body during the pull-out process, during the constant velocity pull-out process, the outer porous are subjected to more deformation, which makes the unpulled part move relatively slowly, so the atoms in the nanoporous move more slowly, and the deformation effect of the structure makes it have more resistance. Therefore, the pull-out force appears to increase.

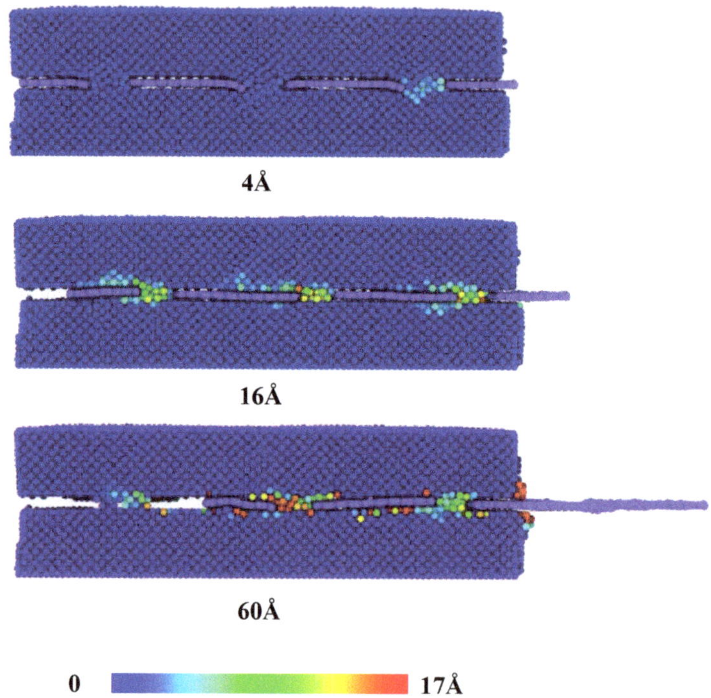

Figure 5. The displacement of copper atoms at the interface of graphene/Cu nanocomposites with various sliding distances.

As mentioned above, the deformation resistance of the nanostructures obtained by depositing copper atoms in nanoporous is the key to enhancing the interfacial mechanical properties of composites. For this purpose, the effect of the number of deposited copper atoms on the pull-out force is investigated in the present work. Figure 6 shows the maximum pull-out force change as a function of the number of deposited copper atoms. A larger maximum pull-out force can be observed as-deposited atoms onto the graphene increase. The reason for this is that more deposited atoms help fill the nanopores and form nanoislands to form diffusion bonds with the bulk copper bonding surface. The nanostructures in nanoporous can increase the deformation resistance effect, consequently leading to improved interfacial mechanical properties. However, when the deposited atoms exceed 1200, the maximum pull-out force decreases instead, mainly because the oversized nanoisland structure is not sufficiently deformed in the bonding, which makes the bonding surface not fully in contact with the graphene surface and reduces the van der Waals force interaction between copper and graphene surface.

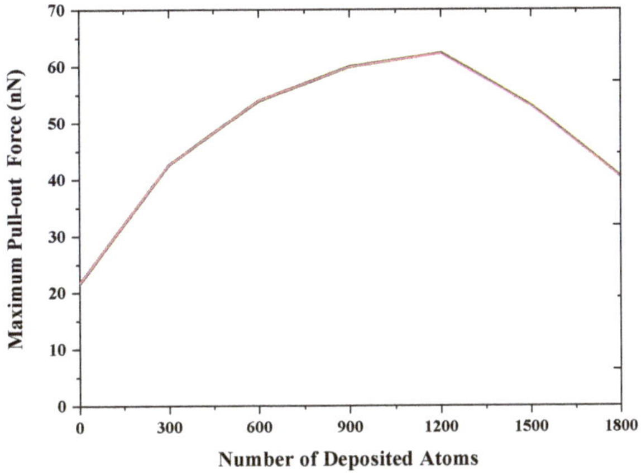

Figure 6. The maximum pull-out force changes as a function of the number of deposited copper atoms for the graphene/Cu composite.

The influence of nanoporous diameters on the reinforcing effect of graphene/Cu nanocomposites is also further studied. We have considered three different nanoporous diameters: small (3.45 Å), medium (8.07 Å), and large (11.93 Å). The number of deposited atoms is 500. As shown in Figure 7, the larger the nanoporous, the greater the maximum pull-out force. The reason is that larger nanoporous helps to form bigger and stronger nanostructures in the nanoporous, which are not deformable and require greater pull-out forces to destroy the nanostructures. However, larger holes make the effective area of graphene reduced, which may affect the ability of graphene to perform thermal and electrical transport functions.

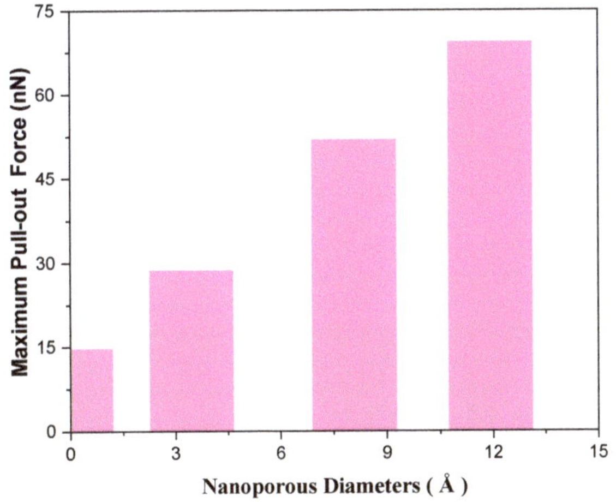

Figure 7. The maximum pull-out force changes vs. nanoporous diameters.

4. Conclusions

In this study, we report a novel Cu-nanoporous graphene-Cu based bonding technology that is of low bonding temperature and high reliability. The process of depositing copper atoms and thermocompression bonding has been numerically investigated by employing MD simulations. Numerical results show that deposition of copper atoms onto nanoporous graphene can help to generate nanoislands on the graphene surface, facilitating atomic diffusion bonding to bulk copper bonding surfaces at low temperatures. Moreover, the resistance of nanostructure in the nanoporous can dramatically improve the interfacial mechanical properties of graphene/Cu nanocomposites. It is worth mentioning that an increase in the number of deposited atoms enhances the maximum pull-out force, but too many atoms deposited will reduce the interfacial strength. While larger nanoporous helps to form bigger and stronger nanostructures in the nanoporous, which are not deformable and require greater pull-out forces to destroy the nanostructures. Decidedly, the research findings of this study provide a feasible and facile route for low-temperature metal surface bonding with a high-performance metal nanocomposites interface layer reinforced by nanoporous graphene.

Author Contributions: Conceptualization, X.S. and M.C.; methodology, X.S.; software, J.Z.; validation, R.Z., W.Z. and X.S.; formal analysis, J.Z.; investigation, X.S.; resources, R.Z.; data curation, W.Z.; writing—original draft preparation, X.S.; writing—review and editing, M.C.; visualization, W.Z.; supervision, X.S.; project administration, X.S.; funding acquisition, X.S. All authors have read and agreed to the published version of the manuscript.

Funding: This research was funded by [Zhongyuan Science and Technology Innovation Leadership Program of China] grant number [214200510014]. And The APC was funded by [Zhongyuan Science and Technology Innovation Leadership Program of China].

Institutional Review Board Statement: Not applicable.

Informed Consent Statement: Not applicable.

Data Availability Statement: The raw/processed data required to reproduce these findings cannot be shared at this time as the data also forms part of an ongoing study.

Conflicts of Interest: The authors declare no conflict of interest.

References

1. Lancaster, A.; Keswani, M. Integrated circuit packaging review with an emphasis on 3D packaging. *Integration* **2018**, *60*, 204–212. [CrossRef]
2. Han, H.; Lee, C.; Kim, Y.; Lee, J.; Kim, R.; Kim, J.; Yoo, B. Cu to Cu direct bonding at low temperature with high density defect in electrodeposited Cu. *Appl. Surf. Sci.* **2021**, *550*, 149337. [CrossRef]
3. Panigrahy, A.K.; Chen, K.-N. Low Temperature Cu–Cu Bonding Technology in Three-Dimensional Integration: An Extensive Review. *J. Electron. Packag.* **2018**, *140*, 010801. [CrossRef]
4. Wang, J.; Wang, Q.; Wu, Z.; Tan, L.; Cai, J.; Wang, D. Plasma combined self-assembled monolayer pretreatment on electroplated-Cu surface for low temperature Cu–Sn bonding in 3D integration. *Appl. Surf. Sci.* **2017**, *403*, 525–530. [CrossRef]
5. Liang, L.; Zhang, J.; Xu, Y.; Zhang, Y.; Wang, W.; Yang, J. The effect of pressure and orientation on Cu-Cu3Sn interface reliability under isothermal ageing and monotonic traction via molecular dynamics investigation. *Mater. Des.* **2018**, *149*, 194–204. [CrossRef]
6. Shie, K.-C.; Gusak, A.M.; Tu, K.N.; Chen, C. A kinetic model of copper-to-copper direct bonding under thermal compression. *J. Mater. Res. Technol.* **2021**, *15*, 2332–2344. [CrossRef]
7. Alian, A.R.; Ju, Y.; Meguid, S.A. Comprehensive atomistic modeling of copper nanowires-based surface connectors. *Mater. Des.* **2019**, *175*, 107812. [CrossRef]
8. Song, X.; Wu, S.; Zhang, R. Computational Study on Surface Bonding Based on Nanocone Arrays. *Nanomaterials* **2021**, *11*, 1369. [CrossRef]
9. Mou, Y.; Cheng, H.; Peng, Y.; Chen, M. Fabrication of reliable Cu-Cu joints by low temperature bonding isopropanol stabilized Cu nanoparticles in air. *Mater. Lett.* **2018**, *229*, 353–356. [CrossRef]
10. Zhang, S.; Xu, X.; Lin, T.; He, P. Recent advances in nano-materials for packaging of electronic devices. *J. Mater. Sci. Mater. Electron.* **2019**, *30*, 13855–13868. [CrossRef]
11. Kim, Y.; Lee, J.; Yeom, M.S.; Shin, J.W.; Kim, H.; Cui, Y.; Kysar, J.W.; Hone, J.; Jung, Y.; Jeon, S.; et al. Strengthening effect of single-atomic-layer graphene in metal–graphene nanolayered composites. *Nat. Commun.* **2013**, *4*, 2114. [CrossRef] [PubMed]
12. An, Z.; Li, J.; Kikuchi, A.; Wang, Z.; Jiang, Y.; Ono, T. Mechanically strengthened graphene-Cu composite with reduced thermal expansion towards interconnect applications. *Microsyst. Nanoeng.* **2019**, *5*, 20. [CrossRef] [PubMed]
13. Peng, W.; Sun, K. Effects of Cu/graphene interface on the mechanical properties of multilayer Cu/graphene composites. *Mech. Mater.* **2020**, *141*, 103270. [CrossRef]
14. Wang, H.; Leong, W.S.; Hu, F.; Ju, L.; Su, C.; Guo, Y.; Li, J.; Li, M.; Hu, A.; Kong, J. Low-Temperature Copper Bonding Strategy with Graphene Interlayer. *ACS Nano* **2018**, *12*, 2395–2402. [CrossRef] [PubMed]

15. Kakanakova-Georgieva, A.; Gueorguiev, G.; Sangiovanni, D.G.; Suwannaharn, N.; Ivanov, I.G.; Cora, I.; Pécz, B.; Nicotra, G.; Giannazzo, F. Nanoscale phenomena ruling deposition and intercalation of AlN at the graphene/SiC interface. *Nanoscale* **2020**, *12*, 19470–19476. [CrossRef]
16. Hidalgo-Manrique, P.; Lei, X.; Xu, R.; Zhou, M.; Kinloch, I.A.; Young, R.J. Copper/graphene composites: A review. *J. Mater. Sci.* **2019**, *54*, 12236–12289. [CrossRef]
17. Montazeri, A.; Panahi, B. MD-based estimates of enhanced load transfer in graphene/metal nanocomposites through Ni coating. *Appl. Surf. Sci.* **2018**, *457*, 1072–1080. [CrossRef]
18. Zhao, S.; Zhang, Y.; Yang, J.; Kitipornchai, S. Significantly improved interfacial shear strength in graphene/copper nanocomposite via wrinkles and functionalization: A molecular dynamics study. *Carbon N. Y.* **2021**, *174*, 335–344. [CrossRef]
19. Hickman, J.; Mishin, Y. Temperature fluctuations in canonical systems: Insights from molecular dynamics simulations. *Phys. Rev. B* **2016**, *94*, 184311. [CrossRef]
20. Hao, H.; Lau, D. Atomistic modeling of metallic thin films by modified embedded atom method. *Appl. Surf. Sci.* **2017**, *422*, 1139–1146. [CrossRef]
21. Long, X.J.; Li, B.; Wang, L.; Huang, J.; Zhu, J.; Luo, S. Shock response of Cu/graphene nanolayered composites. *Carbon N. Y.* **2016**, *103*, 457–463. [CrossRef]
22. Plimpton, S. Fast Parallel Algorithms for Short-Range Molecular Dynamics. *J. Comput. Phys.* **1995**, *117*, 1–19. [CrossRef]
23. Stukowski, A. Visualization and analysis of atomistic simulation data with OVITO–the Open Visualization Tool. *Model. Simul. Mater. Sci. Eng.* **2009**, *18*, 15012. [CrossRef]

nanomaterials

MDPI

Review

Multiscale Mechanical Performance of Wood: From Nano- to Macro-Scale across Structure Hierarchy and Size Effects

Yuri I. Golovin [1,2], Alexander A. Gusev [1,3,4,*], Dmitry Yu. Golovin [1], Sergey M. Matveev [3] and Inna A. Vasyukova [1]

1 Institute "Nanotechnology and Nanomaterials", G.R. Derzhavin Tambov State University, 392000 Tambov, Russia; yugolovin@yandex.ru (Y.I.G.); tarlin@yandex.ru (D.Y.G.); vasyukovaia@gmail.com (I.A.V.)
2 Department of Chemical Enzymology, School of Chemistry, Lomonosov Moscow State University, 119991 Moscow, Russia
3 Research and Educational Center "Sustainable Development of the Forest Complex", Voronezh State Forestry University Named after G.F. Morozov, 394087 Voronezh, Russia; lisovod@bk.ru
4 Department of Functional Nanosystems and High-Temperature Materials, National University of Science and Technology "MISIS", 119991 Moscow, Russia
* Correspondence: nanosecurity@mail.ru; Tel.: +7-91-0756-4546

Abstract: This review describes methods and results of studying the mechanical properties of wood at all scales: from nano- to macro-scale. The connection between the mechanical properties of material and its structure at all these levels is explored. It is shown that the existing size effects in the mechanical properties of wood, in a range of the characteristic sizes of the structure of about six orders of magnitude, correspond to the empirical Hall-Petch relation. This "law" was revealed more than 60 years ago in metals and alloys and later in other materials. The nature, as well as the particular type of the size dependences in different classes of materials can vary, but the general trend, "the smaller the stronger", remains true both for wood and for other cellulose-containing materials. The possible mechanisms of the size effects in wood are being discussed. The correlations between the mechanical and thermophysical properties of wood are described. Several examples are used to demonstrate the possibility to forecast the macromechanical properties of wood by means of contactless thermographic express methods based on measuring temperature diffusivity. The research technique for dendrochronological and dendroclimatological studies by means of the analysis of microhardness and Young's modulus radial dependences in annual growth rings is described.

Keywords: wood; nano-, micro-, meso-, and macro-structure; multiscale mechanical properties; size effects; Hall-Petch law; dendrochronology

Citation: Golovin, Y.I.; Gusev, A.A.; Golovin, D.Y.; Matveev, S.M.; Vasyukova, I.A. Multiscale Mechanical Performance of Wood: From Nano- to Macro-Scale across Structure Hierarchy and Size Effects. *Nanomaterials* **2022**, *12*, 1139. https://doi.org/10.3390/nano12071139

Academic Editors: Mohammad Malikan, Shahriar Dastjerdi and Takuya Kitaoka

Received: 2 March 2022
Accepted: 24 March 2022
Published: 29 March 2022

1. Introduction

Interest in wood and other cellulose-containing materials, composites in particular, had considerably increased by the beginning of the 21st century. The studies of wood nano- and micro-structures have especially intensified in the last decade (Figure 1) [1–5]. Several reasons can be named. Mineral resources (especially various metallic and nonmetallic materials, coal, oil, and natural gas) are being extracted at continually rising rates, and open-cycle processing technologies create ever-growing volumes of industrial and household waste. This poses a threat to the biosphere because of the environmental pollution and increased carbon dioxide concentration in the atmosphere, while only a fraction of the manufactured materials are recycled and reused. The situation is aggravated by a sharp increase in polymer material manufacturing for packaging, which are seldom recycled and mostly non-biodegradable. The surging pressure on the environment requires more and more efforts for its neutralization.

In this regard, a wider use of biogenic materials as well as substituting them for traditional ones seems a promising step. Such cellulose-containing substances as modified wood and various agricultural vegetable wastes, and especially the nano- and micro-cellulose they contain, offer the best potential for numerous applications.

Cellulose is the most common natural linear polymer polysaccharide $(C_6H_{10}O_5)_n$ in the biosphere. The materials formed on its basis provide vast advantages:

1. Unlike the majority of extracted mineral resources, cellulose-containing materials have sustainable and renewable sources, namely forests, field crops, and aqua cultures [1,3–7];
2. These materials are multifunctional; they can be used in construction and industrial manufacturing [5,8], for producing cardboard, paper, packaging [9–12], and textile goods [13,14], in electronics [15], photonics [16], and energetics [17,18], in environmental remediation and wastewater treatment [5,19–22], medicine [23–27], military [28] and household applications, and in many other spheres [1,3–5,29];
3. Wood, cellulose-containing plant materials, and bio-composites are gaining more and more popularity each year. Among their most attractive features we should name their environmental friendliness, biodegradability, after-service "self-destruction" that leaves no toxic products [1–5], and their ability to be modified [30];
4. These materials are perfect for creating a closed carbon cycle, which does not increase the carbon dioxide content in the atmosphere [1–3,6], and it is a well known fact that this gas contributes to the greenhouse effect and to the average annual temperature growth;
5. Nano- and microstructural components in the wood structure (nanocrystals, nanofibrils, cellulose microfibers) possess mechanical properties (tensile strength σ_b, Young's modulus E, etc.) comparable with, and even exceeding the same properties of such high strength construction materials as steels, titanium, and aluminum-based alloys. Additionally, if we take into consideration their lower density ρ (~1.5 g/cm^3 in nanocellulose vs. ~8 g/cm^3 in steels, ~4.5 g/cm^3 in titanium-based, and ~2.8 g/cm^3 in aluminum-based alloys), then we discover that the specific values of strength σ_b/ρ and stiffness E/ρ of nano-/microcellulose can exceed manifold those of steels and alloys;
6. Finally, they are manufacturable, non-toxic, and comparatively inexpensive.

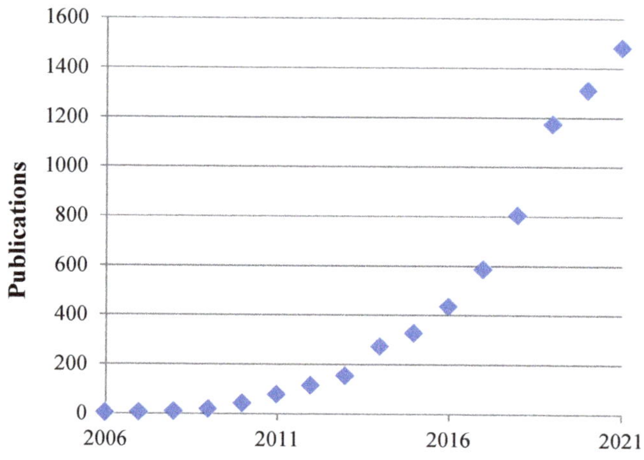

Figure 1. Growth dynamics of the number of research papers on the structure, properties, and applications of the materials comprising natural nano- and micro-fibers. Data obtained from Scopus using the following search parameters "TITLE-ABS-KEY" with keywords «Nanocellulose or micro-cellulose» on 28 February 2022.

Certainly, wood and other cellulose-containing materials have several disadvantages. They require special treatment, as they are flammable and hygroscopic. High humidity makes them lose some of their strength properties, while low humidity causes deformation.

They succumb to rot and unwanted biodegradation. Besides, in their original, state the mechanical properties in every sort of wood strongly depend on the conditions of its growth, usage, and testing humidity, the structure of cell walls and annual growth rings, proportion of young and mature wood, stress condition, size of the sample or stressed area, and also the direction, rate, and duration of load application. The aforementioned considerations have obstructed identifying the universal patterns that form the mechanical properties of different species of wood. Nevertheless, some generalizations can be derived from the literature and from accumulated experience, as outlined below.

In the present review, we explore the methods and results of a multiscale study of the mechanical properties of various wood species, in connection with peculiarities of their nano-, micro-, and meso-scale structural levels of material organization. The analysis of literature data shows that, in a huge range of characteristic sizes of the structural units (about six orders of magnitude), mechanical properties of wood generally follow the Hall-Petch relation, which is well known in material science. The possibilities for non-destructive assessment of the mechanical properties of wood by means of contactless measurement of the temperature diffusivity tensor components are discussed, as well as using the scanning nanoindentation method for evaluating woods' micromechanical characteristics, in order to obtain dendrochronological and dendroclimatological data. The main scopes of the review are presented in Figure 2.

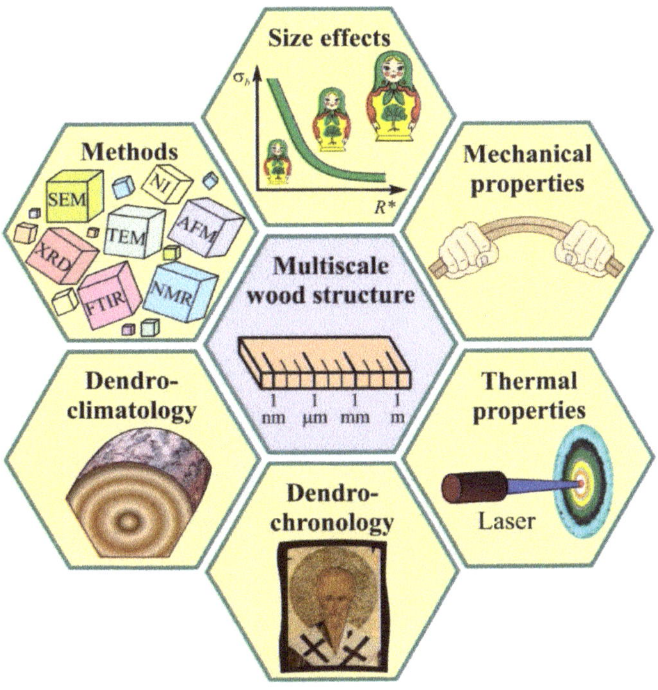

Figure 2. The main scopes of the review.

2. The Hierarchical Structure of Wood

From the point of view of material science, wood is a hierarchically organized natural composite with a complicated structure and a clear heterogeneity and anisotropy of all its properties, as well as an ability to regenerate [5,31–35]. In the wood architecture, one can distinguish, though only tenuously, several size and hierarchical levels (Figure 3), namely atomic–molecular, nano- (nanocrystals, nanofibrils), micro- (microfibers, cell walls), meso- (cells, large vessels, radial rays), and macro-level (annual growth rings, macroscopic structural defects, cracks, etc.) [31–33,35]. They all contribute to forming the complex of physical, chemical, and mechanical properties [31,36]. A large range of the characteristic size of the structural components of wood (about six to eight orders of magnitude) and a wide scope of tasks and questions emerging from the study of this material all require a varied arsenal of research techniques and means to implement them. They will be briefly analyzed in the following section.

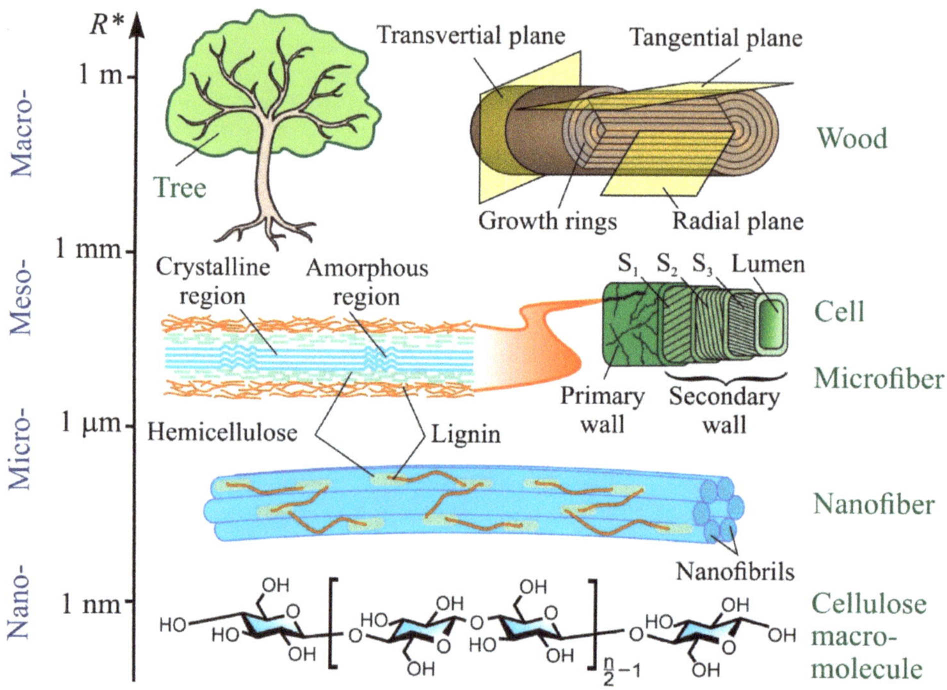

Figure 3. Scale hierarchy in wood structure and its main components. R^* is the characteristic size.

Identification of patterns in the formation of the macro-properties of wood, as derived from its nano-, micro, meso-, and macrostructure, is the most important task in wood science. There are many reasons for the interest in the relations between macromechanical properties of wood and its nano- and microstructures, as well as physical characteristics, thermal characteristics in particular. Let us enumerate the most important ones. Firstly, the relevant patterns help to elucidate the nature and mechanisms of formation of the parameters most significant for practical applications of wood in the macroscale, i.e., its mechanical and thermal properties. Secondly, nanomechanical strength properties, being much higher than those at the micro- and macro-scale, indicate the potential for strengthening, which may approach the ultimate tensile strength of nanocrystalline cellulose (~10 GPa). Thirdly, the increased use of composite and nanocomposite materials in different spheres of engineering, construction, biochemical technologies, and medicine paves the way to replacing traditional metals and alloys with more lightweight and ecologically friendly composites. For example, the bodies of the most recent Boeing and Airbus airplane models consist, by weight, of more than 50% of fiber-reinforced composites. Their popularity in the auto industry, shipbuilding, sports equipment manufacturing, etc. is growing fast. However, glass and basalt fibers used for composite reinforcement, not to mention carbon micro- and nano-fibers, have some adverse properties from the ecological point of view; they are quite expensive and still unable to conquer the wide market for consumer goods. Cellulose fiber is by about an order of magnitude less expensive than fibrous glass while having almost the same mechanical characteristics. Therefore, it is important to understand the nature of strength and damage mechanisms in microcellulose fibers, and to find approaches to improve their strength, thus enhancing the properties of textiles, non-woven materials, and the composites they are used to reinforce. When correlations are revealed between the mechanical characteristics and other physical properties, for example thermal properties, this information will be of great use for developing non-destructive contactless thermophysical methods for evaluating mechanical characteristics, instead of applying labor-intensive destructive techniques. Fourthly and finally, many tree species have a lifespan of several hundred or even thousands of years, with sequoia as an example. In their nano-, micro-, meso-, and macro-structure they accumulate a vast amount of information about the climatic conditions during their growth and about ecological catastrophes they have witnessed. The variations in composition and structure are inevitably reflected in the local physico-mechanical properties of wood. This natural archive can serve as a valuable source of information for climatology and for dating various events in earth's history (dendrochronology).

3. Methods of Studying the Structural and Mechanical Properties of Wood at Various Levels of Scale

The aim of the classic wood science is to discover and describe the dependence of woods' macromechanical, physico-chemical, and service properties on its inner structural characteristics, humidity, and external thermodynamic factors [6,37,38]. Wood type classification and its grading, according to mechanical properties, is an important pragmatic task [39]. Since the end of the last century, more and more attention is being paid to the fine structure of wood at the nanoscale. This interest was brought forward, on the one hand, by the growth in nanotechnology and nanometrology, and on the other hand by realization of what untapped resources are hidden at the nanoscale.

In the recent 15–20 years, numerous modern methods and means traditionally used in solid state physics and material science are being applied for studying the micro-structure and physico-chemical properties of wood [2,6,33,35,40].

Micro-structure is studied by means of transmission and scanning electron microscopy, scanning probes (mostly atomic force), confocal laser, and optical microscopy in various modes. Numerous X-ray methods are used to determine the composition and the parameters of atomic- and micro-structures. The character and degree of order of cellulose molecules in nano-fibers, the angle between the micro-fibers and the long axis of the cell, are determined by X-ray diffractometry and microtomography, as well as small-angle (SAXS) and wide-angle (WAXS) X-ray scattering. Elemental and molecular composition is revealed by spectroscopic methods, such as X-ray fluorescence, various types of spectroscopy such as infrared (IR), Fourier transform IR (FTIR), Raman, Brillouin, nuclear magnetic resonance (NMR), and other analytical methods. Together, they cover a huge spatiotemporal range of structures and events in them, namely more than twelve orders of magnitude in time and about eight orders of magnitude in length (Figure 4) [32]. The comparative analysis of the most widely employed physical methods for studying the molecular, sub-cellular, and cellular structures of wood can be found in most recent reviews [32–34,41].

To study mechanical properties at the nano- and micro-scale, a number of nano-/micro-mechanical testing (SSMT—small scale mechanical testing) methods [42–45] are employed. Atomic force microscopy (AFM) [46–49] and nano-indentometry (NI) [50–59] can be named as the most widely used ones.

They have similar structure flowcharts (see Figure 5) and capabilities [45,50,52]. In both cases, a high precision driver brings a probe, with the radius of its curve being from a few (in AFM) to a few tens (in NI) of nanometers, close to the studied surface and the probe starts interacting with it. The force P and penetration depth h of the probe are measured continuously, and their alteration kinetics are registered throughout the testing cycle (Figure 6a). Most commonly a P–h diagram (similar to a σ–ε diagram created during macro-testing) is built using the obtained data (Figure 6b), and standardized algorithms are applied to calculate about ten various mechanical characteristics, such as Young's modulus, contact stiffness, hardness, fracture toughness, creep rate, etc., at the nano-/micro-scale. In NI, a three-sided diamond Berkovich tip is used, as it is better calibrated from the point of view of the real tip geometry than the one used in AFM, thus providing more accurate and reliable quantitative data. Among the variety of proposed techniques for mechanical characteristics extraction from raw data, the method proposed and developed by Oliver and Pharr [60–62] has become the most widely used and has been incorporated into ISO standard [63], so that this method has been used for processing all NI experiments described in this review.

SSMT methods were used to examine the mechanical properties of individual cellulose nanofibrils and microfibers [64–68], cell walls [32,69–75], layers of early and late wood in annual growth rings [74], and to obtain plenty of other interesting data. However, there are very few papers that compare and analyze several scale levels at once [76–79]. Thus, connection between the properties of individual nano- and micro-structural elements of wood and their influence upon macro-mechanical characteristics cannot yet be traced.

The analysis of the structure and role of annual growth rings in shaping wood macro-properties requires, at the very least, one-dimensional, or better yet, two-dimensional scanning of certain physical characteristics. Three-dimensional imaging can be applied as well. Dating archaeological finds, works of art, climate changes and events, and ecological catastrophes based on the changes in growth ring structure and width is a separate issue. These approaches are known as dendrochronology, dendroecology, and dendroclimatology, respectively. The width of rings, proportion of early wood (EW) to late wood (LW), and changes in their morphology reflect the specific growing conditions each long-lived plant witnessed during each vegetation season.

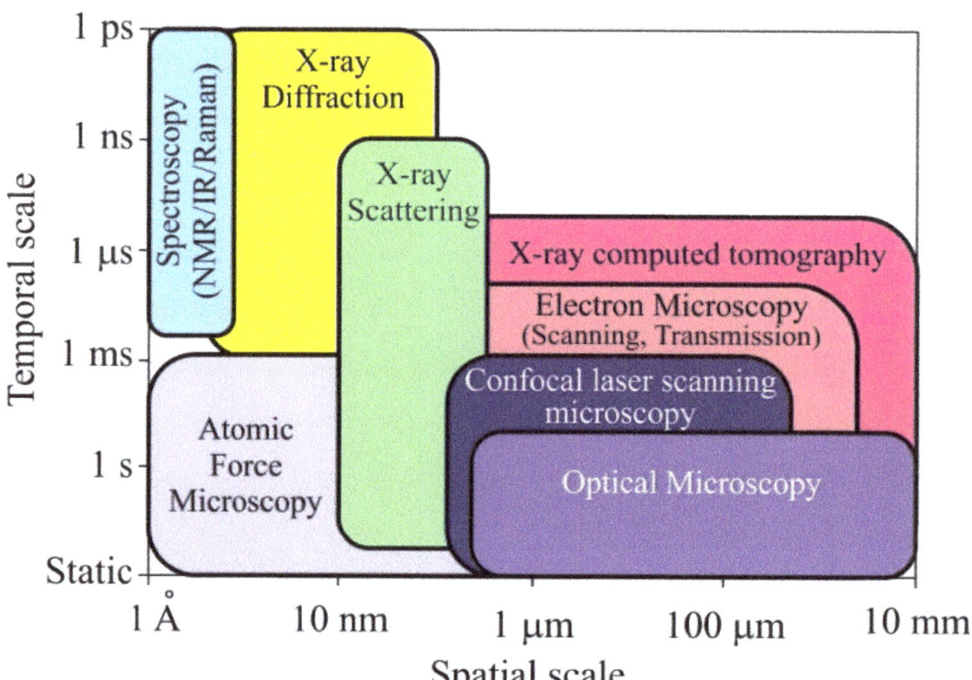

Figure 4. Schematic of the spatiotemporal ranges of the most popular physical methods for studying wood structures. Adapted with permission from Ref. [32]. Copyright 2021, Wiley-VCH.

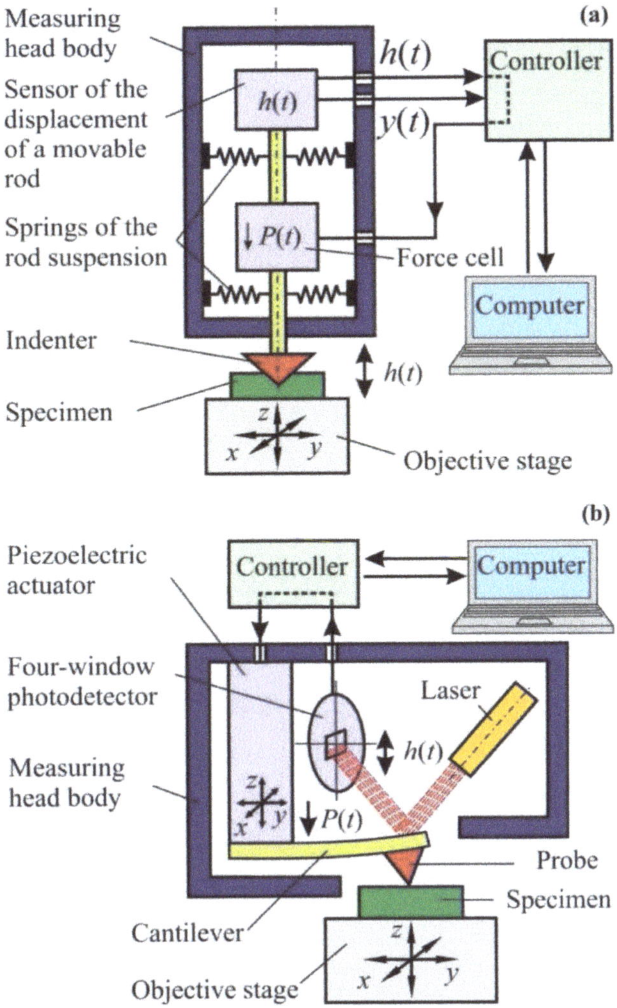

Figure 5. Schematic diagrams of (**a**) the nano-indentometer and (**b**) the atomic-force microscope.

While microstructure and physico-chemical properties of wood are studied with elaborate modern equipment, examination and analysis of annual growth rings for dendrochronological and dendroclimatological applications is carried out using simple optical methods, where primary information is derived from the difference in reflectivity between EW and LW. Quite often, the same approach is employed while assessing wood strength and other service properties. A detailed description of dendrochronological methods developed by the beginning of this century is given in [80]. These methods reveal only geometrical and morphological characteristics of the studied object (annual growth rings width, proportion of EW and LW in them, variations from ring to ring, etc.) and allow for comparison between the data obtained by different methods.

Numerous attempts have been made to improve traditional dendrochronological methods, mostly by modifications introduced to the sample preparation techniques, staining, use of blue light instead of white, application of computer vision technologies, and mathematical data processing (see e.g., [81–88]). However, despite this, the capabilities of the approach based on the analysis of transversal section images and photodensitometry remain severely limited, as the reflective optical properties of wood are variable and their connection with other wood characteristics, such as mechanical and thermal, are either ambiguous or very weak.

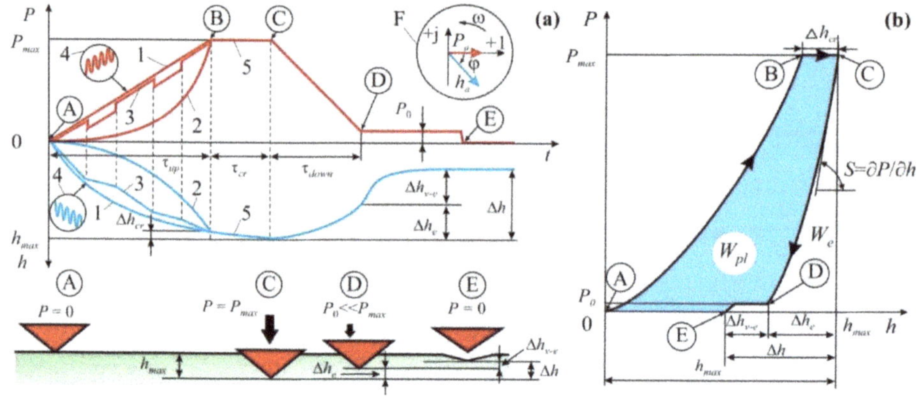

Figure 6. Two methods of representing the data obtained by means of normal nanoindentation: (**a**) as kinetic curves $P(t)$ and $h(t)$; (**b**) as a $P-h$-diagram (P—force, h—indenter displacement). Adapted with permission from Ref. [45]. Copyright 2021, Springer Nature. The circled letters (from A to E) mark the characteristic points on the loading curves and the indenter position relative to the sample surface. Inset F shows the vector diagram depicting correlation in the complex plane between the vectors of oscillating force and the resulting indenter displacement in the CSM method. Five loading regimes are marked from 1 to 5. The indices at P and h mean as follows: *up*—increase; *cr*—creep; *down*—drop; *e*—elasticity; *v−e*—viscoelasticity; *max*—maximum value; W_e—elastic energy; W_{pl}—energy absorbed and dissipated by the sample in a single load–unload cycle.

In order to expand the capabilities of the analysis of mechanical properties in their connection with the architecture of wood ring structures, the following methods have been used: two-dimensional mapping of properties on cross-sections of tree trunks by AFM methods [89–92] and NI [93] scanning, 3D X-ray [94,95] and NMR tomography [96], and synchrotron-based X-ray microscopy [97]. However, these methods are complicated, labor-intensive and require expensive or unique equipment; therefore, they are used only sporadically. The method of X-ray densitometry [98] presents fewer difficulties, but it requires access from both sides to a perfectly flat sample cut exactly perpendicular to the long axis of wood cells.

It should be noted that the mechanical properties of wood and cellulose-containing materials show a significant dependence on the rate of monotonous loading, oscillating load frequency, and duration of load application. They can vary between samples and change over time in a significantly greater range than in similarly structured technogenic composites (e.g., in glass and carbon fiber-reinforced plastics) [2,5,40,41]. Such variability of properties makes identification of common regularities in their formation even more challenging.

To sum up, we should mention that, thus far, the overall links between the properties of all the scales and hierarchical levels—from cellulose nanocrystals (CN) to macro-samples—require additional study. However, there is a considerable volume of information on every individual level. The following sections present the examples of the most representative data from the lowest to the highest scale levels of the structure.

4. Nanocellulose and Elementary Nanofibrils

Cellulose is the most common natural polymer and the major structural component that provides strength to wood and stems of grass, reed, bamboo, etc. [5,40]. Cellulose is a macro-molecular polysaccharide $(C_6H_{10}O_5)_n$, consisting of linear chains of several tens to many hundred n of β-(1→4) linked glucose molecules (Figure 2). In its origin, cellulose can belong to three types: plant, regenerated, and bacterial [5]. The current state of affairs in extraction and functionalization of cellulose nanofibers is described in Handbook [5] and in the most recent reviews [99–102].

Cellulose molecules easily form nanocrystals with a lateral size of 3–10 nm and being 100–300 nm long. These nanocrystals form nanofibrils 5–20 nm in diameter and up to many hundreds of nanometers long. Inside, nanofibril cellulose is present in an amorphous–crystalline state as a series of alternating domains. The amorphous phase, to some extent, reduces the strength of the nanofibril, but makes it more supple and elastic. The most typical structural characteristics of nanocellulose-based formations are presented in Table 1 [67].

Table 1. Structural characteristics of cellulose nanocrystals and individual nanofibrils (CNC and CNF, respectively) [67]. Copyright 2017, Wiley books.

Characteristics	CNCs	CNFs
Length of nanoparticles (nm)	100–500	$\geq 10^3$
Lateral size of nanoparticles (nm)	5–30	10–40
Aspect ratio of nanoparticles	10–50	60–100
Length of crystallites (nm)	70–200	60–150
Lateral size of crystallites (nm)	5–10	3–7
Crystallinity (%)	72–80	50–65
Amorphicity (%)	20–28	35–50
Specific gravity (g cm^{-3})	1.57–1.59	1.54–1.56
Specific volume (cm^3 g^{-1})	0.63–0.64	0.64–0.65
Porosity (cm^3 g^{-1})	0.01–0.05	0.1–0.2

Mechanical properties of nanocrystalline cellulose have been characterized by various methods, including calculations of bond strength inside macro-molecules and between them, computer generated simulations, processing data from IR and Raman spectroscopy, AFM, WAXS, and others [5,40,65,68]. A brief overview of the mechanical characteristics of nanocellulose is given in Table 2 [67]. The variability of data is explained by specific characteristics of the calculation schemes, models, raw data processing algorithms, and also by the difficulty of carrying out direct measurements at the nanoscale. The differences in age, structure, and origin of wood affect the experimental results as well. Besides, the mechanical properties of nanocellulose samples depend on their size significantly. For instance, transversal Young modulus reduction by a factor of 1.6 has been reported in [78] for increasing NC diameter from 2.5 nm to 6.5 nm.

Table 2. Mechanical characteristics of cellulose nanocrystals and individual nanofibrils (CNC and CNF, respectively) [67]. Copyright 2017, Wiley books.

Characteristics	CNCs	CNFs
Modulus axial (GPa)	140–160	30–40
Modulus transversal (GPa)	15–30	10–15
Tensile strength axial (GPa)	8–10	0.8–1
Tensile strength transversal (GPa)	About 1	About 0.1

5. Cellulose Microfibers

The typical hierarchy of wood structure at higher levels continues with nano- and micro-fibers. They are formed by elementary nanofibrils, mainly due to hydrogen bonds. Nanofibrils form strands surrounded by a matrix composed of lignin (an aromatic polymer polyphenol), hemicellulose (low molecular weight branched polysaccharide), pectin (gel-forming polysaccharide), and water [5,40,68]. Cellulose content in the fibers can vary in a wide range. For example, it is 40–60% in the wood fibers of various species and can exceed 96% in cotton fibers [5,23,24,31,68].

Nano- and micro-structures of cellulose materials and their properties strongly depend upon the specifics of interaction between nanocrystals in elementary fibrils and the ordering and binding of the latter in nano- and micro-fibers [103–107]. Mechanical, strength in particular, properties of cellulose nano- and micro-structures are structure sensitive, just as those of most other other organic and non-organic materials. In turn, their morphology and inner structure depend upon plant species, their growth conditions, and cellulose extraction technology. The dominant role in determining fiber properties belongs to the cellulose content in nanofiber, the degree of its crystallinity, and specifics of nanofiber binding at the material. The angle between the nanofibril axis and nanofiber or cell axis has significant impact too. A comprehensive review [104] contains various data concerning morphology, microstructure, and mechanical properties of micro-fibers of various origin (Figure 7) and examples of their application for polymer composite reinforcement.

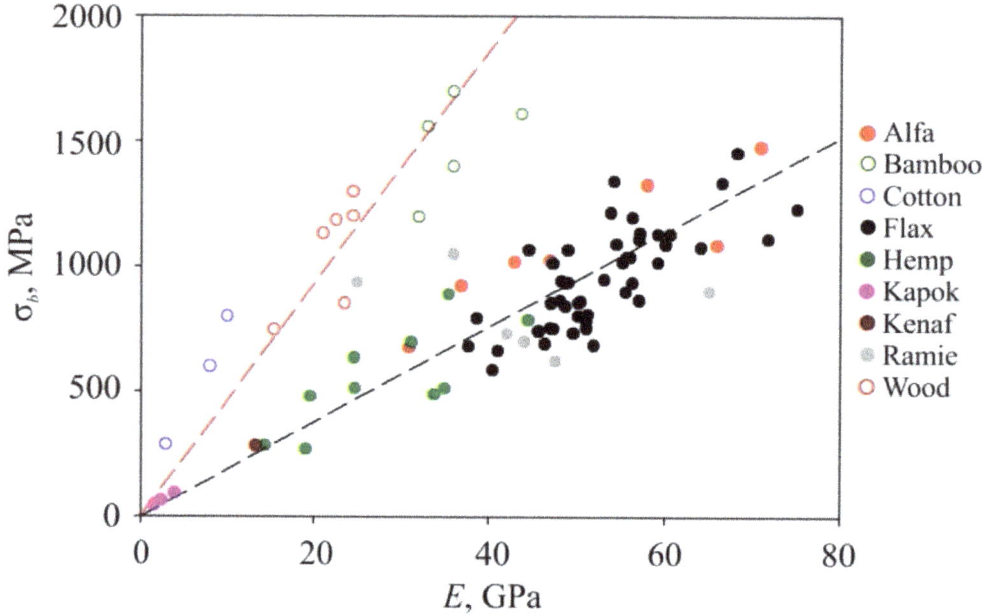

Figure 7. Dependence of cellulose fiber tensile strength upon Young modulus for various plant materials. Adapted with permission from Ref. [104]. Copyright 2018, Elsevier.

The strongest of the studied micro-cellulose fibrils have demonstrated Young modulus E = 75–85 GPa and tensile strength σ_b = 1.6–1.7 GPa, so that the ratio $E/\sigma_b \approx 50$. One of the possible techniques allowing researchers to reach such high mechanical properties is described in [66]. The authors have used the efficient technique of double hydrodynamic ordering of nanocrystals and nanofibrils to produce the fibers with diameter 6–8 μm. Their tensile strength reached 1.1 GPa. Nanofibril cross-linking has increased the fibers' strength up to σ_b = 1.57 GPa.

As follows from fundamental considerations, the theoretical strength of any defect-free material can reach 0.1 E, while the strongest micro-cellulose fibers mentioned above have values around 0.015–0.020 E. Hence, even the strongest studied micro-fibers have the potential of increasing their strength by 3–5 times.

It should be mentioned that the data concerning cellulose micro-fibers mechanical properties differ significantly depending on the measurement technique (see Table 3) [104]. Results obtained using AFM and NI are in agreement with each other regarding the measurement accuracy, despite using different probes and measurement techniques, so that they are just as reliable as the ones obtained using the undebatable method of uniaxial tension. Usually, its tensile strength is two to three times higher than compression strength or hardness [5,40,104,106,107], unlike void-free materials, where the reversed value is quite typical. For example, the Tabor ratio for metals is well known, where hardness is three times higher than the yield stress. We suppose that this difference is due to specific nanofibril behavior when subjected to tensile and compressive stress or in hardness measurements. In the first case, the molecular chains are strained and partially oriented along the fiber axis, which increases their strength. Indentation, on the other hand, used both in NI and AFM, promotes the arising of compression stress that causes micro-fibril bending, and micro-fibril buckling failure, which occurs earlier than their failure in uniaxial tension.

Table 3. Microfibers Young moduli E obtained using different methods for three materials [104].

Sample	Elementary Fibre Tensile Modulus (GPa)	Nanoindentation Modulus (GPa)	AFM Mapping Modulus (GPa)
Eden flax	68.9 ± 24.6	20.4 ± 1.1	21.3 ± 2.2
Bamboo	43.6 ± 0.6	21.3 ± 1.7	21.3 ± 2.9
Tension wood	18–40	14–20	11

For many applications such as aviation, space aeronautics, the automotive industry, sport equipment etc., the most important mechanical characteristics are not absolute but specific ones, i.e., normalized on material density ρ. Figure 8 shows the absolute σ_b, specific strength σ_b/ρ, and Young modulus E for highly oriented cellulose nano- and micro-fibers when compared to those for macroscopic wood and other materials. As it can be seen, the specific strength of defect-free nanocellulose can be manifold higher than that of aluminum or titanium alloys or constructional steels. However, nano- and micro-cellulose materials are inferior to metals at thermal and crack resistance, failure deformation, and other related energy characteristics. Only some polymers such as kevlar and carbon microfibers can contest nanocellulose at specific strengths. Single-wall carbon nanotubes and graphene are manifold superior at specific strength to any other known material.

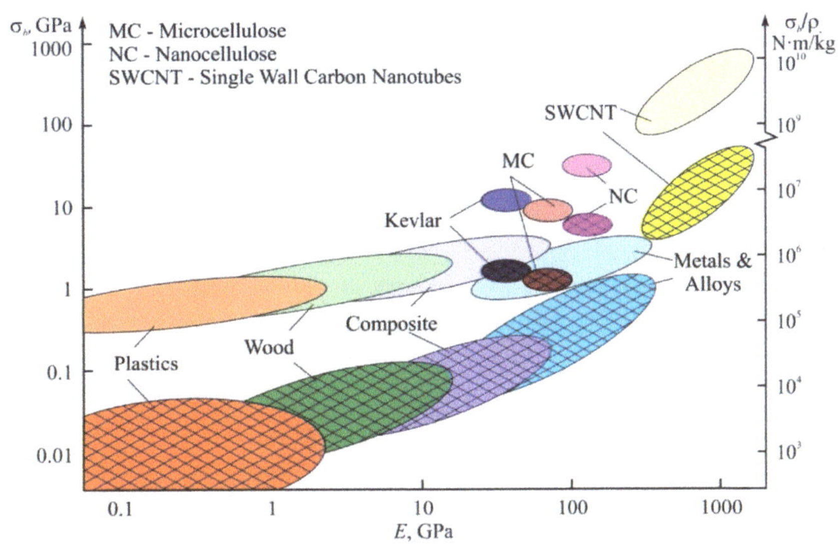

Figure 8. Mechanical properties of nanocellulose (NC) and cellulose microfibers (MC) in comparison to common and perspective constructional materials. Crosshatched areas denote absolute values of tensile strength σ_b, while non-crosshatched ones are the strength normalized over material density ρ.

Lastly, let us mention one more advantage of natural cellulose fibers produced from the wood. It is several times cheaper than the flax fibers and an order of magnitude cheaper than ecologically unsafe glass fibers widely used in composite reinforcement applications [104]. Highly ordered cellulose microfibers have nearly the same strength as glass fibers already and have a good prospects of further strength increases.

6. Cells and Cell Walls

While a tree grows, cellulose microfibers integrating with other components such as lignin, hemicellulose, pectin, water, etc., form walls of cells that are highly elongated in the direction of tree trunk axis. Several layers are discerned within cell wall including primary wall and multilayered secondary walls that usually consists of three layers, named S1, S2, and S3, which differ in the angle μ between cellulose microfibers and cell long axis. The secondary wall provides the main contribution to cell stiffness and mechanical strength. Cell size diminishes, cell wall width increases, and the cross-section of internal capillary reduces while going from early wood (EW) that is a part of the annual ring formed at the first stage of vegetation, to late wood (LW), formed at the second stage.

To study mechanical properties of wood cells, various SSMT methods are used [45], and the most widespread ones are AFM [89,92] and NI [70,71,74,75,93]. Let us present some typical examples of NI application with load P_{max} = 0.1–1 mN to this problem. The authors of [74] studied radial dependence of cell wall longitude Young modulus E_{NI} and nanohardness H_{NI} in two annual rings of common pine (*Pinus sylvestris* L.) wood, corresponding to the ages of 7 and 74 years. As could be seen at Figure 9a, in going from EW to LW, E_{NI} increases by nearly 50%, while H_{NI} (Figure 9b) increases by just 5–7%.

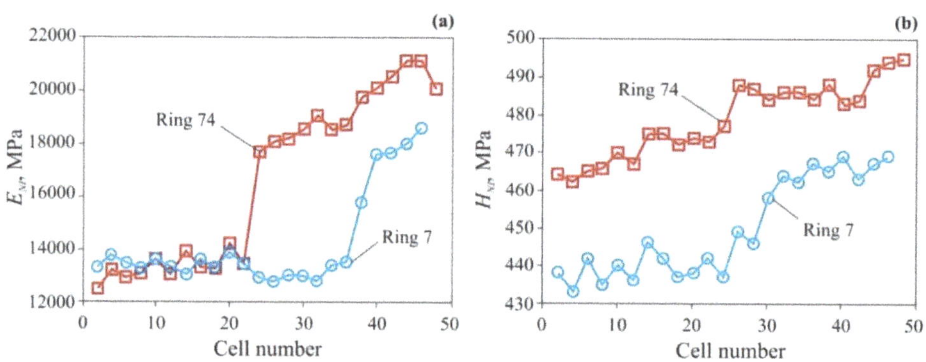

Figure 9. Dependencies of cell wall longitude Young modulus E_{NI} (**a**) and nanohardness H_{NI} (**b**) upon cell sequential number in the annual growth ring for two rings with ages of 7 and 74 years. Adapted with permission from Ref. [74]. Copyright 2020, PAN.

A number of other papers report similar data supporting that cell wall nanohardness varies not too much at different layers, rings, or even tree species. For instance, the following results are reported: H_{NI} = 0.35–0.42 GPa for *Pinus massoniana* Lamb. In [107], H_{NI} = 0.41–0.53 GPa for *Masson pine*, coinciding within the measurement accuracy for EW and LW in [108] and H_{NI} = 0.34–0.54 GPa for *Pinus taeda*, not discerning EW and LW, in [109]. Similar results are reported for H_{NI} in cell walls junction through the middle lamella for *Norway spruce*. Nanohardness in the cell corner middle lamella was estimated to be 0.34 ± 0.16 GPa [110].

The NI technique allows more detailed measuring cell wall elastic properties and determining the main components of elasticity tensor. So, the measured value of Young modulus of secondary wall S2 has been reported to be 26.3 GPa in the longitudinal direction and 4.5 GPa in the lateral one [111].

The most informative experiments are carried out in situ in an electron microscope column using a sharp indenter or flat piston [112]. Simultaneous recording of loading diagrams and obtaining visual information concerning the nano-/micro-structure evolution allow the studying of the micro-mechanisms of deformation and failure [113].

There are a number of papers studying the mechanical properties of micro-pillars cut from cell walls by focused ion beams (FIB). The pillars were subjected to uniaxial compression in a scanning electron microscope column (SEM) [114,115]. It allows simultaneous recording of σ-ε diagram and specific features of micro-pillar deformation.

Such works are not numerous due to high labor content and rather complex and expensive equipment required for sample preparation. However, they provide direct confirmation of quantitative information obtained using NI, allow for obtaining unique information concerning various deformation modes, buckling, and failure mechanisms of cell walls, and verification of various behavior models of wood hierarchic structures under load.

7. Annual Growth Rings

As follows from the data presented in the section above, cell wall Young modulus E_{NI} does not vary significantly with the indent location, be it late or early wood layer, secondary wall, or cell conjugation region with middle lamella. The ring age, weather conditions during its formation, or particular tree species does not change it more than by a factor of 1.5–2. Nanohardness H_{NI} dependence upon these factors is even weaker. Nevertheless, macromechanical properties woods of differing origin can differ manifold reaching up to an order of magnitude or even more. Evidently, this weak correlation between nano- and macro-properties is due to the difference in cell wall thickness, the relative share of late wood, and the number of large tracheides and other wood structure elements which reduce wood macroscopic strength. To close the gap between nano- and macro- scale mechanical properties, the nanoindentation tests were carried out at loads ranging from 5 to 500 mN and reported in the set of papers [116–119], unlike the 0.1–1 mN range usually used in studying cell walls. It extended the deformed region over the whole cell or several cells up to 50–150 μm, as opposed to precise targeting at the cell wall.

The values of Young modulus E_{eff} and microhardness H_{eff} obtained this way can be considered as effective, as long as, being the result of averaging over the cells structure, they incorporate not only mechanical properties of cell walls but also their thickness, material porosity, and microdefects, just as in macroscopic mechanical tests. However, indentation size was at least an order of magnitude less than annual ring width. It allowed obtaining E_{eff} and H_{eff} spatial distributions across several annual rings.

Typical E_{eff} and H_{eff} radial dependencies measured at cross-sections of common pine (*Pinus sylvestris* L.), which represent coniferous trees, are shown at Figure 10, and pedunculate oak (*Quercus robur* L.), which represent hardwood trees, are shown at Figure 11 [116,117]. As could be seen, both species manifest pronounced periodicity of local mechanical properties. Positions of abrupt changes in E_{eff} and H_{eff} coincide with annual growth ring boundaries determined by wood color change using the standard optical method. Changes of E_{eff} and H_{eff} in transition from EW to LW within each annual ring are gradual in pine and abrupt in oak. The linear dependence between E_{eff} and H_{eff} with almost the same slope $m = 0.017 \pm 0.002$ has been observed (see Figures 10b and 11b). In other words, the ductility ratio DR = E_{eff}/H_{eff} is found to be around 60 for both tree species. This value is quite typical for many other species, for example DR in gum-tree lies within the 54 to 68 range, with an average of 61 [120], and in beech it is around 55 [121].

As could be seen at Figures 10a and 11a, E_{eff} and H_{eff} vary not too much within each EW layer and across several EW layers (relative variance of both values is around 10–15%), despite that the weather conditions during layer growth may differ significantly. For instance, the year 2010 was very dry. It affected the ring width w_o, so that it is less than half of an average one, however, E_{eff} and H_{eff} values for EW are almost the same as appropriate values for other years (Figure 10a). The lateral size of EW cells in different rings does not differ significantly either, but cell wall width does. Thus, ring width variation is mainly due to the difference in cell morphology and its quantity in the layer, while the mechanical properties of cell wall material are almost the same.

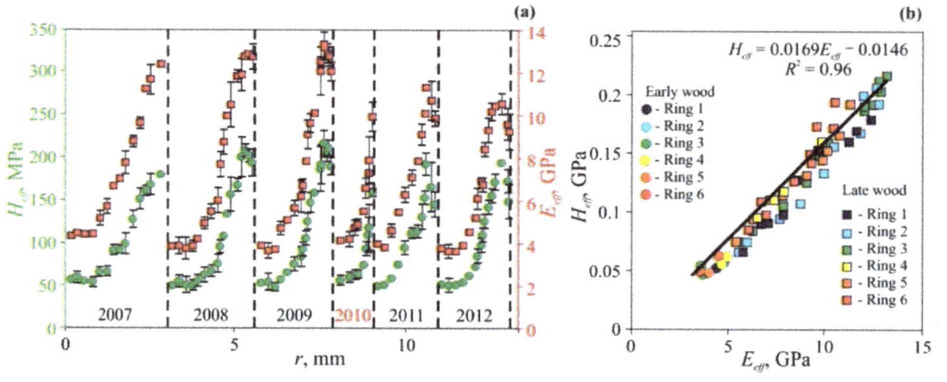

Figure 10. Micromechanical properties of common pine annual growth rings measured at P_{max} = 500 mN [116]. (**a**) Spatial dependencies of H_{eff} and E_{eff} over radial distance r for six successive rings. Ring boundaries are shown using dashed lines. The extraordinarily draughty 2010 year is highlighted by red color. (**b**) Dependence of hardness H_{eff} upon Young modulus E_{eff} for six successive rings.

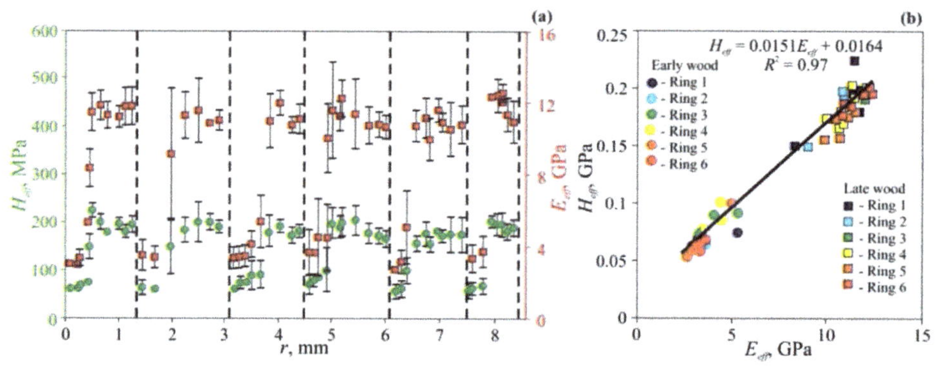

Figure 11. Micromechanical properties of pedunculate oak annual growth rings measured at P_{max} = 500 mN [116]. (**a**) Spatial dependencies of H_{eff} and E_{eff} over radial distance r for six successive rings. Ring boundaries are shown using dashed lines. (**b**) Dependence of hardness H_{eff} upon Young modulus E_{eff} for six successive rings.

As follows from the data presented, the ratio of averaged Young moduli for LW and EW is around 3.1 for pine and 3.5 for oak. The ratio of averaged hardness for LW and EW is close to those measurements, and equal 3.7 for pine and 3.0 for oak. These values stay for average values calculated for individual rings, but in some years, they can differ from the mean substantially. Obviously, such variations are due to anomalous weather conditions during these years, and narrower growth rings corresponding to these years evidence the same. However, year-to-year variation in mechanical properties is much higher than in annual ring widths, which is usually used for climate reconstruction. Thus, E_{eff} and H_{eff} measurements can be a much more sensitive and accurate dendroclimatological method than annual growth rings width measurement.

The values of E_{eff} and H_{eff} measured as described above are two to three times lower than those of cell walls. This is expected as long as the former are affected by material porosity. However, H_{eff} is two to three times higher than macroscopic Brinell hardness [8,76]. It can be formally qualified as size effect (SE) in wood mechanical properties. However, determining the relative contributions of void-free material properties, porosity, and the related differences in deformation mechanisms in such SEs requires separate research.

8. Mechanical Properties of Wood at Macroscopic Scale

The largest part of the literature discussing mechanical properties of the wood concerns properties at the macroscopic scale [4,8,32,35,122]. As long as mechanical properties of the wood demonstrate prominent anisotropy, reference data typically comprise Young modulus, hardness, strength, and other properties for directions along and across fibers separately [35]. The methodological review of generally used approaches and experimental techniques of mechanical testing of the wood can be found in [123]. Standard methods of timber mechanical testing are described in [124]. Methods of timber strength classification can be found in [35,125].

The most general relationships between the mechanical properties and structure of the wood are discussed below. Going from juvenile wood (JW) that is formed during the first 5–20 years of tree growth to mature wood (MW), specific gravity, cell length and wall thickness, percentage of late wood, and strength increase, while fibril angle μ, moisture content W, and annual ring width w decrease. Moisture content W increases by 1% in the range between 10–12% and 50–60% results in a decrease in uniaxial compression strength by 5% and in uniaxial tensile strength by 2–2.5% both along and across fibers [126]. Young modulus across the fibers diminishes with growing W too but at the slower rate of around 1.5% per 1% of W. Typical tensile strength of softwoods (fir, pine, spruce, cedar, etc.) along the fibers is between 45 and 112 GPa, while that of hardwoods (beech, oak, maple, elm, etc.) is between 70 and 120 GPa [8]. Cooling from room temperature to −195 °C results in increase in compressive strength of dry wood with $W = 12\%$ by a factor of 2–2.5, while heating to 50 °C rises it by 10–20% [8]. Holding the wood under load for a long time diminishes its strength. So, it drops by 10–15% for an hour, by 20–25% for a month, and 30–35% for a year when compared to 1 min load [8]. Fracture toughness of macroscopic wood samples range from 250 kPa m$^{1/2}$ for Western white pine to 517 kPa m$^{1/2}$ for Yellow poplar [35]. In addition to the species, growth conditions and moisture content, structure defects significantly affect wood macromechanical properties [8,122,123]. Their variation can reach 15–35% from sample to sample due to such sensitivity. More detailed data concerning mechanical properties for various sample size and testing conditions can be found in [4,8,32,35,122,125,126] and Table 4 in Section 10.

9. Modification and Hardening of Wood and Cellulose

Drastic reduction of wood mechanical characteristics at increasing characteristic size of the samples stimulates development of various modification techniques for both macroscopic wood products [30,127,128] and nano- and micro-cellulose [5,129,130]. Several classes of wood modification techniques could be distinguished, including (a) chemical processing (acetylation, furfurylation, resin impregnation etc.); (b) thermally-based processing; (c) thermo–hydro–mechanical processing (surface densification); (d) microwaves, plasma, and laser light treatment; (e) mineralization; (f) biological treatment [30]. Besides improvement of mechanical properties, wood modification can be aimed at reducing water absorption or susceptibility to rotting and biodegradation, enhancing fire resistance or antiseptic properties, improving dimensional stability or resistance to acids or bases, ultraviolet radiation etc. Getting back to mechanical properties, let us mention some examples of wood modification leading to significant improvement of such properties. So, the widely used Compreg technique, consisting of wood compression before the resin is cured within the material, leads to considerable compressive strength increases that are even higher than wood density increases; tensile strength, and flexural strength increase less than its density increases [30]. As long as wood density can be increased up to a factor of 2–2.5 during this processing, its strength can be increased nearly twofold. At the same time, material toughness drops by 25–50%. The hardness can be raised by Compreg more substantially, and the factor of 10 to 20 has been reported [30]. Another modification technique is described in review [31]. The two-step process involves the partial removal of lignin and hemicellulose from the natural wood via a boiling process in an aqueous mixture of NaOH and Na_2SO_3, followed by hot-pressing, leading to the total collapse of cell walls and the complete densification of the natural wood with highly aligned cellulose nanofibers. The processed wood has a specific strength higher than that of most structural metals and alloys, making it low-cost and high-performance. More detailed information concerning techniques and results of wood modifications aimed at changing the mechanical and other service properties of the wood can be found in books [8,30,127], reviews [128], and original papers [131–136].

10. Size Effects in Wood

All the size effects in materials are usually divided into internal ones that depend on nano- and micro-structures of the object, and the external ones, depending on the size and shape of the sample, loading method, and size of the area under load. There is not much systematized information available regarding both types of SEs at different scale hierarchical levels of wood structure. In our survey, we present the most interesting and typical data. They are summed up in Table 4.

As follows from Table 4, the strength of cellulose nanocrystals, assessed both by calculations and by experimental techniques, is 4.9–10 GPa, while the strength of nanofibrils with a diameter 3–15 nm is close to the lower end of this range [5,40,65–68]. These values exceed the strength of cellulose microfibers 8–12 μm in diameter, which is 0.5–1.65 GPa, by about one order of magnitude [5,40,66,104,106,107]. The typical nanohardness values H_{NI} for cell walls with a thickness of 2–5 μm are about 0.3–0.5 GPa [70–75,107–110], which is 2–3 times less than the strength of cellulose microfibers. As it is shown in Figures 10a and 11a, the effective values of microhardness, H_{eff}, which take into account the porosity in layers EW and LW, are several times lower (from two to four times) than H_{NI}. However, the H_{eff} value is several times higher than Brinell macrohardness HB [76], and bending and uniaxial tensile strength obtained in macrotests [8].

It is evident that the values of effective Young's modulus and hardness at meso- and macro-levels fall so dramatically not only because of the internal reasons defined by molecular and supramolecular structures, but also due to a great influence of nano-/micro-porosity of the material that should be attributed to the number of pores, capillaries, and larger tracheides with a high aspect ratio, ubiquitous in in the wood structure. Their presence results in several considerable differences between the mechanical behavior of wood and non-porous bodies. Firstly, Tabor's rule, according to which the hardness of soft materials exceeds their yield stress or strength by about three times, is almost never met. On the contrary, in most wood species, their macroscopic hardness is several times lower than the yield stress and ultimate tensile strength determined for macroscopic samples. Apparently, in all similar events, the reason is that during indentation and uniaxial compression, the tested wood cell structure loses its stability long before the manifestation of plastic deformation and tensile rupture. These events strongly depend on the direction of the load application in relation to the long axis of the cell due to anisotropy of the mechanical properties of wood. In the longitudinal direction, they are higher by about an order of magnitude than in the transverse direction. It is difficult to provide an explanation for the various possible modes and mechanisms of wood deformation at the nano- and micro- scale because of insufficient experimental data. However, the quantitative information quoted above leads to the conclusion that there exist highly pronounced SEs in wood which cause a sharp decrease in strength/hardness from ~10 GPa in nanocrystalline cellulose to ~0.1 GPa or less in macrovolumes of wood. This means that all cellulose-containing materials have a large strengthening potential that can be realized through optimally organized nano- and micro-structures, and employment of relevant technologies.

Table 4. Some mechanical properties of wood structure components.

Specimen		Young's Modulus, GPa		Tensile Strength, GPa		Stress Strength, GPa		Hardness, GPa		Reference
	*R**	‖	⊥	‖	⊥	‖	⊥	‖	⊥	
CNC (Cellulose Nanocrystals)	5–30 nm	140–160	15–30	8–10	~1					[67]
	3–20 nm	105–168		7.5–9						[103]
	5–70 nm	150–175								[129]
	~10 nm	110–220	10–50	7.5–7.7						[40]
CNF (Cellulose Nanofibril)	10–40 nm	30–40	10–15	0.8–1	~0.1					[67]
CMF (Cellulose Microfibers)	10–70 μm	120–140		0.75–1.08						[36]
	10–50 μm	12–27		0.3–1.4						[40]
	10–30 μm	15–27		0.55–1.3						[104]
	~10 μm	86		1.57						[66]
Cell wall										
Pinus sylvestris L.		17 ± 5						0.46 ± 0.03		[74]
Pinus massoniana								0.38 ± 0.04		[107]
Masson pine								0.47 ± 0.06		[108]
Pinus taeda								0.44 ± 0.1		[109]

Table 4. *Cont.*

Specimen	R^*	Young's Modulus, GPa ‖	Young's Modulus, GPa ⊥	Tensile Strength, GPa ‖	Tensile Strength, GPa ⊥	Stress Strength, GPa ‖	Stress Strength, GPa ⊥	Hardness, GPa ‖	Hardness, GPa ⊥	Reference
EW layers										
Pinus sylvestris L.	~1 mm	4 ± 1						0.05 ± 0.01		[116]
Quercus robur L.		4 ± 1						0.08 ± 0.02		[116]
LW layers										
Pinus sylvestris L.	~1 mm	11 ± 2						0.18 ± 0.04		[116]
Quercus robur L.		12 ± 1						20 ± 0.02		[116]
Bulk wood										
Pine (misc.)		8.5–13.7		0.08–0.12	0.002–0.003	0.04–0.06	0.003–0.007			[8]
Pinus sylvestris L.								0.03–0.04	0.01–0.02	[116]
Pinus sylvestris L.								0.04–0.05		[117]
Oak (misc.)	10–1000 mm	10.3–13.9		0.08–0.16	0.003–0.007	0.04–0.06	0.006–0.009			[8]
Quercus robur L.								0.06–0.07		[117]

Legend: R^*—characteristic size ‖—along the fibers ⊥—across the fibers.

Representation of the strength characteristics as a function of characteristic dimensions of the structure R^* in double-logarithmic coordinates provides a distinctive hockey stick-shaped diagram (Figure 12). The descending part of the curve in the nano- and micro-domain has a slope close to −0.5, a feature it shares with Hall-Petch relation, which is well-known in materials science and described for the first time for polycrystalline metals more than 60 years ago [137–139]:

$$\sigma_y = \sigma_0 + A(d)^{-0.5} \tag{1}$$

where σ_y is the yield stress, d is the crystallite size, σ_0 and A are material constants. In most cases, the value of A that used to be called the Hall-Petch constant turned out to be deformation-dependent [139]. Later it was clarified that external dimensions affect the mechanical properties in the similar way, though the index of power may differ from 0.5 [45].

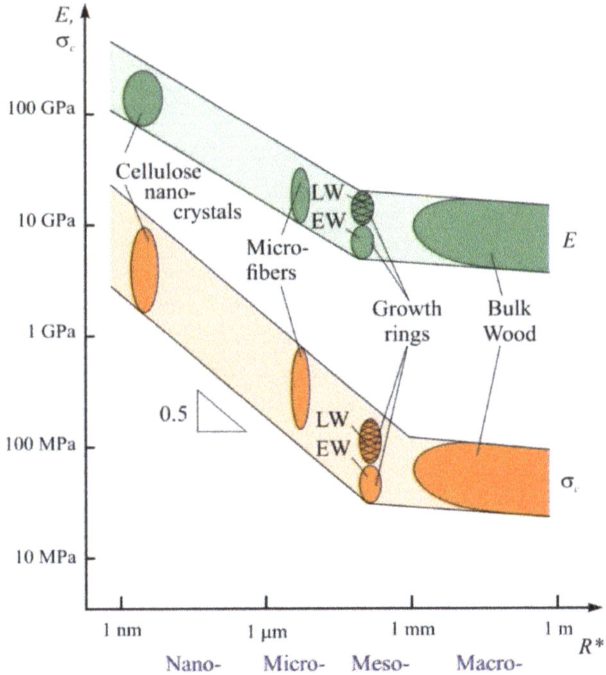

Figure 12. The size dependence of cellulose-containing materials strength. (Data collected by the authors). R^* is the characteristic size.

Relations, similar to (1) were discovered for hardness [140]

$$H = H_0 + A(R^*)^{-0.5} \tag{2}$$

where R^* is the characteristic dimension of the locally deformed area, which, in the process of indentation with a Berkovich tip, is usually taken equal to the indentation depth h_{max}.

The specific internal and external SEs have been studied not only in void-free poreless materials (metals, alloys, rocks, composites, etc.), but also in such porous materials as ceramics, solidified foams [141–144], and organic gels [145].

Due to a discrete character of damage accumulation in porous materials under load, the loading diagrams obtained in NI and in microsample deformation contain deformation jumps. They carry information about elementary events, their rate and statistics as a function of size of the area under load, deformation rate and other experiment conditions [143].

Obviously, the causes for drop in strength/hardness with increase in R^* may differ in different groups of materials. Nevertheless, some similarities can be observed, for example, the principle "the smaller the stronger" works in wood as well, though the index of power for R^* may vary within quite a wide range, namely from 0.2 to 1. Therefore, many authors suggest other dependences to account for SEs, such as $1/R^*$, $\ln(R^*)/R^*$ and others [146,147].

The softening mechanisms turning on with the increase in both internal and external characteristic size of the system require additional study of interrelations between the multilevel nano-/micro-structure of wood and its physico-mechanical properties. However, there are reasons to suppose that micromechanics of thin filaments, walls, partitions, as well as the macromechanical behavior of wood may have a lot in common with those in other highly porous materials [148–154]. Therefore, the general approaches and models developed for the analysis of the latter can be applied to wood as well. Plausibly, such softening depends upon cell wall width, cell morphology, aspect ratio and adhesion, and percentage of the lumens in wood crosscut area.

Some SEs can also be observed at the macro-scale and, to some extent, at meso-scale, though they are less pronounced than that at the nano- or micro-scale. They can be attributed to a growing possibility of large defect (cracks, delaminations, knots, and other wood defects) occurrence in larger objects, and can be described using Weibull statistics [155,156]. However, such a study lies outside the scope of our present review.

11. Nanomechanics in Dendrochronology

Jumps in E_{eff} and H_{eff} at the edges of annual growth rings made it possible to measure their width w_{NI} using scanning NI. Then, the comparison was carried out with the w_0 value determined by an optical method (analysis of image contrast). The image processing method was similar to the one used in the widely used LINTAB equipment. The comparison of the data obtained by these two techniques for measuring tree-ring width is presented in Figure 13; one can see that differences between them do not exceed 2–3% for pine and 4–5% for oak and lime. Mean average deviation for six to seven rings was about 2%. In effect, this means that the scanning indentation method can be used as an alternative to the optical one, or can complement it, providing some additional data on the local mechanical properties.

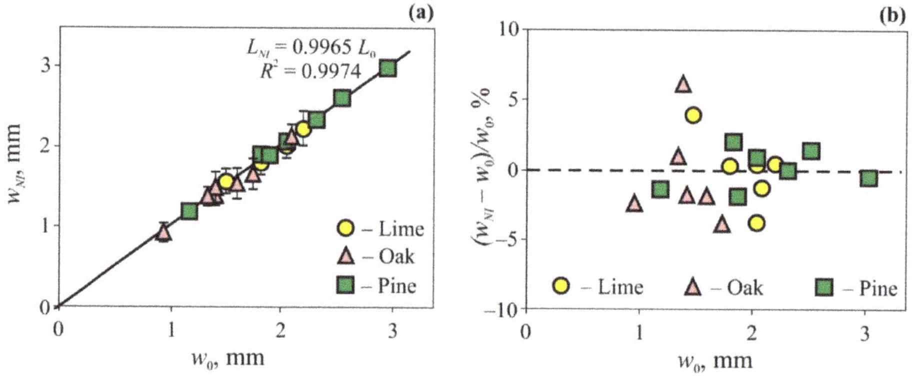

Figure 13. The results of annual growth ring width measurements obtained by nanoindentation w_{NI} and by the optical method w_0 (**a**), and the discrepancies between these methods (**b**) [117].

12. Correlation between Thermal Diffusivity and Mechanical Properties of Wood

Kinetic thermophysical characteristics (thermal conductivity λ and diffusivity a) and mechanical properties of wood (Young's modulus, strength, hardness) depend on the same factors, namely on composition, structure, density, porosity, humidity, and specifics of interconnections between microstructural units [5,8,10]. Moreover, in both groups, the same pattern can be observed: higher wood density is accompanied by higher values of the characteristics mentioned above. So, it seems reasonable to suggest that there is a certain correlation between these two groups of properties. Once revealed, such relations could allow switching from laborious and material extensive destructive mechanical tests to non-destructive contactless measurements of λ or a. Such approaches could be used to estimate relative mechanical properties, and to sort and grade materials and products made of wood, fiber-reinforced composites, etc.

It should be noted that despite a wealth of information concerning the mechanical and thermal properties of natural and modified wood, wood-based layered materials, as well as composites reinforced with artificial and natural organic fibers, they are measured in separate tests executed on different samples. Exceptions are several papers that employ thermal non-destructive testing to estimate structural damage of wood [157,158], defects [159,160], porosity [161], and anisotropy [162,163] of composites and their possible impact on material mechanical properties. A micromechanical model for the prediction of effective thermal conductivity in two- and three-phase composites is proposed in [164].

The dependence of thermal diffusivity tensor a_{ij} on Brinelle hardness HB in common pine (*Pinus sylvestris* L.), pedunculate oak (*Quercus robur* L.), and small-leaf lime (*Tilia cordata Mill.*) wood at various humidity levels was studied in [116,165,166].

The thermal diffusivity tensor components $a_{ij} = \lambda_{ij}/\rho C_p$—where λ_{ij} represents the thermal conductivity tensor components, ρ is the material density, and C_p is the specific thermal capacitance—were measured using an original non-destructive thermal imaging technique described in detail in [167–169]. The method is based upon local stepped heating at small spots on the sample surface by a focused laser beam while continuously monitoring the surface temperature distribution with a thermal camera. Heat propagation in such a setup is close to spherical symmetry in isotropic materials, while in orthotropic materials, the isothermal surfaces are close to three-axis ellipsoids with the axes fully determined by the main components of a_{ij} tensor and the time elapsed since heating onset, provided that the distance to the heating center is at least several times higher than the heating beam radius. Therefore, the a_{ij} values were determined by processing dynamic thermal images obtained on lateral, radial, and transverse crosscuts of wood samples, as described in [167–169].

The values of hardness HB_W at current humidity values W normalized to the hardness HB_{W10} at W = 10% for the lateral and radial faces were statistically indistinguishable (Figure 14); therefore, they were approximated as a single set by the following linear function $(HB_W/HB_{W10}) = 1.59(a_l/a_n) - 1.34$ with the coefficient of determination $R^2 = 0.75$, where a_l and a_n are longitudinal and lateral components of a_{ij} tensor, accordingly. The hardness measured on the face perpendicular to the fibers was found to be independent of tensor a_{ii} components, so that it could be used for method and equipment calibration.

Similar results were obtained for other porous materials. For example, a linear relationship between the coefficient of thermal conductivity λ and compressive strength σ_b in lightweight porous cement composites containing aerogel has been reported in [170]. A decrease in porosity resulted in λ growth from 0.39 to 0.47 W/m·K, and simultaneous σ_b increase from 7.5 to 15 MPa.

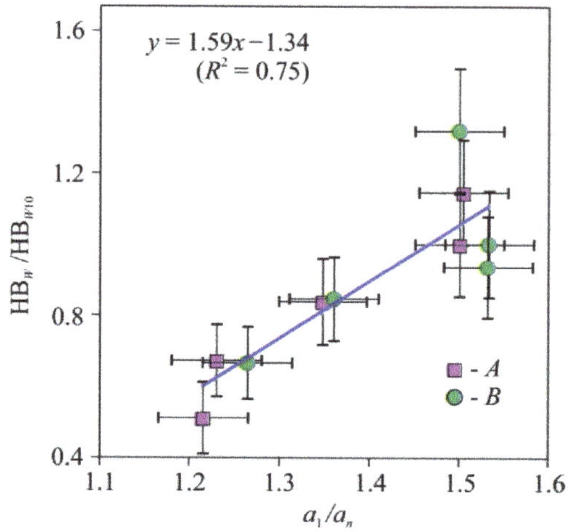

Figure 14. Dependence of the relative macrohardness HB_W/HB_{W10} of pine wood upon thermal diffusivity anisotropy—a_1/a_n. HB—Brinell hardness (12.7 mm sphere diameter, 1 mm indentation depth), a_1, a_n—thermal diffusivity coefficients along and across the fibers. HB_W/HB_{W10} is HB_W values on radial (A) and tangential (B) faces, as shown in Figure 2, normalized to HB_W at humidity $W = 10\%$ (HB_{W10}) [116].

13. Discussion

Despite the great variety of plant domain and trees in particular, their structural composition is similar, and this similarity stays across different hierarchical scale levels. They are composed of the same organic molecules that form nanocrystals and nanofibrils, which in turn form microfibers and then cell walls. The tree trunk contains the cells organized in annual growth rings. It is interesting that not only such a structure is more or less universal, but its individual elements in various species have similar morphology and close mechanical properties. Additionally, the higher the scale level in this hierarchy the lower its mechanical characteristics. This decrease is roughly in accordance with the Hall-Petch relation, and this fact has not been noticed before. Reconciling mechanical strength and physiological functions, nature had to find some compromise. Capillaries, large tracheids, and gum canals weaken wood structure inevitably, so that their specifics determine in large part the macromechanical properties of timber products. Relations between wood mechanical properties and the architecture and features of its structure are studied insufficiently yet. In particular, quantitative relations between the polymerization degree of polysaccharides, microfibers and cell walls structure, morphology and shape of the cells, wood microarchitecture, turgor pressure etc., on one hand, and macroscopic mechanical characteristics such as stiffness, hardness, strength, strain at break, and fracture toughness on the other hand, are unknown. Meanwhile, the turgor of a plant cell, for example, averages around 0.44 MPa and can be as high as 2 MPa [35]. It is essential in providing mechanical stability of any plant. It is known that higher cellulose percentage and crystallinity generally provide higher strength and stiffness to microfibers [8,36,40,67,68], as long as other structural components have much lower mechanical characteristics. In particular, hemicellulose stiffness and strength are several tens times lower, and around two orders of magnitude lower for lignin than that of cellulose [36]. Besides, an increase in moisture content leads to much faster deterioration of their mechanical properties than that in cellulose. However, reliable systematic research into cellulose, hemicellulose, lignin, and pectin percentages impacts on wood mechanical properties at different structural levels are lacking. Nevertheless, the role of chemical composition is very significant, especially accounting for the high variance of the above percentages, even in plants of the same species, and more so for plants of differing species [104].

A better understanding of formation regularities of wood hierarchical structure and their relations with mechanical properties could allow for controlling the latter, as well as manufacturing materials with desired properties. In theory, it could be achieved by means of controlled growth of the plant in an artificial environment, modification of wood structure using more efficient techniques, or reconstruction of the material that could provide it with new design and properties. All these approaches are used to some extent already, but without physical substantiation their efficiency is low. The best results have been achieved at nano- and micro-scales by growing nanofibers, and aligning and cross-linking them to form high strength microfibers. A scientifically valid and optimized design of nanocellulose-based materials can fulfill natural potential of this polymer and allow manufacturing ecologically friendly, high-strength, economically efficient materials.

14. Conclusions

In spite of the fact that people have been using different kinds of wood and wood derivatives for many thousands of years, the variety of cellulose-containing materials is so vast, and their multilevel architecture is so complicated, that the nature of their properties has not been completely revealed yet. As a result, the potential of these natural materials is not fully used. In order to make accurate predictions about wood characteristics, to be able to control them during growth, or to modify them with various treatments, and to preserve them in service, we need a deep comprehension of which elements of woods' structure, their condition, orientation, interconnection, and evolution, as well as which defects and flaws in the nano-, micro-, meso-, and macro-structure, influence particular macroscopic properties. First of all, these properties should be understood as the most common and important key elements of wood microstructure, such as nanocrystalline and amorphous phases of cellulose, microfibrils, cell walls, and annual growth rings.

Various modern research methods provide multiscale data on wood mechanical properties at different structural levels, from nano- to macro-scale. The interrelated information from these levels can offer new approaches to the cultivation of wood with predetermined mechanical properties, for example, with high strength and elastic properties, necessary acoustic characteristics, or low creep rate, and would help to create effective physically substantiated methods of wood strengthening. High specific strength of nanocellulose, exceeding that of almost all the other construction materials, except nanocarbons, encourages designing new extra-strong, ecologically friendly materials that would take advantage of this potential.

The results of scanning nano- and micro-mechanical properties of wood across several successive annual growth rings provides the basis for innovative methods and techniques for dendroclimatological and dendrochronological applications that would complement the existing approaches. Since the average cell size in a wood cross-section is about 30–50 μm, and the average ring width is 1–3 mm, there are approximately 50–100 cells in each ring. Technically, nanoindentation provides a means to measure the mechanical characteristics of each individual cell. Therefore, the temporal resolution limit for NI in dendrochronological applications is close to one week, making it a much more precise method with better temporal resolution than the traditional optical ones.

Author Contributions: Methodology—Y.I.G.; project administration—A.A.G., I.A.V.; supervision—Y.I.G.; data processing—D.Y.G.; writing—draft, review & editing Y.I.G., D.Y.G., S.M.M., A.A.G.; visualization—S.M.M., I.A.V. All authors have read and agreed to the published version of the manuscript.

Funding: The results were obtained using the equipment of the Center for Collective Use of Scientific Equipment of TSU named after G.R. Derzhavin. This research was funded by the Russian Scientific Foundation grant 20-19-00602 (measurement of thermal properties), and grant 21-14-00233 (macro- and micromechanical testing), and by the Ministry of Science and Higher Education of the Russian Federation, the contract 075-15-2021-709, unique identifier of the project RF-2296.61321X0037 (equipment maintenance).

Data Availability Statement: All the data is available within the manuscript.

Conflicts of Interest: The authors declare no conflict of interest.

References

1. Wang, J.; Wang, L.; Gardner, D.J.; Shaler, S.M.; Cai, Z. Towards a cellulose-based society: Opportunities and challenges. *Cellulose* **2021**, *28*, 4511–4543. [CrossRef]
2. Plocher, J.; Mencattelli, L.; Narducci, F.; Pinho, S. Learning from nature: Bio-inspiration for damage-tolerant high-performance fibre-reinforced composites. *Compos. Sci. Technol.* **2021**, *208*, 108669. [CrossRef]
3. Zhu, J.Y.; Agarwal, U.P.; Ciesielski, P.N.; Himmel, M.E.; Gao, R.; Deng, Y.; Morits, M.; Österberg, M. Towards sustainable production and utilization of plant-biomass-based nanomaterials: A review and analysis of recent developments. *Biotechnol. Biofuels* **2021**, *14*, 114. [CrossRef] [PubMed]
4. Pandey, K.K.; Ramakantha, V.; Chauhan, S.S.; Kumar, A.N.A. *Wood is Good: Current Trends and Future Prospects in Wood Utilization*; Springer Nature Singapore Pte Ltd.: Singapore, 2017.
5. Kargarzadeh, H.; Ahmad, I.; Thomas, S.; Dufresne, A. *Handbook of Nanocellulose and Cellulose Nanocomposites*; Wiley-VCH Verlag GmbH & Co. KGaA: Weinheim, Germany, 2017; Volume 1, p. 849. [CrossRef]
6. Ma, T.; Hu, X.; Lu, S.; Liao, X.; Song, Y.; Hu, X. Nanocellulose: A promising green treasure from food wastes to available food materials. *Crit. Rev. Food Sci. Nutr.* **2020**, *62*, 989–1002. [CrossRef] [PubMed]
7. Van Hai, L.; Son, H.N.; Seo, Y.B. Physical and bio-composite properties of nanocrystalline cellulose from wood, cotton linters, cattail, and red algae. *Cellulose* **2015**, *22*, 1789–1798. [CrossRef]
8. General Technical Report FPL-GTR-282. In *Wood handbook—Wood as an Engineering Material*; U.S. Department of Agriculture, Forest Service, Forest Products Laboratory: Madison, WI, USA, 2021. Available online: https://www.fs.usda.gov/treesearch/pubs/62200 (accessed on 1 March 2022).
9. Holik, H. *Handbook of Paper and Board*; Wiley-VCH Verlag GmbH & Co. KGaA: Weinheim, Germany, 2006. [CrossRef]
10. Kirwan, M.J. *Handbook of Paper and Paperboard Packaging Technology*, 2nd ed.; John Wiley & Sons Limited,: Hoboken, NJ, USA, 2012. [CrossRef]
11. Vilásia, G.; Martins, V.; Palezi, S.C.; Alves-Silva, G.F.; Santos, L.G. Biodegradable Packaging Materials and Techniques to Improve Their Performance. In *Food Packaging: The Smarter Way*; Springer: Singapore, 2022; pp. 61–105. [CrossRef]
12. Nechita, P.; Roman, M. Review on Polysaccharides Used in Coatings for Food Packaging Papers. *Coatings* **2020**, *10*, 566. [CrossRef]
13. Bunsell, A.R. *Handbook of Properties of Textile and Technical Fibres*; Elsevier Science: Amsterdam, The Netherlands, 2018.
14. Felgueiras, C.; Azoia, N.G.; Gonçalves, C.; Gama, M.; Dourado, F. Trends on the Cellulose-Based Textiles: Raw Materials and Technologies. *Front. Bioeng. Biotechnol.* **2021**, *9*, 608826. [CrossRef] [PubMed]
15. Thomas, S.; Pottathara, Y.B. *Nanocellulose Based Composites for Electronics*; Elsevier Inc.: Amsterdam, The Netherlands, 2021. [CrossRef]
16. Nishiyama, Y. Retrieving structural information from scattering and attenuation data of transparent wood and (Nano)paper. *J. Bioresour. Bioprod.* **2021**, *6*, 187–194. [CrossRef]
17. Wang, F.; Cheong, J.Y.; Lee, J.; Ahn, J.; Duan, G.; Chen, H.; Zhang, Q.; Kim, I.; Jiang, S. Pyrolysis of Enzymolysis-Treated Wood: Hierarchically Assembled Porous Carbon Electrode for Advanced Energy Storage Devices. *Adv. Funct. Mater.* **2021**, *31*, 2101077. [CrossRef]
18. Zou, Y.; Yang, P.; Yang, L.; Li, N.; Duan, G.; Liu, X.; Li, Y. Boosting solar steam generation by photothermal enhanced polydopamine/wood composites. *Polymer* **2021**, *217*, 123464. [CrossRef]
19. Shak, K.P.Y.; Pang, Y.L.; Mah, S.K. Nanocellulose: Recent advances and its prospects in environmental remediation. *Beilstein J. Nanotechnol.* **2018**, *9*, 2479–2498. [CrossRef] [PubMed]
20. Abdelhamid, H.N.; Mathew, A.P. Cellulose-Based Materials for Water Remediation: Adsorption, Catalysis, and Antifouling. *Front. Chem. Eng.* **2021**, *3*, 790314. [CrossRef]
21. Syeda, H.I.; Yap, P.-S. A review on three-dimensional cellulose-based aerogels for the removal of heavy metals from water. *Sci. Total Environ.* **2021**, *807*, 150606. [CrossRef]
22. Palacios Hinestroza, H.; Urena-Saborio, H.; Zurita, F.; Guerrero de León, A.A.; Sundaram, G.; Sulbarán-Rangel, B. Nanocellulose and Polycaprolactone Nanospun Composite Membranes and Their Potential for the Removal of Pollutants from Water. *Molecules* **2020**, *25*, 683. [CrossRef]
23. Mali, P.; Sherje, A.P. Cellulose nanocrystals: Fundamentals and biomedical applications. *Carbohydr. Polym.* **2021**, *275*, 118668. [CrossRef]
24. Swingler, S.; Gupta, A.; Gibson, H.; Kowalczuk, M.; Heaselgrave, W.; Radecka, I. Recent Advances and Applications of Bacterial Cellulose in Biomedicine. *Polymers* **2021**, *13*, 412. [CrossRef]
25. Raut, H.K.; Das, R.; Liu, Z.; Liu, X.; Ramakrishna, S. Biocompatibility of Biomaterials for Tissue Regeneration or Replacement. *Biotechnol. J.* **2020**, *15*, e2000160. [CrossRef]
26. Abdul Khalil, H.P.S.; Adnan, A.; Yahya, E.B.; Olaiya, N.; Safrida, S.; Hossain, M.S.; Balakrishnan, V.; Gopakumar, D.A.; Abdullah, C.; Oyekanmi, A.; et al. A Review on Plant Cellulose Nanofibre-Based Aerogels for Biomedical Applications. *Polymers* **2020**, *12*, 1759. [CrossRef]
27. Joseph, B.; Sagarika, V.K.; Sabu, C.; Kalarikkal, N.; Thomas, S. Cellulose nanocomposites: Fabrication and biomedical applications. *J. Bioresour. Bioprod.* **2020**, *5*, 223–237. [CrossRef]

28. Norrrahim, M.N.F.; Kasim, N.A.M.; Knight, V.F.; Ujang, F.A.; Janudin, N.; Razak, M.A.I.A.; Shah, N.A.A.; Noor, S.A.M.; Jamal, S.H.; Ong, K.K.; et al. Nanocellulose: The next super versatile material for the military. *Mater. Adv.* **2021**, *2*, 1485–1506. [CrossRef]
29. Ajdary, R.; Tardy, B.L.; Mattos, B.D.; Bai, L.; Rojas, O.J. Plant Nanomaterials and Inspiration from Nature: Water Interactions and Hierarchically Structured Hydrogels. *Adv. Mater.* **2020**, *33*, e2001085. [CrossRef] [PubMed]
30. Sandberg, D.; Kutnar, A.; Karlsson, O.; Jones, D. Wood modification technologies. In *Principles, Sustainability, and the Need for Innovation*; CRC Press (Taylor & Francis Group): Boca Raton, FL, USA, 2021; p. 442.
31. Chen, C.; Kuang, Y.; Zhu, S.; Burgert, I.; Keplinger, T.; Gong, A.; Li, T.; Berglund, L.; Eichhorn, S.J.; Hu, L. Structure–property–function relationships of natural and engineered wood. *Nat. Rev. Mater.* **2020**, *5*, 642–666. [CrossRef]
32. Toumpanaki, E.; Shah, D.U.; Eichhorn, S.J. Beyond What Meets the Eye: Imaging and Imagining Wood Mechanical–Structural Properties. *Adv. Mater.* **2020**, *33*, 2001613. [CrossRef]
33. Donaldson, L.A. Wood cell wall ultrastructure The key to understanding wood properties and behaviour. *IAWA J.* **2019**, *40*, 645–672. [CrossRef]
34. Börjesson, M.; Westman, G. Crystalline Nanocellulose—Preparation, Modification and Properties. In *Cellulose. Fundamental Aspects and Current Trends*; IntechOpen: London, UK, 2015; pp. 159–191.
35. Codjoe, J.M.; Miller, K.; Haswell, E.S. Plant cell mechanobiology: Greater than the sum of its parts. *Plant Cell* **2021**, *34*, 129–145. [CrossRef]
36. Gibson, L.J. The hierarchical structure and mechanics of plant materials. *J. R. Soc. Interface* **2012**, *9*, 2749–2766. [CrossRef]
37. Olorunnisola, A.O. *Design of Structural Elements with Tropical Hardwoods*; Springer International Publishing AG: Gewerbestrasse, Switzerland, 2018. [CrossRef]
38. Jahan, Z.; Niazi, M.B.K.; Gregersen, W. Mechanical, thermal and swelling properties of cellulose nanocrystals/PVA nanocomposites membranes. *J. Ind. Eng. Chem.* **2018**, *57*, 113–124. [CrossRef]
39. Mankowski, P.; Burawska-Kupniewska, I.; Krzosek, S.; Grzeskiewicz, M. Influence of pine (*Pinus sylvestris* L.) growth rings width on the strength properties of structural sawn timber. *BioResources* **2020**, *14*, 9287–9297. [CrossRef]
40. Moon, R.J.; Martini, A.; Nairn, J.; Simonsen, J.; Youngblood, J. Cellulose nanomaterials review: Structure, properties and nanocomposites. *Chem. Soc. Rev.* **2011**, *40*, 3941–3994. [CrossRef]
41. Baghaei, B.; Skrifvars, M. All-Cellulose Composites: A Review of Recent Studies on Structure, Properties and Applications. *Molecules* **2020**, *25*, 2836. [CrossRef] [PubMed]
42. Bhushan, B. *Nanotribology and Nanomechanics: An Introduction*, 2nd ed.; Springer: Berlin/Heiderberg, Germany; New York, NY, USA, 2008.
43. Tiwari, A. *Nanomechanical Analysis of High Performance Materials*; Springer Science & Business Media: Heidelberg, Germany, 2014. [CrossRef]
44. Ranganathan, N.M. *Materials Characterization: Modern Methods and Applications*; CRC Press: Boca Raton, FL, USA, 2015.
45. Golovin, Y.I. Nanoindentation and Mechanical Properties of Materials at Submicro- and Nanoscale Levels: Recent Results and Achievements. *Phys. Solid State* **2021**, *63*, 1–41. [CrossRef]
46. Neugirg, B.R.; Koebley, S.R.; Schniepp, H.C.; Fery, A. AFM-based mechanical characterization of single nanofibres. *Nanoscale* **2016**, *8*, 8414–8426. [CrossRef] [PubMed]
47. Hsueh, C.-H.; Schmauder, S.; Chen, C.-S.; Chawla, K.K. *Handbook of Mechanics of Materials*; Springer Nature: Singapore, 2019.
48. Melelli, A.; Arnould, O.; Beaugrand, J.; Bourmaud, A. The Middle Lamella of Plant Fibers Used as Composite Reinforcement: Investigation by Atomic Force Microscopy. *Molecules* **2020**, *25*, 632. [CrossRef] [PubMed]
49. Charrier, A.; Lereu, A.; Farahi, R.H.; Davison, B.H.; Passian, A. Nanometrology of Biomass for Bioenergy: The Role of Atomic Force Microscopy and Spectroscopy in Plant Cell Characterization. *Front. Energy Res.* **2018**, *6*, 11. [CrossRef]
50. Golovin, Y.I. Nanoindentation and mechanical properties of solids in submicrovolumes, thin near-surface layers, and films: A Review. *Phys. Solid State* **2008**, *50*, 2205–2236. [CrossRef]
51. Golovin, Y.I. Nanoindentation as an instrument for integrated evaluation of physicomechanical properties of materials in submicrovolumes (a review). *Ind. Lab. Mater. Diagn.* **2009**, *75*, 45–59. (In Russian)
52. Golovin, Y.I. *Nanoindentirovaniye i Yego Vozmozhnosti (Nanoindentation and Its Capability)*; Mashinostroenie: Moscow, Russia, 2009. (In Russian)
53. Fischer-Cripps, A.C. *Nanoindentation*; Springer: New York, NY, USA, 2011.
54. Oyen, M.L. *Handbook of Nanoindentation with Biological Applications*; Pan Stanford Publishing Pte. Ltd.: Singapore, 2011.
55. Nemecek, J. *Nanoindentation in Materials Science*; InTech: London, UK, 2012.
56. TTze, W.T.Y.; Wang, S.; Rials, T.G.; Pharr, G.M.; Kelley, S.S. Nanoindentation of wood cell walls: Continuous stiffness and hardness measurements. *Compos. Part A Appl. Sci. Manuf.* **2007**, *38*, 945–953. [CrossRef]
57. Tiwari, A.; Natarajan, S. *Applied Nanoindentation in Advanced Materials*; John Wiley & Sons: New York, NY, USA, 2017. [CrossRef]
58. Qian, L.; Zhao, H. Nanoindentation of Soft Biological Materials. *Micromachines* **2018**, *9*, 654. [CrossRef]
59. Perepelkin, N.V.; Borodich, F.M.; Kovalev, A.E.; Gorb, S.N. Depth-Sensing Indentation as a Micro- and Nanomechanical Approach to Characterisation of Mechanical Properties of Soft, Biological, and Biomimetic Materials. *Nanomaterials* **2019**, *10*, 15. [CrossRef]
60. Oliver, W.C.; Pharr, G.M. An improved technique for determining hardness and elastic modulus using load and displacement sensing indentation experiments. *J. Mater. Res.* **1992**, *7*, 1564–1583. [CrossRef]

61. Oliver, W.C.; Pharr, G.M. Measurement of hardness and elastic modulus by instrumented indentation: Advances in understanding and refinements to methodology. *J. Mater. Res.* **2004**, *19*, 3–20. [CrossRef]
62. Oliver, W.C.; Pharr, G.M. Nanoindentation in materials research: Past, present, and future. *MRS Bull.* **2010**, *35*, 897–907. [CrossRef]
63. ISO/TC 164/SC 3/WG 4. *Revision of ISO 14577—Metallic Materials—Instrumented Indentation Test for Hardness and Materials Parameters*; ISO: Geneva, Switzerland, 1981.
64. Hitam, C.; Jalil, A. Recent advances on nanocellulose biomaterials for environmental health photoremediation: An overview. *Environ. Res.* **2021**, *204*, 111964. [CrossRef] [PubMed]
65. Eichhorn, S.J.; Dufresne, A.; Aranguren, M.I.; Marcovich, N.E.; Capadona, J.R.; Rowan, S.; Weder, C.; Thielemans, W.; Roman, M.; Renneckar, S.; et al. Review: Current international research into cellulose nanofibres and nanocomposites. *J. Mater. Sci.* **2010**, *45*, 1–33. [CrossRef]
66. Mittal, N.; Ansari, F.; Gowda, V.K.; Brouzet, C.; Chen, P.; Larsson, P.T.; Roth, S.V.; Lundell, F.; Wågberg, L.; Kotov, N.A.; et al. Multiscale Control of Nanocellulose Assembly: Transferring Remarkable Nanoscale Fibril Mechanics to Macroscale Fibers. *ACS Nano* **2018**, *12*, 6378–6388. [CrossRef]
67. Ioelovich, M. Characterization of various kinds of nanocellulose. In *Handbook of Nanocellulose and Cellulose Nanocomposites*; Kargarzadeh, H., Ahmad, I., Thomas, S., Dufresne, A., Eds.; Wiley-VCH Verlag GmbH & Co.: Weinheim, Germany, 2017; pp. 51–100. [CrossRef]
68. Huang, J.; Dufresne, A.; Lin, N. *Nanocellulose: From Fundamental to Advanced Materials*; Wiley-VCH, Verlag GmbH & Co.: Wienheim, Germany, 2019.
69. Burgert, I.; Keplinger, T. Plant micro- and nanomechanics: Experimental techniques for plant cell-wall analysis. *J. Exp. Bot.* **2013**, *64*, 4635–4649. [CrossRef]
70. Eder, M.; Arnould, O.; Dunlop, J.W.C.; Hornatowska, J.; Salmén, L. Experimental micromechanical characterisation of wood cell walls. *Wood Sci. Technol.* **2012**, *47*, 163–182. [CrossRef]
71. Konnerth, J.; Eiser, M.; Jager, A.; Bader, T.K.; Hofstetter, K.; Follrich, J.; Ters, T.; Hansmann, C.; Wimmer, R. Macro- and micro-mechanical properties of red oak wood (*Quercus rubra* L.) treated with hemicellulases. *Holzforschung* **2010**, *64*, 447–453. [CrossRef]
72. Nelson, N.; Stubbs, C.J.; Larson, R.; Cook, D.D. Measurement accuracy and uncertainty in plant biomechanics. *J. Exp. Bot.* **2019**, *70*, 3649–3658. [CrossRef]
73. Keckes, J.; Burgert, I.; Frühmann, K.; Müller, M.; Kölln, K.; Hamilton, M.; Burghammer, M.; Roth, S.V.; Stanzl-Tschegg, S.; Fratzl, P. Cell-wall recovery after irreversible deformation of wood. *Nat. Mater.* **2003**, *2*, 810–813. [CrossRef] [PubMed]
74. Mania, P.; Nowicki, M. Nanohardness and elasticity of cell walls of Scots pine (*Pinus sylvestris* L.) juvenile and mature wood. *Bull. Pol. Acad. Sci. Tech. Sci.* **2020**, *68*, 1237–1241. [CrossRef]
75. Jakes, J.E.; Stone, D.S. Best Practices for Quasistatic Berkovich Nanoindentation of Wood Cell Walls. *Forests* **2021**, *12*, 1696. [CrossRef]
76. Vincent, M.; Tong, Q.; Terziev, N.; Daniel, G.; Bustos, C.; Escobar, W.G.; Duchesne, I. A comparison of nanoindentation cell wall hardness and Brinell wood hardness in jack pine (*Pinus banksiana* Lamb.). *Wood Sci. Technol.* **2013**, *48*, 7–22. [CrossRef]
77. Normand, A.C.; Charrier, A.M.; Arnould, O.; Lereu, A.L. Influence of force volume indentation parameters and processing method in wood cell walls nanomechanical studies. *Sci. Rep.* **2021**, *11*, 5739. [CrossRef]
78. Pakzad, A.; Simonsen, J.; Heiden, P.A.; Yassar, R.S. Size effects on the nanomechanical properties of cellulose I nanocrystals. *J. Mater. Res.* **2011**, *27*, 528–536. [CrossRef]
79. Smithers, E.; Luo, J.; Dyson, R.J. Mathematical principles and models of plant growth mechanics: From cell wall dynamics to tissue morphogenesis. *J. Exp. Bot.* **2019**, *70*, 3587–3600. [CrossRef]
80. Vaganov, E.A.; Hughes, M.K.; Shashkin, A.V. Growth dynamics of conifer tree rings. In *Images of Past and Future Environments*; Springer: Berlin/Heidelberg, Germany, 2006.
81. Meko, D.M.; Friedman, J.M.; Touchan, R.; Edmondson, J.R.; Griffin, E.R.; Scott, J.A. Alternative standardization approaches to improving streamflow reconstructions with ring-width indices of riparian trees. *Holocene* **2015**, *25*, 1093–1101. [CrossRef]
82. Gärtner, H.; Cherubini, P.; Fonti, P.; von Arx, G.; Schneider, L.; Nievergelt, D.; Verstege, A.; Bast, A.; Schweingruber, F.H.; Büntgen, U. A Technical Perspective in Modern Tree-ring Research - How to Overcome Dendroecological and Wood Anatomical Challenges. *J. Vis. Exp.* **2015**, *97*, e52337. [CrossRef]
83. Zhang, X.; Li, J.; Liu, X.; Chen, Z. Improved EEMD-based standardization method for developing long tree-ring chronologies. *J. For. Res.* **2019**, *31*, 2217–2224. [CrossRef]
84. Arbellay, E.; Jarvis, I.; Chavardès, R.D.; Daniels, L.; Stoffel, M. Tree-ring proxies of larch bud moth defoliation: Latewood width and blue intensity are more precise than tree-ring width. *Tree Physiol.* **2018**, *38*, 1237–1245. [CrossRef]
85. Balzano, A.; Novak, K.; Humar, M.; Čufar, K. Application of confocal laser scanning microscopy in dendrochronology. *Les/Wood* **2019**, *68*, 5–17. [CrossRef]
86. Samusevich, A.; Lexa, M.; Vejpustková, M.; Altman, J.; Zeidler, A. Comparison of methods for the demarcation between earlywood and latewood in tree rings of Norway spruce. *Dendrochronologia* **2020**, *60*, 125686. [CrossRef]
87. Zhang, W.; Zhao, T.; Su, X.; Wu, B.; Min, Z.; Tian, Y. A Tree Ring Measurement Method Based on Error Correction in Digital Image of Stem Analysis Disk. *Forests* **2021**, *12*, 464. [CrossRef]

88. Björklund, J.; von Arx, G.; Fonti, P.; Stridbeck, P.; De Mil, T.; Neycken, A.; Seftigen, K. The utility of bulk wood density for tree-ring research. *Dendrochronologia* **2021**, *69*, 125880. [CrossRef]
89. Rongpipi, S.; Ye, D.; Gomez, E.D.; Gomez, E. Progress and Opportunities in the Characterization of Cellulose—An Important Regulator of Cell Wall Growth and Mechanics. *Front. Plant Sci.* **2019**, *9*, 1894. [CrossRef]
90. Khalili, A.A.; Ahmad, M.R. A Review of Cell Adhesion Studies for Biomedical and Biological Applications. *Int. J. Mol. Sci.* **2015**, *16*, 18149–18184. [CrossRef]
91. Cascione, M.; de Matteis, V.; Rinaldi, R.; Leporatti, S. Atomic force microscopy combined with optical microscopy for cells investigation. *Microsc. Res. Tech.* **2016**, *80*, 109–123. [CrossRef]
92. Garcia, R. Nanomechanical mapping of soft materials with the atomic force microscope: Methods, theory and applications. *Chem. Soc. Rev.* **2020**, *49*, 5850–5884. [CrossRef]
93. Davì, V.; Chevalier, L.; Guo, H.; Tanimoto, H.; Barrett, K.; Couturier, E.; Boudaoud, A.; Minc, N. Systematic mapping of cell wall mechanics in the regulation of cell morphogenesis. *Proc. Natl. Acad. Sci. USA* **2019**, *116*, 13833–13838. [CrossRef]
94. Moghaddam, M.S.; Bulcke, J.V.D.; Wålinder, M.; Claesson, P.M.; Van Acker, J.; Swerin, A. Microstructure of chemically modified wood using X-ray computed tomography in relation to wetting properties. *Holzforschung* **2017**, *71*, 119–128. [CrossRef]
95. Thomas, J.; Collings, D.A. Imaging spiral grain in pinus radiate with X-ray microtomography. In *Wood Is Good: Current Trends and Future Prospects in Wood Utilization*; Springer Nature Singapore Pte Ltd.: Singapore, 2017; pp. 29–36. [CrossRef]
96. Kang, X.; Kirui, A.; Widanage, M.C.D.; Mentink-Vigier, F.; Cosgrove, D.J.; Wang, T. Lignin-polysaccharide interactions in plant secondary cell walls revealed by solid-state NMR. *Nat. Commun.* **2019**, *10*, 347. [CrossRef] [PubMed]
97. Alves, E.E.N.; Rodriguez, D.R.O.; Rocha, P.D.A.; Vergütz, L.; Junior, L.S.; Hesterberg, D.; Pessenda, L.C.R.; Tomazello-Filho, M.; da Costa, L.M. Synchrotron-based X-ray microscopy for assessing elements distribution and speciation in mangrove tree-rings. *Results Chem.* **2021**, *3*, 100121. [CrossRef]
98. Jacquin, P.; Longuetaud, F.; Leban, J.-M.; Mothe, F. X-ray microdensitometry of wood: A review of existing principles and devices. *Dendrochronologia* **2017**, *42*, 42–50. [CrossRef]
99. Dhali, K.; Ghasemlou, M.; Daver, F.; Cass, P.; Adhikari, B. A review of nanocellulose as a new material towards environmental sustainability. *Sci. Total Environ.* **2021**, *775*, 145871. [CrossRef]
100. Nasution, H.; Yahya, E.B.; Khalil, H.P.S.A.; Shaah, M.A.; Suriani, A.B.; Mohamed, A.; Alfatah, T.; Abdullah, C.K. Extraction and Isolation of Cellulose Nanofibers from Carpet Wastes Using Supercritical Carbon Dioxide Approach. *Polymers* **2022**, *14*, 326. [CrossRef]
101. Tortorella, S.; Buratti, V.V.; Maturi, M.; Sambri, L.; Franchini, M.C.; Locatelli, E. Surface-Modified Nanocellulose for Application in Biomedical Engineering and Nanomedicine: A Review. *Int. J. Nanomed.* **2020**, *ume 15*, 9909–9937. [CrossRef]
102. Thakur, V.; Guleria, A.; Kumar, S.; Sharma, S.; Singh, K. Recent advances in nanocellulose processing, functionalization and applications: A review. *Mater. Adv.* **2021**, *2*, 1872–1895. [CrossRef]
103. Thomas, B.; Raj, M.C.; Athira, K.B.; Rubiah, M.H.; Joy, J.; Moores, A.; Drisko, G.L.; Sanchez, C. Nanocellulose, a Versatile Green Platform: From Biosources to Materials and Their Applications. *Chem. Rev.* **2018**, *118*, 11575–11625. [CrossRef]
104. Bourmaud, A.; Beaugrand, J.; Shah, D.U.; Placet, V.; Baley, C. Towards the design of high-performance plant fibre composites. *Prog. Mater. Sci.* **2018**, *97*, 347–408. [CrossRef]
105. Lee, H.-R.; Kim, K.; Mun, S.C.; Chang, Y.K.; Choi, S.Q. A new method to produce cellulose nanofibrils from microalgae and the measurement of their mechanical strength. *Carbohydr. Polym.* **2018**, *180*, 276–285. [CrossRef] [PubMed]
106. Patera, A.; Bonnin, A.; Mokso, R. Micro- and Nano-Scales Three-Dimensional Characterisation of Softwood. *J. Imaging* **2021**, *7*, 263. [CrossRef] [PubMed]
107. Zarna, C.; Opedal, M.T.; Echtermeyer, A.T.; Chinga-Carrasco, G. Reinforcement ability of lignocellulosic components in biocomposites and their 3D printed applications—A review. *Compos. Part C Open Access* **2021**, *6*, 100171. [CrossRef]
108. Wu, Y.; Wu, X.; Yang, F.; Zhang, H.; Feng, X.; Zhang, J. Effect of Thermal Modification on the Nano-Mechanical Properties of the Wood Cell Wall and Waterborne Polyacrylic Coating. *Forests* **2020**, *11*, 1247. [CrossRef]
109. Huang, Y.H.; Fei, B.H.; Yu, Y.; Wang, S.Q.; Shi, Z.Q.; Zhao, R.J. Modulus of elasticity and hardness of compression and oppo-site wood cell walls of masson pine. *Bioresources* **2012**, *7*, 3028. [CrossRef]
110. Gindl, W.; Gupta, H.S.; Schöberl, T.; Lichtenegger, H.; Fratzl, P. Mechanical properties of spruce wood cell walls by nanoindentation. *Appl. Phys. A* **2004**, *79*, 2069–2073. [CrossRef]
111. Jäger, A.; Hofstetter, K.; Buksnowitz, C.; Gindl-Altmutter, W.; Konnerth, J. Identification of stiffness tensor components of wood cell walls by means of nanoindentation. *Compos. Part A Appl. Sci. Manuf.* **2011**, *42*, 2101–2109. [CrossRef]
112. Yu, Q.; Legros, M.; Minor, A. In situTEM nanomechanics. *MRS Bull.* **2015**, *40*, 62–70. [CrossRef]
113. Yu, Y.; Fei, B.; Zhang, B.; Yu, X. Cell-wall mechanical properties of bamboo investigated by in-situ imaging nanoindentation. *Wood Fiber Sci.* **2007**, *39*, 527–535.
114. Raghavan, R.; Adusumalli, R.-B.; Buerki, G.; Hansen, S.; Zimmermann, T.; Michler, J. Deformation of the compound middle lamella in spruce latewood by micro-pillar compression of double cell walls. *J. Mater. Sci.* **2012**, *47*, 6125–6130. [CrossRef]
115. Klímek, P.; Sebera, V.; Tytko, D.; Brabec, M.; Lukeš, J. Micromechanical properties of beech cell wall measured by micropillar compression test and nanoindentation mapping. *Holzforschung* **2020**, *74*, 899–904. [CrossRef]

116. Golovin, Y.I.; Tyurin, A.I.; Golovin, D.Y.; Samodurov, A.A.; Matveev, S.M.; Yunack, M.A.; Vasyukova, I.A.; Zakharova, O.V.; Rodaev, V.V.; Gusev, A.A. Relationship between Thermal Diffusivity and Mechanical Properties of Wood. *Materials* **2022**, *15*, 632. [CrossRef] [PubMed]
117. Golovin, Y.I.; Tyurin, A.I.; Gusev, A.A.; Matveev, S.M.; Golovin, D.Y.; Samodurov, A.A.; Vasyukova, I.A.; Yunack, M.A.; Kolesnikov, E.A.; Zakharova, O.V. Scanning nanoindentation as an instrument of studying local mechanical properties distribution in wood and a new technique for dendrochronology. *Zhurnal Tekhnicheskoy Fiz.* **2022**, *92*, 575–587. (In Russia) [CrossRef]
118. Golovin, Y.I.; Tyurin, A.I.; Gusev, A.A.; Matveev, S.M.; Golovin, D.Y.; Vasyukova, I.A. Distribution of mechanical properties in annual growth rings of deciduous trees measured using scanning nanoindentation. *Tech. Phys. Lett.* **2022**, *in press*.
119. Golovin, Y.I.; Tyurin, A.I.; Golovin, D.Y.; Samodurov, A.A.; Vasyukova, I.A. Nanoindentation as a Tool for High-Resolution Dendrochronology. *Sov. Phys. J.* **2021**, *63*, 2041–2042. [CrossRef]
120. Carrillo-Varela, I.; Valenzuela, P.; Gasitua, W.; Mendoca, R.T. An evaluation of fiber biometry and nanomechanical proper-ties of different eucalyptus species. *Bioresources* **2019**, *14*, 6433. [CrossRef]
121. Stanzl-Tschegg, S.; Beikircher, W.; Loidl, D. Comparison of mechanical properties of thermally modified wood at growth ring and cell wall level by means of instrumented indentation tests. *Holzforschung* **2009**, *63*, 443–448. [CrossRef]
122. Borgström, E. Design of timber structures. In *Structural Aspects of Timber Construction-1*; Swedish Forest Industries Federation: Stockholm, Sweden, 2016.
123. Shah, D.U.; Reynolds, T.; Ramage, M.H. The strength of plants: Theory and experimental methods to measure the mechanical properties of stems. *J. Exp. Bot.* **2017**, *68*, 4497–4516. [CrossRef]
124. ASTM. *Standard Methods for Testing Small Clear Specimens of Timber*; ASTM D143–94; ASTM International: West Conshohocken, PA, USA, 2000. [CrossRef]
125. Porteous, J.; Kermani, A. Timber as a Structural Material. In *Structural Timber Design to Eurocode 5*; Blackwell: Padstow, UK, 2008; pp. 1–49. [CrossRef]
126. Hoffmeyer, P. Strength under long term loading. In *Timber Engineering*; Thelandersson, S., Larsen, H.J., Eds.; Wiley: Chichester, UK, 2003; pp. 131–152.
127. Jones, D.; Sandberg, D.; Goli, G.; Todaro, L. *Wood Modification in Europe: A State-of-the-Art about Processes, Products and Applications*; Firenze University Press: Firenze, Italy, 2020.
128. Sandberg, D.; Kutnar, A.; Mantanis, G. Wood modification technologies—A review. *iForest Biogeosci. For.* **2017**, *10*, 895–908. [CrossRef]
129. Trache, D.; Tarchoun, A.F.; Derradji, M.; Hamidon, T.S.; Masruchin, N.; Brosse, N.; Hussin, M.H. Nanocellulose: From Fundamentals to Advanced Applications. *Front. Chem.* **2020**, *8*, 392. [CrossRef]
130. Thybring, E.; Fredriksson, M. Wood Modification as a Tool to Understand Moisture in Wood. *Forests* **2021**, *12*, 372. [CrossRef]
131. Huang, C.; Gong, M.; Chui, Y.; Chan, F. Mechanical behaviour of wood compressed in radial direction-part I. New method of determining the yield stress of wood on the stress-strain curve. *J. Bioresour. Bioprod.* **2020**, *5*, 186–195. [CrossRef]
132. Huang, C.; Chui, Y.; Gong, M.; Chana, F. Mechanical behaviour of wood compressed in radial direction: Part II. Influence of temperature and moisture content. *J. Bioresour. Bioprod.* **2020**, *5*, 266–275. [CrossRef]
133. Ding, L.; Han, X.; Jiang, S. Impregnation of poplar wood with multi-functional composite modifier and induction of in-situ polymerization by heating. *J. Wood Chem. Technol.* **2021**, *41*, 220–228. [CrossRef]
134. Han, X.; Wang, Z.; Ding, L.; Chen, L.; Wang, F.; Pu, J.; Jiang, S. Water molecule-induced hydrogen bonding between cellulose nanofibers toward highly strong and tough materials from wood aerogel. *Chin. Chem. Lett.* **2021**, *32*, 3105–3108. [CrossRef]
135. Han, X.; Wu, W.; Wang, J.; Tian, Z.; Jiang, S. Hydrogen-Bonding-Aided Fabrication of Wood Derived Cellulose Scaffold/Aramid Nanofiber into High-Performance Bulk Material. *Materials* **2021**, *14*, 5444. [CrossRef]
136. Tian, F.; Chen, L.; Xu, X. Dynamical mechanical properties of wood-high density polyethylene composites filled with recycled rubber. *J. Bioresour. Bioprod.* **2021**, *6*, 152–159. [CrossRef]
137. Armstrong, R.W. 60 Years of Hall-Petch: Past to Present Nano-Scale Connections. *Mater. Trans.* **2014**, *55*, 2–12. [CrossRef]
138. Cordero, Z.; Knight, B.E.; Schuh, C.A. Six decades of the Hall–Petch effect – a survey of grain-size strengthening studies on pure metals. *Int. Mater. Rev.* **2016**, *61*, 495–512. [CrossRef]
139. Voyiadjis, G.Z.; Yaghoobi, M. *Size Effects in Plasticity: From Macro- to Nano-*; Academic Press: New York, NY, USA, 2019.
140. Voyiadjis, G.Z.; Yaghoobi, M. Review of Nanoindentation Size Effect: Experiments and Atomistic Simulation. *Crystals* **2017**, *7*, 321. [CrossRef]
141. Toivola, Y.; Stein, A.; Cook, R. Depth-sensing indentation response of ordered silica foam. *J. Mater. Res.* **2004**, *19*, 260–271. [CrossRef]
142. Maréchal, M.; Estrada, E.D.C.; Moulin, G.; Almeida, G.; Lv, P.; Cuvelier, G.; Bonazzi, C. New insulating and refractory mineral foam: Structure and mechanical properties. *Mater. Sci. Eng. A* **2020**, *780*, 139153. [CrossRef]
143. Kasyap, S.S.; Senetakis, K. Application of Nanoindentation in the Characterization of a Porous Material with a Clastic Texture. *Materials* **2021**, *14*, 4579. [CrossRef] [PubMed]
144. Devillard, J.; Adrien, J.; Roux, S.; Meille, S.; Maire, E. Highlighting the role of heterogeneity on the indentation hardness of foamed gypsum. *J. Eur. Ceram. Soc.* **2020**, *40*, 3795–3805. [CrossRef]
145. Sarkisyan, V.; Sobolev, R.; Frolova, Y.; Vorobiova, I.; Kochetkova, A. A Study of the Quantitative Relationship between Yield Strength and Crystal Size Distribution of Beeswax Oleogels. *Gels* **2022**, *8*, 39. [CrossRef]

146. Lu, S.; Kan, Q.; Zaiser, M.; Li, Z.; Kang, G.; Zhang, X. Size-dependent yield stress in ultrafine-grained polycrystals: A multiscale discrete dislocation dynamics study. *Int. J. Plast.* **2021**, *149*, 103183. [CrossRef]
147. Li, Y.; Bushby, A.J.; Dunstan, D.J. The Hall–Petch effect as a manifestation of the general size effect. *Proc. R. Soc. A* **2016**, *472*, 20150890. [CrossRef]
148. Coussy, O. *Mechanics and Physics of Porous Solids*; John Wiley & Sons, Ltd.: Chichester, UK, 2010. [CrossRef]
149. MacMinn, C.W.; Dufresne, E.R.; Wettlaufer, J.S. Large Deformations of a Soft Porous Material. *Phys. Rev. Appl.* **2016**, *5*, 044020. [CrossRef]
150. Bouterf, A.; Maire, E.; Roux, S.; Hild, F.; Brajer, X.; Gouillart, E.; Boller, E. Analysis of compaction in brittle foam with multiscale indentation tests. *Mech. Mater.* **2018**, *118*, 22–30. [CrossRef]
151. Cook, R.F. Model for instrumented indentation of brittle open-cell foam. *MRS Commun.* **2018**, *8*, 1267–1273. [CrossRef]
152. Zhang, Y.; Zhang, Y.; Tang, L.; Liu, Z.; Jiang, Z.; Liu, Y.; Zhou, L.; Zhou, X. Uniaxial compression constitutive equations for saturated hydrogel combined water-expelled behavior with environmental factors and the size effect. *Mech. Adv. Mater. Struct.* **2021**, *28*, 1–12. [CrossRef]
153. Guével, A.; Rattez, H.; Veveakis, E. Morphometric description of strength and degradation in porous media. *Int. J. Solids Struct.* **2022**, *241*, 111454. [CrossRef]
154. Karimi, M.; Massoudi, M.; Walkington, N.; Pozzi, M.; Dayal, K. Energetic formulation of large-deformation poroelasticity. *Int. J. Numer. Anal. Methods Géoméch.* **2022**, *46*, 910–932. [CrossRef]
155. Weibull, W. A Statistical Distribution Function of Wide Applicability. *J. Appl. Mech.* **1951**, *18*, 293–297. [CrossRef]
156. Kasyap, S.S.; Li, S.; Senetakis, K. Investigation of the mechanical properties and the influence of micro-structural characteristics of aggregates using micro-indentation and Weibull analysis. *Constr. Build. Mater.* **2020**, *271*, 121509. [CrossRef]
157. Catena, A.; Catena, G. Overview of Thermal Imaging for Tree Assessment. *Arboric. J.* **2008**, *30*, 259–270. [CrossRef]
158. Vidal, D.; Pitarma, R. Infrared Thermography Applied to Tree Health Assessment: A Review. *Agriculture* **2019**, *9*, 156. [CrossRef]
159. Palumbo, D.; Cavallo, P.; Galietti, U. An investigation of the stepped thermography technique for defects evaluation in GFRP materials. *NDT E Int.* **2018**, *102*, 254–263. [CrossRef]
160. Addepalli, S.; Zhao, Y.; Roy, R.; Galhenege, W.; Colle, M.; Yu, J.; Ucur, A. Non-destructive evaluation of localised heat damage occurring in carbon composites using thermography and thermal diffusivity measurement. *Measurement* **2018**, *131*, 706–713. [CrossRef]
161. Meola, C.; Toscano, C. Flash Thermography to Evaluate Porosity in Carbon Fiber Reinforced Polymer (CFRPs). *Materials* **2014**, *7*, 1483–1501. [CrossRef]
162. Guo, J.; Gao, X.; Toma, E.; Netzelmann, U. Anisotropy in carbon fiber reinforced polymer (CFRP) and its effect on induction thermography. *NDT E Int.* **2017**, *91*, 1–8. [CrossRef]
163. Adamczyk, W.; Ostrowski, Z.; Ryfa, A. Development of a non-destructive technique for measuring thermal conductivity of material with small anisotropy based on application of the reduced order technique. *Measurement* **2020**, *165*, 108078. [CrossRef]
164. Lages, E.N.; Marques, S.P.C. Prediction of effective thermal conductivity of multiphase composites with periodic microstructures using an expanded micromechanical model. *Int. J. Therm. Sci.* **2021**, *171*, 107226. [CrossRef]
165. Golovin, Y.I.; Tyurin, A.I.; Golovin, D.Y.; Samodurov, A.A.; Vasyukova, I.A. The Relationship between the Mechanical Properties of Anisotropic Materials and Their Thermophysical Characteristics Using the Example of Pine Wood. *Tech. Phys. Lett.* **2021**, *47*, 92–95. [CrossRef]
166. Golovin, Y.I.; Golovin, D.Y.; Samodurov, A.A.; Tyurin, A.I.; Kabanov, D.A. Correlation between the Mechanical and Thermal Properties of Common Pine Wood (*Pínus sylvéstris* L.). *Bull. Russ. Acad. Sci. Phys.* **2021**, *85*, 723–727. [CrossRef]
167. Golovin, D.Y.; Divin, A.G.; Samodurov, A.A.; Tyurin, A.I.; Golovin, Y.I. Temperature diffusivity measurement and nondestructive testing requiring no extensive sample preparation and using stepwise point heating and IR thermography. In *Failure Analysis*; Huang, Z.-M., Hemeda, S., Eds.; InTech: London, UK, 2019; pp. 124–160. [CrossRef]
168. Golovin, D.Y.; Divin, A.G.; Samodurov, A.A.; Tyurin, A.I.; Golovin, Y.I. A New Rapid Method of Determining the Thermal Diffusivity of Materials and Finished Articles. *J. Eng. Phys. Thermophys.* **2020**, *93*, 234–240. [CrossRef]
169. Golovin, D.Y.; Tyurin, A.I.; Samodurov, A.A.; Divin, A.G.; Golovin, Y.I. *Dinamicheskiye Termograficheskiye Metody Nerazrushayushchego Ekspress-Kontrolya (Dynamic thermographic methods of NDT)*; Technosfera: Moscow, Russia, 2019. (In Russian)
170. Adhikary, S.K.; Rudžionis, Ž.; Tučkutė, S. Characterization of novel lightweight self-compacting cement composites with incorporated expanded glass, aerogel, zeolite and fly ash. *Case Stud. Constr. Mater.* **2022**, *16*, e00879. [CrossRef]

nanomaterials

Article

Wear of Functionally Graded Coatings under Frictional Heating Conditions

Vladimir B. Zelentsov, Polina A. Lapina * and Boris I. Mitrin

Research and Education Center "Materials", Don State Technical University, 344000 Rostov-on-Don, Russia; vbzelen@gmail.com (V.B.Z.); boris.mitrin@gmail.com (B.I.M.)
* Correspondence: polina_azarova86@mail.ru

Abstract: Multilayered and functionally graded coatings are extensively used for protection against wear of the working surfaces of mechanisms and machines subjected to sliding contact. The paper considers the problem of wear of a strip made of a functionally graded material, taking into account the heating of the sliding contact from friction. Wear is modeled by a moving strip along the surface of a hard abrasive in the form of a half-plane. With the help of the integral Laplace transform with respect to time, the solutions are constructed as convolutions from the law of the introduction of an abrasive into the strip and the original in the form of a contour integral of the inverse Laplace transform. The study of the integrands of contour quadratures in the complex plane allowed determination of the regions of stable solutions to the problem. Unstable solutions of the problem lead to the concept of thermoelastic instability of the contact with friction and formed regions of unstable solutions. The solutions obtained made it possible to determine a formula for the coefficient of functionally graded inhomogeneity of the coating material and to study its effect on the occurrence of thermoelastic instability of the contact taking friction into account, as well as on its main characteristics: temperature, displacement, stress and wear of the functionally graded material of the coating. The effects of the abrasive speed, contact stresses and temperature on wear of the coating with the functionally graded inhomogeneity of the material by the depth were investigated.

Keywords: functionally graded material; thermoelasticity; sliding contact; wear; heating from friction; thermoelastic instability

Citation: Zelentsov, V.B.; Lapina, P.A.; Mitrin, B.I. Wear of Functionally Graded Coatings under Frictional Heating Conditions. *Nanomaterials* **2022**, *12*, 142. https://doi.org/10.3390/nano12010142

Academic Editor: Mohammad Malikan

Received: 13 December 2021
Accepted: 28 December 2021
Published: 31 December 2021

1. Introduction

Design of multilayered materials increased rapidly in the 1980–1990s, thus enabling the introduction of such materials with properties variable by depth in a number of industries such as construction, transport and other fields. Overall, their necessity in such fields was due to the increasing requirements for strength, wear and tear of details and parts. Recently, methods and technologies for the manufacturing of protective coatings from functionally graded materials (FGM) to micro- and nanometer technological levels have been developed: the centrifugal method [1–7], the technique of pulsed-laser deposition [8,9], the technique of magnetron sputtering [9], the plasma-spray technique [10,11], electrophoretic deposition and anodizing [12,13], micro-arc oxidation [14] and others [15].

FGM coatings are widely used to protect friction surfaces from wear. Generally, the tribological properties of FGM coatings are investigated by experimental methods, such as disc–pad [1,5], ball–disc [3,9,11], pin-on-disc [4,13,15], ring–pad [8], pin–plate [10] and other tests [7,14]. The coating wear as a result of testing is determined by examining the wear track with a microscope or a profilometer [1–3,9,10,13] or by weighing the sample before and after the test [4–8,11,15]. In this case, tests are carried out at one value of the experimental load or for a limited set (3–5 options).

The need to optimize the design of FGM coatings, to predict wear of the working surfaces of mechanisms and machines, to diagnose and prevent abnormal situations necessitates mathematical modeling of the wear process. In modern microelectronics, semiconductor materials are used in the form of thin plates with a thickness not exceeding 20–30 μm down to 100 nm, and the presence of a functionally graded inhomogeneity of the material (FGIM) by depth. The process of thinning such plates in practice is carried out by grinding, polishing and washing. Mathematical modeling of the grinding process of the workpiece material is associated with the elucidation of the degree of influence of the properties of the FGIM of the workpiece material on the grinding process; on the nature of the workpiece heating; on the possibility of the occurrence of thermoelastic instability of the sliding contact; on acceleration/deceleration the grinding process, etc.

A lot of research is devoted to mathematical modeling of the contact of bodies with FGM coatings [16–26]. However, insufficient attention has been paid to the problem of modeling of wear or grinding of FGM coatings. There has been little research conducted in this direction [27].

Today, when solving problems of material wear, Archard's relations are most often used [28]. Dow and Burton were among the first investigators who used the Archard relation in the study of wear under conditions of heat release from friction [29], where the conditions for the occurrence of thermoelastic instability of blade sliding along the surface of a half-space were investigated using the method of small perturbations. Aleksandrov and Annakulova considered contact problems taking into account heat release from friction and wear of the coating [30], as well as the problem of mutual wear of coatings [31]. An attempt to develop a thermodynamic model to describe thermomechanical phenomena at a contact, taking into account friction and wear, was undertaken in [32,33]. Beginning with [34,35] a new direction of the development of the model of contact of two elastic bodies considering friction, wear and heat release, based on the principle of virtual energy and the basic laws of thermodynamics, has emerged. The finite element implementation of such a model in a two-dimensional statement was conducted in [36]. In [37–41] the integral Laplace transform was used with the solution in the form of functional series over the poles of the integrands of the contour quadratures of the inverse Laplace transform to solve contact thermoelasticity problems regarding wear. The solution method allows one to establish the parameters of the boundary of thermoelastic instability of a contact with a friction to study the properties of the obtained solutions. In [42], using the method of integral transforms, the contact problem of sliding an elastic coating over the surface of another coating with friction, wear and heat release from friction was reduced to solving a differential equation, and the conditions of thermoelastic stability of such a system were considered. Quasi-static and dynamic uncoupled contact problems of thermoelasticity regarding friction and wear of a rod were considered in [43]. The conditions for the thermoelastic instability emergence during mutual wear of surfaces made of different materials were considered in [44,45]. Due to the large number of parameters in the problems of thermofrictional contact and wear, one-dimensional quasi-static problems were often considered. The connectivity of the deformation fields and temperature in the listed works was neglected and the problems of uncoupled thermoelasticity were considered. The coupled problem of thermoelasticity on wear of a coating taking into account frictional heat release in a quasi-static formulation was considered in [46].

In the present paper, we consider a problem of wear of an elastic coating in the form of a strip made of FGM by a hard abrasive in the form of a half-plane, which slides along the strip at a constant speed, heating it due to friction. The influence of the FGIM of the strip on the process of wear (grinding), heating from friction, the conditions for the occurrence of thermoelastic instability of the sliding contact are investigated.

2. Problem Statement

To investigate the effect of the FGIM of the coating on its wear we consider a quasi-static contact problem of sliding of a rigid heat-insulated abrasive, half-plane I $(h \leq x < \infty)$, sliding with constant velocity V, over the upper surface $(x = h)$ of the elastic thermally conductive coating with thickness $h (0 \leq x \leq h)$. The lower surface of the coating is perfectly bonded to a rigid substrate, half-plane II $(-\infty < x < 0)$. The coating shear modulus $\mu(x)$ varies with depth $0 \le x \le h$. During abrasive (half-plane I) sliding, the coating wear takes place, which can also be thought of as the abrasive grinding of the coating surface. The frictional heat generated at the contact interface flows into the coating. From the initial time moment, the abrasive (half-plane I) slides along the y axis and deforms the upper $(x = h)$ surface of the coating in the negative direction of the x axis according to the indentation law $\Delta(t)$. Until the initial time moment, the coating was at rest, and its temperature was equal to zero. The scheme of the contact problem is shown in Figure 1.

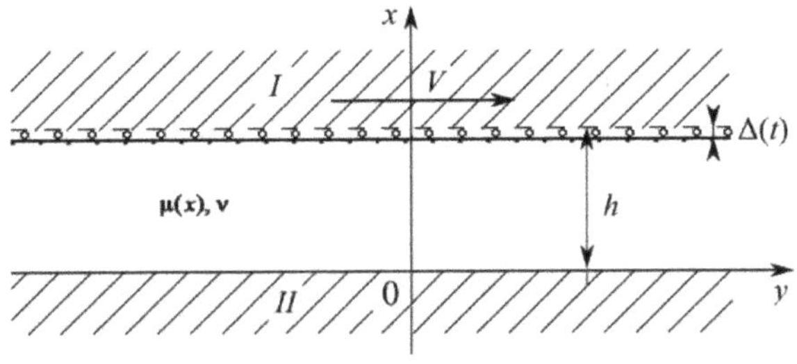

Figure 1. Scheme of the contact problem.

In the described problem formulation, the temperature, stresses and displacements distributions in the coating do not depend on the horizontal coordinate y and are functions only of the vertical coordinate x and time t [37–41]. In the case of a quasi-static formulation without body forces, the stress state of the coating is described by the differential equations of equilibrium

$$\frac{\partial \sigma_{xx}}{\partial x} = 0, \ \frac{\partial \sigma_{xy}}{\partial x} = 0, \quad 0 \le x \le h, \quad t > 0 \tag{1}$$

where $\sigma_{xx}(x,t)$, $\sigma_{xy}(x,t)$ are the normal and shear components of the stresses in the coating.

The heat equation has the form

$$\frac{\partial^2 T}{\partial x^2} - \frac{1}{\kappa}\frac{\partial T}{\partial t} = 0 \quad 0 \le x \le h, \ t > 0 \tag{2}$$

where $T(x,t)$ is the coating temperature, κ is the thermal diffusivity.

The Duhamell–Neumann law takes place [47]

$$\sigma_{xx} = \frac{2(1-\nu)}{1-2\nu}\mu(x)\left(\frac{\partial u}{\partial x} - \frac{1+\nu}{1-\nu}\alpha T\right), \quad \sigma_{xy} = \mu(x)\frac{\partial w}{\partial x}, \tag{3}$$

where $u(x,t)$, $w(x,t)$ are the vertical and horizontal components of the displacements in the coating, $\mu(x)$, ν, α are the shear modulus, Poisson's ratio, coefficient of linear heat expansion of the coating material.

The differential equations of linear uncoupled elasticity are represented by the system of equilibrium Equation (1) and heat Equation (2), which together describe thermoelastic state of the coating.

Boundary conditions for Equation (1) are as follows ($t > 0$):

$$u(h,t) = -\Delta(t) + u_w(t) \tag{4}$$

$$\sigma_{xy}(h,t) = -f\sigma_{xx}(h,t) \tag{5}$$

$$u(0,t) = 0 \tag{6}$$

$$w(0,t) = 0 \tag{7}$$

where f is the coefficient of friction, $u_w(t)$ is the abrasive (half-plane I) displacement due to the coating wear. To find the solution, we use the abrasive wear model [23] in integral form

$$u_w(t) = -fVK^* \int_0^t \sigma_{xx}(h,\tau)d\tau \qquad t > 0 \tag{8}$$

where $\sigma_{xx}(h,t)$ is the compressive normal stress on the contact interface, K^* is the proportionality coefficient between the work of friction forces and the volume of removed material.

The boundary conditions for the heat Equation (2) are ($t > 0$)

$$K\frac{\partial T(h,t)}{\partial x} = Q(t) \tag{9}$$

$$K\frac{\partial T(0,t)}{\partial x} = k(T(0,t) - T(0,0)) \tag{10}$$

where K is the thermal conductivity of the coating material, k is the heat transfer coefficient through the coating–substrate interface, $Q(t) = fV(-\sigma_{xx}(h,t))$ is the amount of frictional heat originatingfrom the contact interface [48]. From (9) it follows that all the heat at the contact is due to friction.

Initial conditions for displacements and temperature are equal to zero:

$$u(x,0) = w(x,0) = T(x,0) = 0, \quad 0 \le x \le h, \tag{11}$$

Thus, the solution of the considered quasi-static thermoelastic contact problem for the elastic FGM coating wear (or grinding) by a certain depth with a hard abrasive in the form of a half-plane, taking into account heating from friction, is derived through the solution of the initial boundary value problem, including the system of the differential equations of elasticity (1) and heat conduction (2) with boundary (4)–(10) and initial (11) conditions. Note that vertical displacements $u(x,t)$, normal stresses $\sigma_{xx}(x,t)$ and temperature $T(x,t)$ are found separately from the horizontal displacements $w(x,t)$. The horizontal displacements $w(x,t)$ are determined from (1), (5), (7) knowing the normal stresses' distribution $\sigma_{xx}(h,t)$.

3. Exact Solution for Arbitrary μ(x)

We proceed to the solution of the formulated problem introducing the Laplace integral transform [49]

$$T^L(x,p) = \int_0^\infty T(x,t)e^{-pt}dt,\ T(x,t) = \frac{1}{2\pi i}\int_{-i\infty+c}^{i\infty+c} T^L(x,p)e^{pt}dp \quad \mathrm{Re}p < c,\ c > 0 \tag{12}$$

To obtain the temperature distribution $T(x,t)$ in the coating, we apply the Laplace transform to heat Equation (2). As a result, the Laplace image $T^L(x,p)$ of the coating temperature has the form

$$T^L(x,p) = A_1 \mathrm{sh}\sqrt{\frac{p}{\kappa}}x + A_2 \mathrm{ch}\sqrt{\frac{p}{\kappa}}x \tag{13}$$

where A_1 and A_2 are arbitrary constants.

The vertical displacements $u(x,t)$ are determined from the first equilibrium Equation (1) and taking into account for the first relation in (3). We use the Laplace transform (12) and obtain the Laplace image $u^L(x,p)$ of the vertical displacements

$$u^L(x,p) = \frac{1+v}{1-v}\alpha\frac{1}{\sqrt{\frac{p}{\kappa}}}\left(A_1 \mathrm{ch}\sqrt{\frac{p}{\kappa}}x + A_2 \mathrm{sh}\sqrt{\frac{p}{\kappa}}x\right) - A_3 B(x) + A_4 \tag{14}$$

where A_1, A_2 are from (13) and A_3, A_4 are additional arbitrary constants. The function $B(x)$ is defined through $\mu(x)$ as follows

$$B(x) = \int_0^x \frac{d\xi}{\mu(\xi)}, \quad 0 \le x \le h \tag{15}$$

where $\mu(x)$ is a continuous function and does not vanish $\mu(x) \neq 0$ for any $x \in [0, h]$.

To determine constants A_k, $k = 1 - 4$ in (13), (14) we apply the Laplace transform to boundary conditions (4), (6), (9) and (10), which results in

$$u^L(h, p) = -\Delta^L(p) + u_w^L(p) \tag{16}$$

$$u^L(0, p) = 0 \tag{17}$$

$$K\frac{dT^L(h, p)}{dx} = -fV\sigma_{xx}^L(h, p) \tag{18}$$

$$K\frac{dT^L(0, p)}{dx} = kT^L(0, p) \tag{19}$$

where

$$u_w^L(p) = -fVK^* \frac{\sigma_{xx}^L(h, p)}{p} \tag{20}$$

$$\sigma_{xx}^L(x, p) = \frac{2(1-v)}{1-2v}\mu_1\left(\frac{du^L(x, p)}{dx} - \frac{1+v}{1-v}\alpha T^L(x, p)\right) \tag{21}$$

where $\mu_1 = \mu(h)$ is the value of the shear modulus at the upper coating boundary and $\Delta^L(p)$ is the Laplace image of the function $\Delta(t)$, representing half-plane I displacement towards the coating.

Substituting (13), (14), (20), (21) into boundary conditions (16)–(19), we obtain the linear system for the determination of the constants A_k, $k = 1 - 4$. Solving this system, we obtain the Laplace images of the temperature, displacement and stress distribution in the coating as follows

$$T^L(x, p) = \frac{1-\nu}{1+\nu}\frac{\hat{V}}{\alpha h}\Delta^L(p)\frac{hB'(h)}{B(h)}\frac{N_T(x, z)}{R(z)} \tag{22}$$

$$N_T(x, z) = \sqrt{z}\left(\mathrm{Bi}\,\mathrm{sh}\sqrt{z}\frac{x}{h} + \sqrt{z}\,\mathrm{ch}\sqrt{z}\frac{x}{h}\right) \tag{23}$$

$$u^L(x, p) = -\Delta^L(p) \cdot \frac{N_u^0(x, z)}{R(z)} \tag{24}$$

$$N_u^0(x, z) = zr(h, z)\frac{B(x)}{B(h)} - \hat{V}\frac{hB'(h)}{B(h)}(r(x, z) - \mathrm{Bi}) \tag{25}$$

$$\sigma_{xx}^L(x, p) = \frac{2(1-v)}{1-2v}\mu(x)\Delta^L(p)\frac{hB'(x)}{B(h)} \cdot \frac{N_\sigma^0(z)}{R(z)} \tag{26}$$

$$N_\sigma^0(z) = zr(h, z) \tag{27}$$

where

$$R(z) = zr(h, z) - \hat{V}\eta((1 - k_w)r(h, z) - \mathrm{Bi}) \tag{28}$$

$$r(x, z) = \mathrm{Bi}\,\mathrm{ch}\sqrt{z}\frac{x}{h} + \sqrt{z}\,\mathrm{sh}\sqrt{z}\frac{x}{h} \tag{29}$$

$$z = \frac{p}{\kappa}h^2, \ \mathrm{Bi} = \frac{kh}{K}, \ k_w = \frac{1-v}{1+v}\frac{KK^*}{\alpha\kappa}, \ \hat{V} = \frac{fV\alpha}{K}\frac{2\mu_1(1+v)h}{1-2v}.$$

For the image $u_w^L(p)$ of the coating wear, we obtain

$$u_w^L(p) = k_w \hat{V} \frac{hB'(h)}{B(h)} \Delta^L(p) \frac{r(h,z)}{R(z)} \tag{30}$$

We apply the inverse Laplace transform to the Laplace images $T^L(x,p)$, $u^L(x,p)$, $\sigma_{xx}^L(x,p)$ and obtain the problem solution in the convolution form ($t > 0$)

$$T(x,t) = \frac{1-v}{1+v} \frac{\hat{V}}{\alpha h} \cdot \frac{hB'(h)}{B(h)} \int_0^t \Delta(\tau) f_T(x, t-\tau) d\tau \tag{31}$$

$$f_T(x,t) = \frac{1}{2\pi i} \int_\Gamma \frac{N_T(x,z)}{t_\kappa R(z)} e^{z\tilde{t}} dz \tag{32}$$

$$u(x,t) = -\int_0^t \Delta(\tau) f_u^0(x, t-\tau) d\tau \tag{33}$$

$$f_u^0(x,t) = \frac{1}{2\pi i} \int_\Gamma \frac{N_u^0(x,z)}{t_\kappa R(z)} e^{z\tilde{t}} dz \tag{34}$$

$$\sigma_{xx}(x,t) = -\frac{2(1-v)}{(1-2v)B(h)} \int_0^t \Delta(\tau) f_\sigma^0(x, t-\tau) d\tau \tag{35}$$

$$f_\sigma^0(x,t) = \frac{1}{2\pi i} \int_\Gamma \frac{N_\sigma(x,z)}{t_\kappa R(z)} e^{z\tilde{t}} dz \tag{36}$$

where $\tilde{t} = \frac{t}{t_\kappa}$, $t_\kappa = \frac{h^2}{\kappa}$.

To obtain (35) from (26) we take into account that $\mu(x)B'(x) \equiv 1$. Coating surface wear $u_{\mathrm{w}}(t)$ is determined by inverting $u_{\mathrm{w}}^L(p)$ in (30)

$$u_{\mathrm{w}}(t) = k_{\mathrm{w}} \hat{V} \frac{hB'(h)}{B(h)} \int_0^t \Delta(\tau) f_w(x, t-\tau) d\tau \tag{37}$$

$$f_{\mathrm{w}}(x,t) = \frac{1}{2\pi i} \int_\Gamma \frac{r(h,z)}{t_\kappa R(z)} e^{z\tilde{t}} dz \tag{38}$$

To find out the existence conditions of integrals in (32), (34), (36), (38) we analyze integrands for $0 \le x \le h$ for large values of the integration variable z ($\arg z = \pi/2$):

$$\begin{aligned} & N_T(x,z)R^{-1}(z) = O\left(z^{-1/2}\right) \ |z| \to \infty \\ & N_u^0(x,z)R^{-1}(z) = \tfrac{B(x)}{B(h)} + O\left(z^{-1/2}\right) \ |z| \to \infty \\ & N_\sigma^0(x,z)R^{-1}(z) = 1 + O\left(z^{-1/2}\right) \ |z| \to \infty \\ & r(h,z)R^{-1}(z) = O(z^{-1}) \ |z| \to \infty \end{aligned} \tag{39}$$

From (39) it follows that the integrands in (34), (36) do not decay at infinity (at $|z| \to \infty$), and the corresponding integrals are divergent and understood in the generalized sense [50]. After regularization of integrals (34), (36) and separation of the generalized part, expressions for the displacements $u(x,t)$ and stresses $\sigma_{xx}(x,t)$ can be written as follows ($t > 0$)

$$u(x,t) = -\frac{B(x)}{B(h)} \Delta(t) - \int_0^t \Delta(\tau) f_u(x, t-\tau) d\tau \qquad 0 \le x \le h,\ t > 0 \tag{40}$$

$$f_u(x,t) = \frac{1}{2\pi i}\int\limits_{\Gamma} \frac{N_u(x,z)}{t_\kappa R(z)} e^{z\tilde{t}} dz \tag{41}$$

$$N_u(x,z) = N_u^0(x,z) - \frac{B(x)}{B(h)} R(z) \tag{42}$$

$$\sigma_{xx}(x,t) = -\frac{2(1-v)}{(1-2v)B(h)}\left(\Delta(t) - \int\limits_0^t \Delta(\tau) f_\sigma(x,t-\tau)d\tau\right) \; 0 \le x \le h, t > 0 \tag{43}$$

$$f_\sigma(x,t) = \frac{1}{2\pi i}\int\limits_{\Gamma} \frac{N_\sigma(x,z)}{t_\kappa R(z)} e^{z\tilde{t}} dz \tag{44}$$

$$N_\sigma(x,z) = N_\sigma^0(x,z) - R(z) \tag{45}$$

where $\tilde{t}$ and t_κ are previously identified, and the contour of integration $\Gamma = \{z : -i\infty + dt_\kappa, +i\infty + dt_\kappa\}$ is a straight line in the complex plane of integration variable z, which is parallel to the imaginary axis and a distance from equal to the value dt_κ. The value of d is chosen so that the contour of integration passes to the right of all of the isolated singular points of integrands.

Integrands in Equations (40) and (43) at $0 \le x \le h$ are meromorphic functions and decay at large z $(-\pi < \arg z < \pi)$

$$\begin{aligned} N_u(x,z)R^{-1}(z) &= O\left(z^{-1/2}\right) |z| \to \infty \\ N_\sigma(x,z)R^{-1}(z) &= O\left(z^{-1/2}\right) |z| \to \infty \end{aligned} \tag{46}$$

These properties of integrands allow us to apply the methods of complex analysis for calculation of the integrals and stability analysis. In quadratures (32), (38), a regularization is carried out to obtain integrands decreasing at infinity, after which they are investigated by the same methods as in (40), (43).

4. Poles of the Integrands

To investigate the stability of the solutions obtained in the previous subsection, we need to study effect of the problem parameters on the integrand poles in integrals (32), (38), (41), (44). The poles of the integrands are zeros of the equation

$$R(z) = zr(h,z) - \eta\hat{V}((1-k_w)r(h,z) - \mathrm{Bi}) = 0, \; |\arg z| < \pi, \; |z| < \infty \tag{47}$$

in the complex plane $z = \xi + i\eta$, where $R(z)$ and $r(x,z)$ at $x = h$ are from (28) and (29), $\eta = hB'(h)/B(h)$ is the coefficient of the FGIM strip. Equation (47) contains four dimensionless parameters of the problem (k_w, $\hat{V}$, Bi, η), which itself contains dimensional parameters, described after (29). In this case, the dimensionless parameter η, the numerical value of which characterizes the FGM coating, consists of $B'(h) = \mu^{-1}(h)$ and $B(h)$ from (15). From the mean value theorem, it follows for $B(h)$ that there is a point $c \in [0,h]$ satisfying the equation

$$B(h) = \mu^{-1}(c)h, \quad c \in [0,h] \tag{48}$$

Then, the parameter η can be represented as the averaged value of the shear modulus in the coating $\mu(c)c \in [0,h]$ divided by the value of $\mu(h)$ at the upper boundary of the coating

$$\eta = \frac{\mu(c)}{\mu(h)}, \quad c \in [0,h] \tag{49}$$

which is its mechanical meaning.

Locating zeros ζ_k of Equation (47) $R(\zeta_k) = 0, k = 0,1,2,\ldots$, their movements in the complex plane $\zeta = \xi + i\eta$ depending on the change in the dimensionless parameters of the problem is the main goal of solving Equation (47).

Equation (47) is solved using numerical methods and complex analysis methods [51]. Analysis of $R(z)$ (47) zeros is similar to [52–54] and is performed for $\hat{V} \in [0, \infty)$ at fixed values of dimensionless parameters k_w, η, Bi. Assuming $\hat{V} = 0$, we find the following equation

$$zr(h, z) = z(\mathrm{Bich}\sqrt{z} + \sqrt{z}\mathrm{sh}\sqrt{z}) = 0 \tag{50}$$

to determine the initial estimates $\zeta_k^0 = \zeta_k(0), k = 0, 1, 2, \ldots$ of zeros $\zeta_k(\hat{V}), k = 0, 1, 2, \ldots$ of Equation (47). In general case, Equation (50) does not allow convenient analytic solutions. However, at Bi $= 0$ Equation (50) possesses analytic solution, and the initial estimations ζ_k^0, $k = 0, 1, 2, \ldots$ of zeros ζ_k, $k = 0, 1, 2, \ldots$ of Equation (47) are found from

$$\zeta_k^0 = -(\pi k)^2, \quad k = 0, 1, 2, \ldots \tag{51}$$

And do not depend on k_w, $\hat{T}$. At Bi $= \infty$, from (50) we obtain another initial estimations ζ_k^0, $k = 0, 1, 2, \ldots$ as follows

$$\zeta_k^0 = -\pi^2(k + 1/2)^2, \quad k = 0, 1, 2, \ldots \tag{52}$$

Asymptotic estimation of ζ_k^0 for large numbers of k has the form of (51). From Formulas (51) and (52) it follows that the initial estimations ζ_k^0, $k = 0, 1, 2, \ldots$ of function $R(z)$ of zeros from (47) are located on the negative part of the real axis or at the origin. As $\hat{V}$ changes from 0 to ∞ at fixed k_w, Bi, η the first two poles ζ_0 and ζ_1 can be located: I—on the real axis $\mathrm{Re}(\zeta_0, \zeta_1) < 0$, $\mathrm{Im}(\zeta_0, \zeta_1) = 0$ at $0 < \hat{V} < \hat{V}_{\mathrm{I}}$; II—in the vertical strip $l_- < \mathrm{Re}(\zeta_0, \zeta_1) < 0$, $|\mathrm{Im}(\zeta_0, \zeta_1)| < \infty$ at $\hat{V}_{\mathrm{I}} < \hat{V} < \hat{V}_{\mathrm{II}}$; III—in the vertical strip $0 < \mathrm{Re}(\zeta_0, \zeta_1) < l_+$, $|\mathrm{Im}(\zeta_0, \zeta_1)| < \infty$ at $\hat{V}_{\mathrm{II}} < \hat{V} < \hat{V}_{\mathrm{III}}$; IV—on the positive part of the real axis $\mathrm{Re}(\zeta_0, \zeta_1) > 0$, $\mathrm{Im}(\zeta_0, \zeta_1) = 0$ at $\hat{V}_{\mathrm{III}} < \hat{V} < \infty$.

The numbers I, II denote regions (we will call them domains of stability), where $\mathrm{Re}(\zeta_0, \zeta_1) < 0$ at $0 < \hat{V} < \hat{V}_{\mathrm{II}}$, while III, IV denote the regions (we will call them domains of instability), where $\mathrm{Re}(\zeta_0, \zeta_1) > 0$ at $\hat{V}_{\mathrm{II}} < \hat{V} < \infty$. Figure 2 shows some examples of the poles $\zeta_0(\hat{V})$ and $\zeta_1(\hat{V})$ trajectories when $\hat{V}$ changes from 0 to ∞ at fixed Bi = 100 for three different values of η = 0.5, 1.0, 2.0 (green, red, blue colors) and following values of k_w= 0.5 (curve ***1***), 0.9 (***2***), 1.0 (***3***), 1.1 (***4***), 1.35 (***5***), 5.0 (***6***). Solid diamonds indicate locations of $\zeta_0(\hat{V})$ and $\zeta_1(\hat{V})$ at $\hat{V}$= 0, while empty ones correspond to $\hat{V} \to \infty$. The point of the trajectory marked as a crossed-out square is a point, after crossing which, with increasing $\hat{V}$, the real poles $\zeta_0(\hat{V})$ and $\zeta_1(\hat{V})$ become a pair of complex conjugated poles, and vice versa. Note that even small change in the parameter k_w, which is proportional to the ratio of the wear factor K^* and the thermal expansion coefficient α, leads to significant variation of $\zeta_0(\hat{V})$ and $\zeta_1(\hat{V})$ trajectories and, to a lesser extent, trajectories of other $\zeta_k(\hat{V}), k = 2, 3, 4, \ldots$. If wear is extensive ($k_w > 1$), both ζ_0, ζ_1 and all others ζ_k, $k = 2, 3, 4, \cdots$ are located in regions I, II (curves ***4–6*** on Figure 2). When thermal expansion of the coating material prevails over wear ($0 < k_w < 1$), the poles ζ_0 and ζ_1 move to the right complex half-plane to regions III, IV (curves ***1–3*** on Figure 2).

One significant property of the poles from regions II, III is noted; they are complex conjugated, i.e., $\zeta_1 = \overline{\zeta_0}$ and $\zeta_0 = \overline{\zeta_1}$.

Wear causes a dramatic effect on the poles' behavior. In quasi-static sliding contact problems with frictional heating but without wear, the poles ζ_0 and ζ_1 remain on the real axis for any $\hat{V} \in [0, \infty)$ [52,53]. In corresponding problems accounting for wear, the poles ζ_0 and ζ_1 can have a non-zero imaginary part, and in this case become complex conjugates $\zeta_1 = \overline{\zeta_0}$, $\zeta_0 = \overline{\zeta_1}$.

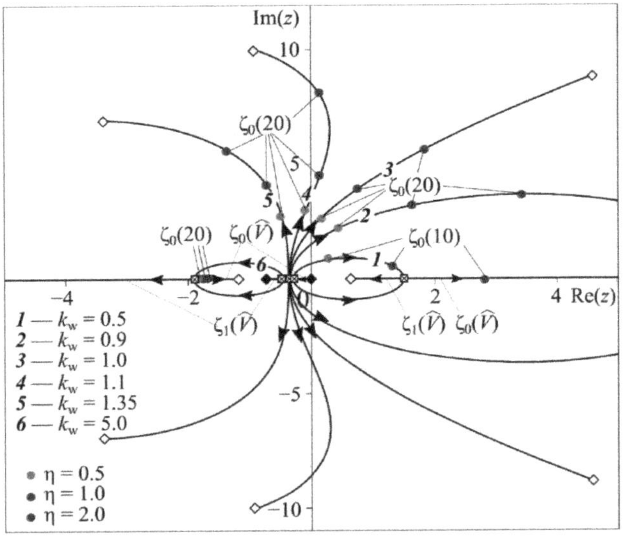

Figure 2. Movement of zeros of $R(z)$ (47) in the complex plane when $\hat{V}$ changes from 0 to ∞.

5. Formulas for Exact Solution

When the poles ζ_k, $k = 0, 1, 2, \ldots$ are known, to calculate the integrals in (32), (41), (44) we calculate the sum of the residues at the points $z = \zeta_k$. For all simple poles we obtain

$$\frac{1}{2\pi i}\int_{\Gamma} \frac{N_a(x,z)}{t_\kappa R(z)} e^{z\tilde{t}} dz = \sum_{k=0}^{\infty} C_a(x,\zeta_k) e^{\zeta_k \tilde{t}}, \tag{53}$$

$$C_a(x,z) = \frac{N_a(x,z)}{t_\kappa R'(z)} \tag{54}$$

By replacing symbolic index a in (53), (54) with T, u or σ, Equation (53) can be used to calculate integral in (32), (41), (44), respectively. If ζ_k and ζ_{k+1}, $k = 0, 1, 2, \ldots$ are complex conjugates ($\zeta_{k+1} = \overline{\zeta_k}$), then

$$C_a(x,z) e^{z\tilde{t}} = 2\mathrm{Re}\frac{N_a(x,z)}{t_\kappa R'(z)} e^{z\tilde{t}} \tag{55}$$

and summation in (53) can be entered by even numbers $k = 2n$, $n = 0, 1, 2, \ldots$ for complex conjugate ζ_k, $k = 0, 1, 2, \ldots$. Using (54) for (32), (41), (44) we obtain

$$f_a(x,t) = \frac{1}{2\pi i}\int_{\Gamma} \frac{N_a(x,z)}{t_\kappa R(z)} e^{z\tilde{t}} dz = \sum_{k=0}^{\infty} C_a(x,\zeta_k) e^{\zeta_k \tilde{t}} a = T, u, \sigma \tag{56}$$

and write out the problem solution in form of series

$$T(x,t) = \frac{1-v}{1+v}\frac{\hat{V}}{\alpha}\frac{B'(h)}{B(h)} \sum_{k=0}^{\infty} C_T(x,\zeta_k) D(\zeta_k,t) \quad 0 \le x \le h,\ t > 0 \tag{57}$$

$$u(x,t) = -\frac{B(x)}{B(h)}\Delta(t) + \sum_{k=0}^{\infty} C_u(x,\zeta_k) D(\zeta_k,t) \quad 0 \le x \le h,\ t > 0 \tag{58}$$

$$\sigma_{xx}(x,t) = -\frac{2(1-v)}{(1-2v)B(h)}\left(\Delta(t) - \sum_{k=0}^{\infty} C_\sigma(x,\zeta_k) D(\zeta_k,t)\right) \quad 0 \le x \le h,\ t > 0 \tag{59}$$

where $C_a(x,z)$ is calculated either by (54) or (55), and $D(z,t)$ is calculated as follows

$$D(z,t) = \int_0^t \Delta(\tau)\exp(z(t-\tau)/t_\kappa)d\tau \qquad t > 0 \tag{60}$$

Calculating $f_w^0(t)$ in (38) by relation

$$f_w^0(t) = \frac{1}{2\pi i}\int_\Gamma \frac{r(h,z)}{t_\kappa R(z)} e^{z\tilde{t}} dz = \sum_{k=0}^{\infty} B_w(\zeta_k) e^{\zeta_k \tilde{t}}, \quad C_w = \frac{r(h,z)}{t_\kappa R'(z)} \tag{61}$$

and substituting it into (37), we obtain a formula for calculation of the coating wear

$$u_w(t) = k_w \hat{V} \frac{hB'(h)}{B(h)} \sum_{k=0}^{\infty} C_w(\zeta_k) D(\zeta_k, t), \quad t > 0 \tag{62}$$

The horizontal displacements $w(x,t)$ are determined from (1), (5), (7) and after integrating (1) take the form

$$w(x,t) = -fB(x)\sigma_{xx}(h,t) \qquad 0 \le x \le h, \ t > 0 \tag{63}$$

6. Domains of Stability and Instability

Analysis of Formulas (57)–(59) for $T(x,t)$, $u(x,t)$, $\sigma_{xx}(x,t)$ shows that, in case of $\mathrm{Re}(\zeta_k) < 0$, $k = 0,1,2,\ldots$ the solution is stable and tends towards a stationary state with increasing t. However, if at least one of ζ_k, $k = 0,1,2,\ldots$ has $\mathrm{Re}(\zeta_k) > 0$, then the magnitude of the solution grows indefinitely when $t \to \infty$ and oscillates with the frequency $\mathrm{Im}(\zeta_k) \neq 0$, which indicates instability of the problem. If the indentation law $\Delta(t)$ is a bounded function

$$m < \Delta(t) < M \qquad m, M > 0, \qquad 0 < t < \infty$$

then the following estimate for the integral in (60) takes place

$$|D(\zeta_k, t)| \ge m\left|\frac{1 - e^{\zeta_k \tilde{t}}}{\zeta_k}\right| \text{ at } \mathrm{Re}(\zeta_k) > 0 \quad k = 0,1,2,\ldots$$

Trajectories of the poles $\zeta_k(\hat{V})$, $k = 0,1,2,\ldots \hat{V} \in [0,\infty)$, lying in the left complex half-plane ($\mathrm{Re}(\zeta_k) < 0$), correspond to the stable solution, and therefore we refer to domains I, II as the domains of stable solutions. Domains III, IV, lying in the right complex half-plane ($\mathrm{Re}(\zeta_k) > 0$, $k = 0,1$), can be referred to as the domains of unstable solutions, because in domain III $\lim_{t\to\infty} T(h,t)$ and $\lim_{t\to\infty} \sigma_{xx}(h,t)$ do not exist (because $\mathrm{Im}(\zeta_k) \neq 0$, $k = 0,1$), and in domain IV we have $\lim_{t\to\infty} T(h,t) = \lim_{t\to\infty} \sigma(h,t) = \infty$ (because $\mathrm{Im}(\zeta_k) = 0$, $k = 0,1$).

Figure 3 presents domains of stability (I, II) and instability (III, IV) in plane ($\hat{V}$, k_w) and boundaries between them for Bi = 100 and different values of η = 0.1; 0.25; 0.5; 1.0; 5.0; 10.0. These plots show that, at fixed k_w, $\hat{V}$, η the parameter Bi affects the position of the boundaries of stability and instability. The point of intersection of boundaries I–IV, lying on axis $\hat{V}$ (Figure 3), shifts depending on Bi and has a coordinate $\hat{V}_* = 2\mathrm{Bi}(2+\mathrm{Bi})^{-1}\eta^{-1}$ ($k_w = 0$). At $\eta = 1$ this result coincides with [41]. The formula for $\hat{V}_*$ shows the effect of the parameter η (Figure 3) on the boundaries of domains I–IV.

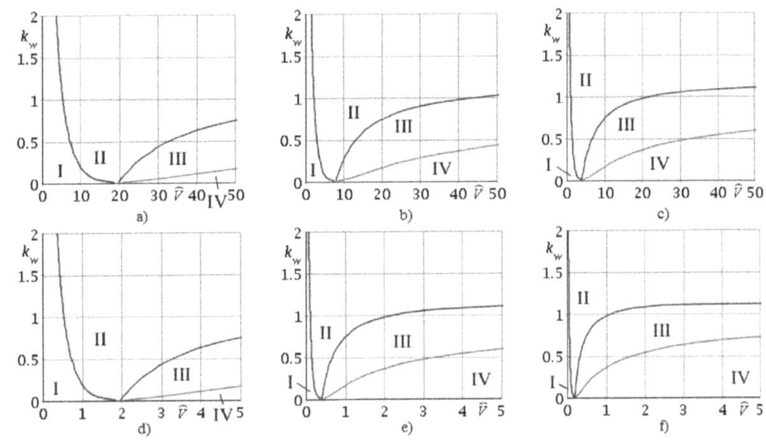

Figure 3. Domains of stability (I, II) and instability (III, IV) of the problem solution at Bi = 100 for different values of η: (**a**) 0.1, (**b**) 0.25, (**c**) 0.5, (**d**) 1.0, (**e**) 5.0, (**f**) 10.0.

Note that the dimensionless parameter η, which characterizes inhomogeneity of the FGM of the coating, significantly affects the boundary between domains of stable (II) and unstable (III) solutions for any k_w. These boundaries are presented in Figure 4 in detail. One can observe that with decrease in η, the stable solution region increases.

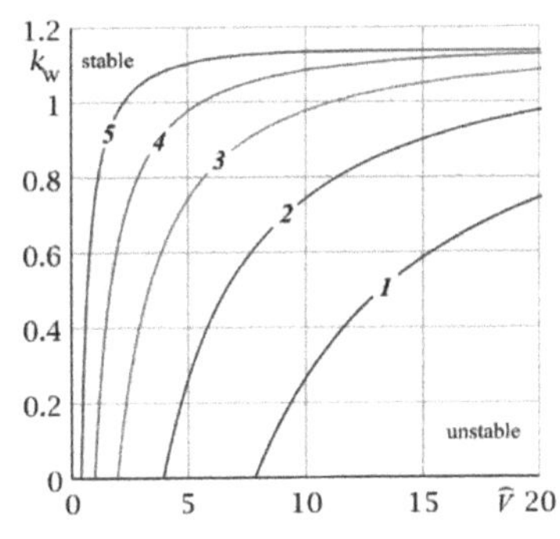

Figure 4. Boundaries between stability and instability domains of the problem solution in the $(\hat{V}, k_w)$ plane at Bi = 100 for different values of η: ***1***—0.25, ***2***—0.5, ***3***—1.0, ***4***—2.0, ***5***—5.0.

7. Features of Wear of FGM Coating

In Section 5 we obtained the exact formulas for the main characteristics of the problem: temperature $T(x,t)$ (57), displacements $u(x,t)$ (58), stresses $\sigma_{xx}(x,t)$ (59) and coating wear $u_w(t)$ (62) on sliding contact. These functions depend on the shear modulus $\mu(x)$ variation by the coating depth $0 \leq x \leq h$. To illustrate the solution, we assume that the shear modulus $\mu(x)$ varies according to a power (parabolic) law

$$\mu(x) = \mu_0\left(a\left(\frac{x}{h}\right)^2 + b\frac{x}{h} + c\right) \tag{64}$$

$$a = 2\left(\frac{\mu_1}{\mu_0} - 2\frac{\mu_{1/2}}{\mu_0} + 1\right),\ b = -\left(\frac{\mu_1}{\mu_0} - 4\frac{\mu_{1/2}}{\mu_0} + 3\right),\ c = 1$$

where $\mu_0 = \mu(0)$, $\mu_1 = \mu(h)$, $\mu_{1/2} = \mu(h/2)$. This means that the shear modulus is equal to μ_0 at the coating–substrate interface ($x = 0$), to μ_1 at the contact interface ($x = h$), and to $\mu_{1/2}$ in the middle of the coating. If $\mu_{1/2} = (\mu_1 + \mu_0)/2$ then $a = 0$ and the law (64) becomes linear.

Calculating integral (15) of the function in (64), we obtain the expression for $B(x)$

$$B(x) = \frac{h}{\mu_1}\begin{cases} \frac{2\chi}{\sqrt{-D}}\operatorname{arctg}\frac{\sqrt{-D}}{\theta(x)} & D < 0 \\ \frac{\chi}{\sqrt{D}}\ln\left|\frac{1+\theta_-(x)}{1-\theta_+(x)}\right| & D > 0 \\ -4\frac{c}{b}\frac{\chi}{\theta(x)} & D = 0 \end{cases} \tag{65}$$

where $D = b^2 - 4ac$, $\chi = \frac{\mu_1}{\mu_0}$, $\theta(x) = 2c + b\frac{x}{h}$, $\theta_\pm(x) = \frac{2a}{b\pm\sqrt{D}}\cdot\frac{x}{h}$.

Equation (65) allows one to determine other characteristics depending on $\mu(x)$

$$B'(x) = \frac{1}{\mu(x)} = \frac{1}{\mu_0\left(a\left(\frac{x}{h}\right)^2 + b\frac{x}{h} + c\right)} \tag{66}$$

$$B'(h) = \frac{1}{\mu_0\chi} \tag{67}$$

$$B(h) = \frac{h}{\mu_1}\begin{cases} \frac{2\chi}{\sqrt{-D}}\operatorname{arctg}\frac{\sqrt{-D}}{2c+b} & D < 0 \\ \frac{\chi}{\sqrt{D}}\ln\left|\frac{1+b_-^1}{1-b_+^1}\right| & D > 0 \\ -4\chi\frac{c}{b(2c+b)} & D = 0 \end{cases} \tag{68}$$

where $D = b^2 - 4ac$, $b_\pm^1 = \frac{2a}{b\pm\sqrt{D}}$.

Then, from (67), (68) we obtain an expression for the parameter η, which characterizes the inhomogeneity of the FGM coating. In case of the parabolic variation of $\mu(x)$ at $0 \le x \le h$, according to (64), it has the form

$$\eta = \frac{hB'(h)}{B(h)} = \frac{h}{\mu_1 B(h)} \tag{69}$$

The law $\Delta(t)$ of indentation of a hard abrasive (half-plane I) into the coating is given by the formula

$$\Delta(t) = \Delta_0\left(\left(e^{\varepsilon t} - 1\right)H(t_0 - t) + H(t - t_0)\right),\ t > 0 \tag{70}$$

where $H(t)$ is the Heaviside step function. Formula (70) assumes that the time section of indentation $0 < t \le t_0$ is active and, when $t > t_0$ it is passive, since $\Delta(t) = \Delta_0$ at $t > t_0$.

The nature of the loss of thermoelastic stability of the main parameters of the contact (temperature $T(h,t)$, contact stresses $\sigma_{xx}(h,t)$, coating wear $u_w(t)$) is studied in detail in Section 6 depending on the value of the parameter η. The boundaries of the thermoelastic stability region on the set of parameter values ($\hat{V}$, k_w) at fixed values η and Bi are also indicated there.

Let us study the effect of the parameter η of the considered thermoelastic problem on wear by a hard abrasive (half-plane I) of an elastic strip made from aluminum with a graded content of alumina (Al_2O_3) on the main characteristics of the contact: temperature $T(h,t)$ from (57), contact stresses $p(t) = -\sigma_{xx}(h,t)$ from (59), coating wear $u_w(t)$ from (62) and wear rate $\dot{u}_w(t)$. This FGM strip is characterized by an increased shear modulus $\mu_1 = \mu(h) = 125.0$ GPa at the contact and shear modulus at the interface with the substrate $\mu_0 = \mu(0) = 25.0$ GPa, $\nu = 0.34$, $\kappa = 88.1 \times 10^{-6}$ mm^2/s, $\alpha = 22.9 \times 10^{-6}$ 1/K, $K = 209.3$ Вт/(м·K), $f = 0.47$, $h = 20$ mm, $\Delta_0 = 5$ mm, $t_0 = 5$ s, $V = 2.5$ mm/s, $\varepsilon = \ln 2$. We consider three different values $\mu_{1/2} = \mu(h/2)$, which, together with their corresponding values η, are presented in Table 1.

Table 1. Parameters of the shear modulus variation with the coating depth.

Curve No.	$\mu_{1/2}$, GPa	η
1	50	0.361287
2	75	0.497067
3	100	0.619564

Variation of the shear modulus $\mu(x)$ by the x-coordinate is illustrated in Figure 5.

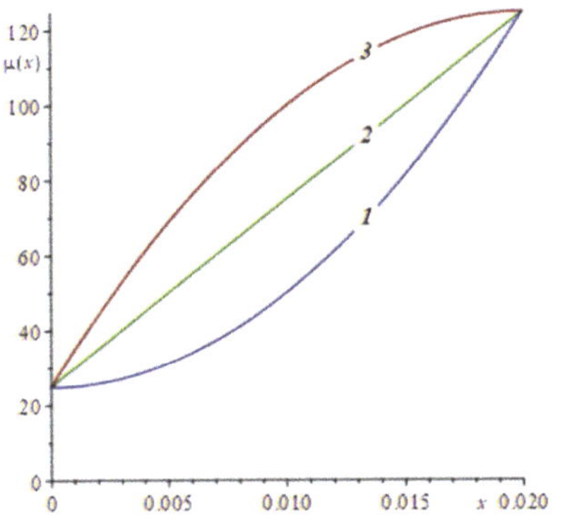

Figure 5. Variation of the shear modulus $\mu(x)$ by the coating depth at the considered values of $\mu_{1/2}$: *1*—50, *2*—75, *3*—100 GPa.

The wear of the coating surface at the depth Δ_0 ends up at $t = t_w$, when the coating wear $u_w(t)$ equals Δ_0, and the contact stress turns to zero ($p(t) = -\sigma_{xx}(h,t) = 0$). We call t_w the time of the coating wear by amount Δ_0. Assuming the wear factor $K^* = 1.0 \times 10^{-11}$ m^2/N, we obtain the values of dimensionless parameters $k_w = 0.511$ and Bi $= 10^5$ using Formula (29).

Table 2 gives the coating wear time, together with maximum values of contact pressure $p(t)$ and temperature $T(h,t)$ depending on the shear modulus in the middle of the coating $\mu_{1/2}$ and sliding velocity V.

Table 2. Values of main contact characteristics depending on shear modulus in the middle of coating $\mu_{1/2}$ and sliding velocity V in case of parabolic variation of the shear modulus $\mu(x)$.

$\mu_{1/2}$, GPa	t_w, s		$\max_{t\in(0,t_w]} p(t)$, GPa		$\max_{t\in(0,t_w]} T(h,t)$, K	
	$V = 2.5$ mm/s	$V = 5.0$ mm/s	$V = 2.5$ mm/s	$V = 5.0$ mm/s	$V = 2.5$ mm/s	$V = 5.0$ mm/s
50	192.3	124.6	0.142	0.144	4.600	8.996
75	128.9	30.2	0.196	0.201	6.292	12.279
100	93.3	18.6	0.246	0.253	7.749	15.153

The effect of the parameter η on the main contact characteristics is illustrated in Figure 6a–d, presenting plots of $T(h,t)$, $p(t)$, $u_w(t)$, $\dot{u}_w(t)$. Curves denoted ***1***, ***2***, ***3*** are plotted for $\mu_{1/2}$ = 50, 75, 100 GPa, respectively.

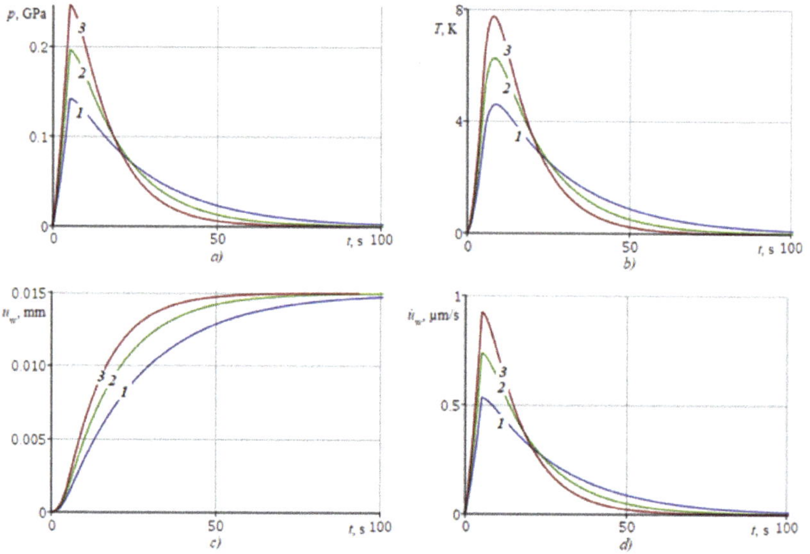

Figure 6. Dependence of the main contact characteristics by time: (**a**) contact pressure $p(t) = -\sigma_{xx}(h,t)$, (**b**) contact temperature $T(h,t)$, (**c**) coating wear $u_w(t)$, (**d**) coating wear rate $\dot{u}_w(t)$, for different values of shear modulus in the middle of the coating $\mu_{1/2}$: ***1***—50, ***2***—75, ***3***—100 GPa.

Figure 6 shows that the coating wear is accompanied by an increase in contact stress $p(t)$ and temperature $T(h,t)$. The growth of the contact stress $p(t)$ is explained by the proportionality of the stresses $\sigma_{xx}(x,t)$ to the shear modulus $\mu(x)$ according to (3). The temperature $T(h,t)$ is proportional to the contact stresses $\sigma_{xx}(x,t)$, according to (9), and, hence, to the shear modulus $\mu(x)$. Since curve **3** indicates a greater stiffness of FGM than curve 1 (Figure 5), the values of $p(t)$ and $T(h,t)$ for curve **3** will exceed the values of these characteristics for curve 1 (Figure 6). The wear of the coating demonstrating higher stiffness (curve **3**) will occur faster than the wear of the less stiff coating (curve ***1***), for the same reason according to (8), and also because the vertical movement of abrasive I carried out forcibly according to the law for $\Delta(t)$. Hence, it follows that it is possible to experimentally determine the depth-variable shear modulus of the FGM coating under wear conditions in accordance with the kinematic law for $\Delta(t)$ of abrasive I, using pressure and temperature sensors at the contact.

For the preparation of materials with grinding, the power supply of the abrasive is often used due to the contact pressure $p(t)$, which must be kept at a certain level. This restriction makes it possible to calculate the kinematic law for $\Delta(t)$ with the condition of the limitation of $p(t)$. Let the formula for the contact stresses be given by the following expression

$$p(t) = p_0\left(\left(e^{\delta t} - 1\right)H(t_0 - t) + H(t - t_0)\right) \tag{71}$$

where $\delta = t_0^{-1} \ln 2$, and t_0 characterizes the time of reaching the stationary mode at $t > t_0$, at which $p(t) = p_0$. To express $\Delta(t)$ in terms of $p(t)$, we use formula (26) in the form

$$p^L(p) = -\frac{2(1-v)}{1-2v}\Delta^L(p)\frac{h}{B(h)} \cdot \frac{N_\sigma^0(z)}{R(z)} \tag{72}$$

$\Delta^L(p)$ is determined from (72) by the following formula

$$\Delta^L(p) = -\frac{1-2v}{2(1-v)}p^L(p)\frac{B(h)}{h} \cdot \frac{R(z)}{N_\sigma^0(z)} \tag{73}$$

After inverting the previous formula, taking into account (27), we obtain the following formula for determining $\Delta(t)$ from $p(t)$

$$\Delta(t) = -\frac{1-2v}{2(1-v)}\frac{B(h)}{h}\int\limits_0^t p(\tau)\cdot f_p(t-\tau)d\tau \quad (74)$$

$$f_p(t) = \frac{1}{2\pi i}\int\limits_\Gamma \frac{R(z)}{t_\kappa z\, r(h,z)}e^{z\tilde{t}}dz \quad (75)$$

where $\tilde{t} = \frac{t}{t_\kappa}$, $t_\kappa = \frac{h^2}{\kappa}$, $R(z)$ and $r(h,z)$ are taken from (28), (29), and $B(h)$ is from the expression (15).

The temperature at the contact $T(h,t)$ with limited $p(t)$ will also be limited, and wear is determined according to (62).

Similarly, it is possible to determine $\Delta(t)$ so that the temperature is limited, for example, by the following formula

$$T(t) = T_0\Big(\Big(e^{\theta t}-1\Big)H(t_0-t)+H(t-t_0)\Big) \quad (76)$$

where $\theta = t_0^{-1}\ln 2$, and t_0 is the time, when $T(t)$ becomes T_0 at $t > t_0$. Using formula (22) for $T^L(x,p)$ according to the above scheme, we obtain

$$T^L(p) = \frac{1-\nu}{1+\nu}\frac{\hat{V}}{\alpha h}\Delta^L(p)\frac{hB'(h)}{B(h)}\frac{N_T(h,z)}{R(z)} \quad (77)$$

$\Delta^L(p)$ is determined from (77) by the following formula

$$\Delta^L(p) = \frac{1+\nu}{1-\nu}\frac{\alpha h}{\hat{V}}T^L(p)\frac{B(h)}{hB'(h)}\frac{R(z)}{N_T(h,z)} \quad (78)$$

After inverting (78), we obtain

$$\Delta(t) = \frac{1+\nu}{1-\nu}\frac{\alpha h}{\hat{V}}\frac{B(h)}{hB'(h)}\int\limits_0^t T(\tau)\cdot f_T(t-\tau)d\tau \quad (79)$$

$$f_T(t) = \frac{1}{2\pi i}\int\limits_\Gamma \frac{R(z)}{t_\kappa N_T(h,z)}e^{z\tilde{t}}dz \quad (80)$$

where $\tilde{t} = \frac{t}{t_\kappa}$, $t_\kappa = \frac{h^2}{\kappa}$.

When $\Delta(t)$ is taken from the expression (79), not only will the temperature on the contact be limited, but also the contact stress $p(t)$. The wear is determined through $\Delta(t)$ from (79) according to (62).

8. Discussion

The present study was aimed at solving the problem of wear-grinding of a FGM coating with a depth-varying shear modulus. According to the results of the studies presented in the paper, it becomes feasible to solve the inverse problem of determining the law of distribution of the depth-varying shear modulus of the FGM l of the coating. It will be necessary to find out what the minimum set of information is for the unambiguous determination of the shear modulus varying by depth. The solution to this problem is important for the design of FGM coatings with special properties (anti-friction, anti-corrosion, wear-resistant, etc.).

9. Conclusions

The considered thermoelastic contact problem of sliding wear by a hard abrasive material at a constant speed over the surface of an elastic coating in the form of a strip made of FGM with an arbitrary shear modulus varying with the depth of the coating made it possible to establish:

- A dimensionless parameter η of the FGIM of the coating, which characterizes the FGM of the coating and its presence in all formulas of the main parameters of the contact with friction;
- Boundaries of the regions of thermoelastic instability in the space of dimensionless parameters of the problem: k_w, $\hat{V}$, Bi, η;
- Features of the kinematic scheme of wear, which consists in setting an arbitrary law of upsetting of a hard abrasive on the surface of a strip made of FGM and leading to an uncontrolled increase in temperature and contact stresses;
- A special grinding scheme, consisting in the development of such a law of displacement of a hard abrasive into the surface of a strip made of FGM, with the help of which the growth of both the temperature at the contact and the contact stresses is limited.

Author Contributions: Problem statement and development of the solution method, V.B.Z.; construction of the solution, V.B.Z. and P.A.L.; calculation, B.I.M.; analysis of the solution, B.I.M. and P.A.L.; conclusions, V.B.Z.; writing—original draft preparation, V.B.Z. and B.I.M.; writing—translation and editing, B.I.M. and P.A.L.; supervision, V.B.Z. All authors have read and agreed to the published version of the manuscript.

Funding: This research was funded by the Government of the Russian Federation, grant number 14.Z50.31.0046.

Institutional Review Board Statement: Not applicable.

Informed Consent Statement: Not applicable.

Data Availability Statement: Data are available on request.

Conflicts of Interest: The authors declare no conflict of interest.

Nomenclature

$w(x,t)$	horizontal component of displacements in the coating
$\sigma_{xx}(x,t)$	normal component of stresses in the coating
$\sigma_{xy}(x,t)$	shear component of stresses in the coating
$T(x,t)$	the coating temperature
h	thickness of the coating
V	sliding velocity of the abrasive (half-plane I)
$\mu(x)$	shear modulus of the coating material
ν	Poisson's ratio of the coating material
α	coefficient of linear heat expansion of the coating material
κ	the thermal diffusivity of the coating material
f	the coefficient of friction
K^*	the proportionality coefficient between the work of friction forces and the volume of removed material
K	the thermal conductivity of the coating material
k	the heat transfer coefficient through the coating–substrate interface
$u_w(t)$	coating wear
$Q(t)$	the amount of frictional heat originated at the contact interface
$\Delta(t)$	the law of indentation of a hard abrasive (half-plane I) into the coating

References

1. Watanabe, Y.; Nakamura, T. Microstructures and wear resistances of hybrid Al–(Al_3Ti+Al_3Ni) FGMs fabricated by a centrifugal method. *Intermetallics* **2001**, *9*, 33–43. [CrossRef]
2. El-Hadad, S.; Sato, H.; Watanabe, Y. Wear of Al/Al_3Zr functionally graded materials fabricated by centrifugal solid-particle method. *J. Mater. Process. Technol.* **2010**, *210*, 2245–2251. [CrossRef]
3. Melgarejo, Z.H.; Suárez, O.M.; Sridharan, K. Wear resistance of a functionally-graded aluminum matrix composite. *Scr. Mater.* **2006**, *55*, 95–98. [CrossRef]
4. Gomes, J.R.; Ribeiro, A.R.; Vieira, A.C.; Miranda, A.S.; Rocha, L.A. Wear mechanisms in functionally graded aluminium matrix composites: Effect of the presence of an aqueous solution. *Mater. Sci. Forum.* **2005**, *492–493*, 33–38. [CrossRef]
5. Sato, H.; Murase, T.; Fujii, T.; Onaka, S.; Watanabe, Y.; Kato, M. Formation of a wear-induced layer with nanocrystalline structure in Al–Al_3Ti functionally graded material. *Acta Mater.* **2008**, *56*, 4549–4558. [CrossRef]
6. Vieira, A.C.; Sequeira, P.D.; Gomes, J.R.; Rocha, L.A. Dry sliding wear of Al alloy/SiCp functionally graded composites: Influence of processing conditions. *Wear* **2009**, *267*, 585–592. [CrossRef]
7. Radhika, N.; Raghu, R. The mechanical properties and abrasive wear behavior of functionally graded aluminum/AlB_2 composites produced by centrifugal casting. *Part. Sci. Technol.* **2017**, *35*, 575–582. [CrossRef]
8. Hu, F.Y.; Cui, A.Y.; Hu, B.; Fu, P.F.; Zhao, P.Z.; Lu, C.L. Microstructure and wear behavior of Ti-matrix functional gradient layer. *Adv. Mater. Res.* **2011**, *311–313*, 2093–2098. [CrossRef]
9. Miyoshi, K.; Pohlchuck, B.; Street, K.; Zabinski, J.; Sanders, J.H.; Voevodin, A.; Wu, R. Sliding wear and fretting wear of diamondlike carbon-based, functionally graded nanocomposite coatings. *Wear* **1999**, *225–229*, 65–73. [CrossRef]
10. Cetinel, H.; Uyulgan, B.; Tekmen, C.; Ozdemir, I.; Celık, E. Wear properties of functionally gradient layers on stainless steel substrates for high temperature applications. *Surf. Coat. Technol.* **2003**, *174–175*, 1089–1094. [CrossRef]
11. Prchlik, L.; Sampath, S.; Gutleber, J.; Bancke, G.; Ruff, A.W. Friction and wear properties of WC-Co and Mo-Mo2C based functionally graded materials. *Wear* **2001**, *249*, 1103–1115. [CrossRef]
12. Askari, E.; Mehrali, M.; Metselaar, I.H.S.C.; Kadri, N.A.; Rahman, M.M. Fabrication and mechanical properties of Al_2O_3/SiC/ZrO_2 functionally graded material by electrophoretic deposition. *J. Mech. Behav. Biomed. Mater.* **2012**, *12*, 144–150. [CrossRef]
13. Siahpoosh, H.; Abbasi, A.; Aliofkhazraei, M.; Khajavi, B.; Majidi, H. Effect of pulse frequency on tribological behaviour of functionally graded Ni/Al_2O_3 nanocomposite coatings. *Procedia Mater. Sci.* **2015**, *11*, 498–502. [CrossRef]
14. Yan, Y.; Han, Y.; Huang, J. Formation of Al_2O_3–ZrO_2 composite coating on zirconium by micro-arc oxidation. *Scr. Mater.* **2008**, *59*, 203–206. [CrossRef]
15. Nai, S.M.L.; Gupta, M.; Lim, C.Y.H. Synthesis and wear characterization of Al based, free standing functionally graded materials: Effects of different matrix compositions. *Compos. Sci. Technol.* **2003**, *63*, 1895–1909. [CrossRef]
16. Guler, M.A.; Erdogan, F. The frictional sliding contact problems of rigid parabolic and cylindrical stamps on graded coatings. *Int. J. Mech. Sci.* **2007**, *49*, 161–182. [CrossRef]
17. Altenbach, H.; Eremeyev, V.A. Eigen-vibrations of plates made of functionally graded material. *Comput. Mater. Contin.* **2009**, *9*, 153–177.
18. Selvadurai, A.P.S.; Katebi, A. Mindlin's problem for an incompressible elastic half-space with an exponential variation in the linear elastic shear modulus. *Int. J. Eng. Sci.* **2013**, *65*, 9–21. [CrossRef]
19. Tokovyy, Y.; Ma, C.-C. An analytical solution to the three-dimensional problem on elastic equilibrium of an exponentially-inhomogeneous layer. *J. Mech.* **2015**, *31*, 545–555. [CrossRef]
20. Aizikovich, S.M.; Vasil'ev, A.S.; Volkov, S.S. The axisymmetric contact problem of the indentation of a conical punch into a half-space with a coating inhomogeneous in depth. *J. Appl. Math. Mech.* **2015**, *79*, 500–505. [CrossRef]
21. Alinia, Y.; Beheshti, A.; Guler, M.A.; El-Borgi, S.; Polycarpou, A.A. Sliding contact analysis of functionally graded coating/substrate system. *Mech. Mater.* **2016**, *94*, 142–155. [CrossRef]
22. Ma, J.; El-Borgi, S.; Ke, L.L.; Wang, Y.S. Frictional contact problem between a functionally graded magnetoelectroelastic layer and a rigid conducting flat punch with frictional heat generation. *J. Therm. Stress.* **2016**, *39*, 245–277. [CrossRef]
23. Volkov, S.S.; Vasiliev, A.S.; Aizikovich, S.M.; Seleznev, N.M.; Leontieva, A.V. Stress-strain state of an elastic soft functionally-graded coating subjected to indentation by a spherical punch. *PNRPU Mech. Bull.* **2016**, *4*, 20–34.
24. Selvadurai, A.P.S.; Katebi, A. The Boussinesq–Mindlin problem for a non-homogeneous elastic halfspace. *Z. Angew. Math. Phys.* **2016**, *67*, 68. [CrossRef]
25. Kudish, I.I.; Volkov, S.S.; Vasiliev, A.S.; Aizikovich, S.M. Lubricated point heavily loaded contacts of functionally graded materials. Part 1. Drycontacts. *Math. Mech. Solids* **2018**, *201823*, 1061–1080. [CrossRef]
26. Vasiliev, A.S.; Volkov, S.S.; Aizikovich, S.M. Approximated analytical solution of a problem on indentation of an electro-elastic half-space with inhomogeneous coating by a conductive punch. *Dokl. Phys.* **2018**, *63*, 18–22. [CrossRef]
27. Singh, S.; Singh, R. Wear modelling of Al-Al_2O_3 functionally graded material prepared by FDM assisted investment castings using dimensionless analysis. *J. Manuf. Process.* **2015**, *20*, 507–514. [CrossRef]
28. Archard, J.F. Contact and rubbing of flat surfaces. *J. Appl. Phys.* **1953**, *24*, 981–988. [CrossRef]
29. Dow, T.A.; Burton, R.A. The role of wear in the initiation of thermoelastic instabilities of rubbing contact. *J. Lubr. Technol.* **1973**, *95*, 71–75. [CrossRef]

30. Alexandrov, V.M.; Annakulova, G.K. A contact problem of thermo-elasticity with wear and heat release caused by friction. *Trenie i Iznos* **1990**, *11*, 24–28.
31. Alexandrov, V.M.; Annakulova, G.K. Interaction between coatings of a body with deformation, wear, and heat release due to friction. *Trenie i Iznos* **1992**, *13*, 154–160.
32. Zmitrowicz, A. A thermodynamical model of contact, friction and wear: I governing equations. *Wear* **1987**, *114*, 135–168. [CrossRef]
33. Zmitrowicz, A. A thermodynamical model of contact, friction and wear: II constitutive equations for materials and linearized theories. *Wear* **1987**, *114*, 169–197. [CrossRef]
34. Strömberg, N.; Johansson, L.; Klarbring, A. Derivation and analysis of a generalized standard model for contact, friction and wear. *Int. J. Solids Struct.* **1996**, *33*, 1817–1836. [CrossRef]
35. Andrews, K.T.; Shillor, M.; Wright, S.; Klarbring, A. A dynamic thermoviscoelastic contact problem with friction and wear. *Int. J. Eng. Sci.* **1997**, *35*, 1291–1309. [CrossRef]
36. Strömberg, N. Finite element treatment of two-dimensional thermoelastic wear problem. *Comput. Methods Appl. Mech. Eng.* **1999**, *177*, 441–455. [CrossRef]
37. Evtushenko, A.A.; Pyr'ev, Y.A. Influence of wear on the development of thermoelastic instability of a frictional contact. *Izv. Ross. Akad. Nauk Mekhanika Tverd. Tela* **1997**, *1*, 114–121.
38. Pyr'ev, Y.A.; Grilitsky, D.V. The non-stationary problem of the frictional contact for a cylinder taking into account heat release and wear. *Appl. Math. Tech. Phys.* **1996**, *37*, 99–104.
39. Awrejcewicz, J.; Pyryev, Y. Thermoelastic contact of a rotating shaft with a rigid bush in conditions of bush wear and stick-slip movements. *Int. J. Eng. Sci.* **2002**, *40*, 1113–1130. [CrossRef]
40. Zelentsov, V.B.; Mitrin, B.I.; Lubyagin, I.A. Effect of wear on frictional heating and thermoelastic instability of sliding contact. *Comput. Contin. Mech.* **2016**, *9*, 430–442. [CrossRef]
41. Zelentsov, V.B.; Mitrin, B.I.; Lubyagin, I.A. Wear resistance of coating materials under the frictional heating conditions. *J. Frict. Wear* **2017**, *38*, 265–271. [CrossRef]
42. Belyakov, N.S.; Nosko, A.P. Thermoelastic problem of friction of plane-parallel layers with allowance for nonstationarity of thermal processes. *J. Frict. Wear* **2010**, *31*, 317–325. [CrossRef]
43. Gu, R.J.; Kuttler, K.L.; Shillor, M. Frictional wear of a thermoelastic beam. *J. Math. Anal. Appl.* **2000**, *242*, 212–236. [CrossRef]
44. Papangelo, A.; Ciavarella, M. The effect of wear on ThermoElastic Instabilities (TEI) in bimaterial interfaces. *Tribol. Int.* **2020**, *142*, 105977. [CrossRef]
45. Papangelo, A.; Ciavarella, M. Can wear completely suppress thermoelastic instabilities? *J. Tribol.* **2020**, *142*, 051501. [CrossRef]
46. Zelentsov, V.B.; Mitrin, B.I.; Lubyagin, I.A.; Kudish, I.I. Diagnostics of wear thermoelastic instability based on sliding contact parameter monitoring. *IMA J. Appl. Math.* **2019**, *84*, 345–365. [CrossRef]
47. Kovalenko, A.D. *Thermoelasticity: Basic Theory and Applications*; Wolters-Noordhoff: Groningen, The Netherlands, 1970.
48. Bowden, F.P.; Tabor, D. *The Friction and Lubrication of Solids*; Oxford University Press: London, UK, 1950.
49. Ditkin, V.A.; Prudnikov, A.P. *Operational Calculus*; Vishaya Shkola: Moscow, Russia, 1975. (In Russian)
50. Brychkov, Y.A.; Prudnikov, A.P. *Integral Transformations of Generalized Functions*; Gordon & Breach Science Publishers, CRC-Press: New York, NY, USA, 1989.
51. Hurwitz, A.; Courant, R. *Theory of Functions*; Nauka Publishers: Moscow, Russia, 1968. (In Russian)
52. Zelentsov, V.B.; Mitrin, B.I.; Aizikovich, S.M.; Ke, L.L. Instability of solution of the dynamic sliding frictional contact problem of coupled thermoelasticity. *Mater. Phys. Mech.* **2015**, *23*, 14–19.
53. Zelentsov, V.B.; Mitrin, B.I.; Aizikovich, S.M. Dynamic and quasi-static instability of sliding thermoelastic frictional contact. *J. Frict. Wear* **2016**, *37*, 213–220. [CrossRef]
54. Zelentsov, V.B.; Mitrin, B.I. Thermoelastic instability in the quasi-static coupled thermoelasticity problem dealt with the sliding contact with frictional heating. *Mech. Solids* **2019**, *54*, 58–69. [CrossRef]

nanomaterials

MDPI

Article

Endowing Acceptable Mechanical Properties of Segregated Conductive Polymer Composites with Enhanced Filler-Matrix Interfacial Interactions by Incorporating High Specific Surface Area Nanosized Carbon Black

Huibin Cheng [1], Xiaoli Sun [1], Baoquan Huang [1], Liren Xiao [2], Qinghua Chen [1,2,3], Changlin Cao [1,*] and Qingrong Qian [1,2,3,*]

1 College of Environmental Science and Engineering, Fujian Normal University, Fuzhou 350007, China; QBX20190099@yjs.fjnu.edu.cn (H.C.); sunxiaoli@fjnu.edu.cn (X.S.); qbh811@fjnu.edu.cn (B.H.); cqhuar@fjnu.edu.cn (Q.C.)
2 Engineering Research Center of Polymer Green Recycling of Ministry of Education, Fuzhou 350007, China; xlr1966@fjnu.edu.cn
3 Fujian Key Laboratory of Pollution Control & Resource Reuse, Fuzhou 350007, China
* Correspondence: caochlin3@fjnu.edu.cn (C.C.); qrqian@fjnu.edu.cn (Q.Q.)

Abstract: Tuning the high properties of segregated conductive polymer materials (CPCs) by incorporating nanoscale carbon fillers has drawn increasing attention in the industry and academy fields, although weak interfacial interaction of matrix-filler is a daunting challenge for high-loading CPCs. Herein, we present a facile and efficient strategy for preparing the segregated conducting ultra-high molecular weight polyethylene (UHMWPE)-based composites with acceptable mechanical properties. The interfacial interactions, mechanical properties, electrical properties and electromagnetic interference (EMI) shielding effectiveness (SE) of the UHMWPE/conducting carbon black (CCB) composites were investigated. The morphological and Raman mapping results showed that UHMWPE/high specific surface area CCB (h-CCB) composites demonstrate an obviously interfacial transition layer and strongly interfacial adhesion, as compared to UHMWPE/low specific surface area CCB (l-CCB) composites. Consequently, the high-loading UHMWPE/h-CCB composite (beyond 10 wt% CCB dosage) exhibits higher strength and elongation at break than the UHMWPE/l-CCB composite. Moreover, due to the formation of a densely stacked h-CCB network under the enhanced filler-matrix interfacial interactions, UHMWPE/h-CCB composite possesses a higher EMI SE than those of UHMWPE/l-CCB composites. The electrical conductivity and EMI SE value of the UHMWPE/h-CCB composite increase sharply with the increasing content of h-CCB. The EMI SE of UHMWPE/h-CCB composite with 10 wt% h-CCB is 22.3 dB at X-band, as four times that of the UHMWPE/l-CCB composite with same l-CCB dosage (5.6 dB). This work will help to manufacture a low-cost and high-performance EMI shielding material for modern electronic systems.

Keywords: interfacial interaction; conducting carbon black network; mechanical property; electromagnetic interference shielding

Citation: Cheng, H.; Sun, X.; Huang, B.; Xiao, L.; Chen, Q.; Cao, C.; Qian, Q. Endowing Acceptable Mechanical Properties of Segregated Conductive Polymer Composites with Enhanced Filler-Matrix Interfacial Interactions by Incorporating High Specific Surface Area Nanosized Carbon Black. *Nanomaterials* **2021**, *11*, 2074. https://doi.org/10.3390/nano11082074

Academic Editors: Mohammad Malikan and Ana María Díez-Pascual

Received: 16 July 2021
Accepted: 11 August 2021
Published: 16 August 2021

1. Introduction

With the rapid advancement in electronic information technology, electronic equipment and communication base stations have become an indispensable part of people's daily lives. Meanwhile, this originated from man-made electronic devices accompanied by high-energy electromagnetic wave radiation pollution and electromagnetic interference have become critical problems that need to be addressed [1,2]. Current electromagnetic interference (EMI) shielding materials mainly include metal and their alloy materials, but these materials have many shortcomings, such as high density, high cost, low corrosions and a complicated manufacturing process [3]. Therefore, the electrically conductive polymer composites (CPCs) containing conductive fillers were identified as one of the excellent candidates for EMI shielding materials due to their excellent merits, such as low density, regulable electrical conductivity and corrosion resistance, etc. [4–7]. To effectively reduce the impact of electromagnetic wave radiation on interfering with the equipment and human health, especially, people paid widespread attention to the design and fabrication of lightweight and highly electrical conducting carbon fillers filled CPCs. In recent years, to obtain higher EMI shielding effectiveness (SE) value, CPCs must require sufficient conductive loss and interface polarized loss. Additionally, a reasonable layout of the conductive network is necessary [8–10]. The construction of stable, efficient and integral conductive paths can be helpful for conductive loss. Unfortunately, highly efficient conductive pathways always require high content filler loading, which inevitably affects the mechanical properties and processibility of materials [11,12]. Thus, how to further obtain an efficient and dense conductive network at low content loading for EMI shielding polymer composites still remains a daunting challenge [13].

The segregated conductive pathway has proven to outperform in enlarging conductive loss mechanism, particularly for lightweight polymer-based EMI shielding materials with segregated structure, multi-element composite structure, laminated structure and multi-interface structure [13–17]. Unfortunately, the segregated conductive pathways may cause low enhancement efficiency of mechanical properties. The choice of the polymer matrices and electrically conductive fillers, and the interface microstructure of polymer-filler are very important for solving undesirable mechanical properties and manufacturing high-performance CPCs. Currently, tremendous efforts in overcoming the disperse and weak interfacial interaction of nanocarbon filler, such as carbon nanotube (CNT) [18–22], graphite (G) [23], carbon black [21,24] and graphene nanoplatelets (GNPs) [25,26], etc. Numerous works have attempted to improve the interface interaction of CPCs by the synergistic effect of different dimensional filler and surface modification of filler [27,28]. Li [23] et al. reported hybrid graphite/carbon black (CB)/ultrahigh molecular weight polyethylene (UHMWPE) composite exhibited a high electrical conductivity and superb EMI SE. The mechanical enhancement is attributed to the synergistic effect of the nano-sized CB particles and large-aspect-ratio graphite platelets, providing a stronger interfacial adhesion between UHMWPE domains. Nah, et al. [29] developed a highly scalable EMI shielding material comprising nanostructured carbon black and poly (methyl methacrylate) (PMMA) nanocomposites via solution mixing, followed by the compression molding method. The obtained PMMA nanocomposite with 10 wt% filler loading possessed a superior EMI SE value of ~28 dB at X band. Moreover, inspired by the microstructure of the bio-structural materials [30–32], people learn from nacre to construct unique structural polymer composites—that is, design the orderly morphology of multi-layered structural polymer matrix and the alignment of inorganic fillers. The thermal conductivity, electrical properties and mechanical properties of polymer/filler nanocomposites were improved via interface engineering [33,34]. However, among these methods in enhancing the filler-matrix, interfacial interactions is relatively complicated for the practical production process. Thus, for CPCs with strong interface adhesion and small defects, whether they can simply employ low-cost and high-electrically conductive nanostructured carbon fillers for developing EMI shielding materials has become a very perspective research hotspot [35,36].

In this work, UHMWPE with unique rheological property serves as a matrix of segregated conductive polymer composites, and nanosized carbon black (CCB) with high and low specific surface area acts as conductive fillers. A comparative study of CPCs with a dense high specific surface area conductive carbon black (h-CCB) packing network and a low specific surface area conductive carbon black (l-CCB) loose network based on an UHMWPE matrix was fabricated through mechanical mixing and compression molding methods. We found that h-CCB particles with the multi-nanopores and intensive polar interaction can not only be used as a conducting agent to enhance the electrical conductivity of UHMWPE, but also as transition layer to improve the filler-matrix interfacial interactions, which in turn effectively suppress interfacial defects between the adjacent

UHMWPE and h-CCB domains, and greatly form a highly efficient dense conductive network in the composite. The achieved high-loading UHMWPE-based CPCs exhibited a significant improvement of EMI shielding performance and mechanical properties. Furthermore, to meet the requirement of actual applications, the influence of h-CCB and l-CCB content on phase morphology, interfacial bonding mechanism, thermal properties and EMI SE of high loading UHMWPE/CCB composites beyond 10 wt% CCB dosage were investigated comprehensively.

2. Materials and Methods

2.1. Materials

CCB (F900A) with a high specific surface area (h-CCB) of 380 m^2/g were purchased from Tianjin Ebory Chemical Co., Ltd., (Tianjin, China). CCB (Orion N990) with a low specific surface area (l-CCB) of 8 m^2/g was obtained from Shenzhen Securities Yoshida Chemical Co., Ltd., (Shenzhen, China). Comparison of the specific surface area and pore size distribution of two kinds of CCB nanoparticles were given in Figure S1 and Table S1. UHMWPE, under the trade of SLL-2, with an average molecular weight of 3.00×10^6 g/mol, was supplied by Shanghai Lianle Chemical Industry Science and Technology Co., Ltd., (Shanghai, China). All raw materials were used as received without further modification.

2.2. Preparation of UHMWPE/CCB Composites

The preparation schematic illustration of segregated UHMWPE/CCB composites is presented in Figure 1. At first, h-CCB particles and UHMWPE powder were dried in a vacuum oven at 80 °C for 6 h. Then, UHMWPE powder and high specific surface area h-CCB particles with compositions of 99.5/0.5, 99/1, 97/3, 95/5, 93/7, 90/10 and 85/15 were premixed by adopting the crusher high-speed mechanical mixing (800Y, Yongkang Platinum Ou Hardware Products Co., Ltd., Zhejiang, China) at a rotation speed of 34,000 r/min for 50 s, obtaining the high specific surface area CCB-coated UHMWPE mixtures. Afterward, the obtained compounds were compression molded at 200 °C for 20 min on a flat vulcanizing machine (ZG-80 T, Guangdong Zhenggong Electromechanical Equipment Technology Co., Ltd., Guangdong, China), followed by cold compression molded to room temperature at a pressure of 17 MPa. The final obtained specimens were denoted as UHMWPE/h-CCB_x composites, in which x means that the weight fraction of h-CCB nanoparticles. For comparison, l-CCB nanoparticles modified UHMWPE composites with the same weight ratios were prepared via high-speed mechanical mixing and compression molding. The samples were remarked as UHMWPE/l-CCB_x composites.

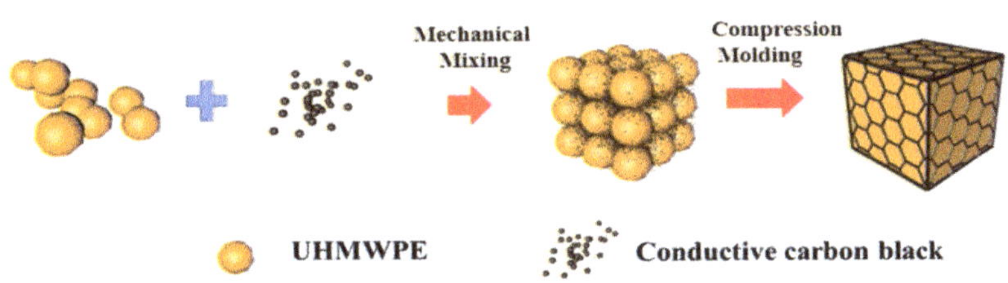

Figure 1. Preparation schematic illustration of segregated UHMWPE/CCB composites.

2.3. Characterization

The morphologies of CCB nanoparticles and composites were investigated using a cold field emission scanning electron microscope (FE-SEM, Regulus 8100, Hitachi, Japan) at an accelerating voltage of 10 kV. All samples were fractured in liquid nitrogen and sputter-coated with gold before observation.

The electrical conductivity was measured by a high-resistance meter (ZC-90G, Shanghai Taiou Electronics Co., Ltd., Shanghai, China) higher than 10^{-5} S/m, four-point probes resistivity measurement (RTS-9, Guangzhou Four Probe Technology Co., Ltd., Guangzhou,

China) below 10^{-5} S/m. The electromagnetic interference shielding performance of the samples depended on scattering parameters, which corresponded to the reflection and transmission of transverse electromagnetic waves, was performed by a Vector network analysis (N5244A, Agilent Technologies, SCC, CA, USA) with the wave-guide method at X-band frequency range (8.2–12.4 GHz) according to ASTM D5568-08, the resultant scattering parameters were used to calculate EMI SE values.

The sheets were cut into dumbbell-shaped specimens for the evaluation of the mechanical properties using a universal testing machine (CMT4104, Shenzhen Sans Material Inspection Co., Ltd., Shenzhen, China). The measurements were repeated five times to get the average values.

The DSC analyses for the samples were performed on a TA Instruments Q20. The samples were first heated up to 200 °C, and then annealed for 5 min to eliminate thermal history, following by cooling down to 40 °C and then reheated up to 200 °C, at heating and cooling rates of 10 °C/min. Thermal gravimetric (TGA) analysis of the samples was performed by TA instrument Q50 from room temperature to 700 °C at a heating rate of 10 °C/min and a nitrogen gas flow of 90 mL/min.

The specific surface area of CCB particles was tested using a BELSOBP-miniII automatic surface area Brunauer-Emmet-Teller (BET) analyzer (BELSOBP-miniII, MicrotracBEL, Osaka, Japan). The specific surface area was calculated according to the Brunauer–Emmett–Teller (BET) method.

To order to indicate intuitively the interface interactions between UHMWPE and CCB particles with various structures, the Raman mapping images of UHMWPE/h-CCB and UHMWPE/l-CCB composites were performed on a micro-laser confocal Raman spectrometer (Thermo Scientific DXR2xi, Waltham, MA, USA). The test condition of Raman was that the laser power is 1.5 mW, and the total exposure is 30 times with 0.025 s exposure time for each spectrum, Raman mapping images with an area of 50 μm^2 in step sizes of 1 m. The imaging data is fitted through the Raman software "OMNICxi", which gives each point a corresponding Raman characteristic peak spectrum.

3. Results

3.1. Characterization of h-CCB, and l-CCB Nanoparticles

We investigated the microstructure and properties of two kinds of CCB nanoparticles. These particles are often agglomerated into large aggregates due to the strong Van der Waals force between particles, leading to a high-percolation threshold value (15–20 wt%) for achieving the interconnected conductive networks [37]. This is mainly due to the zero-dimensional shape of the filler, which is hard to construct highly efficient conductive networks [38]. Figure 2 depicts the morphology and specific surface area of two kinds of nanoscale h-CCB or l-CCB particles. In Figure 2a,b, compared to l-CCB materials, we observed that the particle size of h-CCB is smaller than l-CCB, and most of h-CCB particles aggregate each other, it shows that h-CCB possesses a high specific surface area and strong polar interaction between particles. Figure 2c presents the N_2 adsorption/desorption isotherms of two kinds of different specific surface area CCB nanoparticles; it suggests that the large specific surface area of h-CCB particles was attributed to its abundant porous structure, and the abundant mesoporous structure and high specific surface area of h-CCB particles enable the combined forces of numerous Van der Waals forces to become much stronger, which is beneficial for forming the strong interface interactions of mechanical interlocking between h-CCB and UHMWPE. Figure 2d indicates that Raman spectrum of h-CCB and l-CCB particles is mainly composed of D and G peaks at 1348 cm^{-1} and 1587 cm^{-1}, in which I_D/I_G value presented the level of the different graphitization and defects structure. Moreover, h-CCB particles present an evident 2G peak value in the Raman spectrum, the I_D/I_G value of h-CCB is lower than l-CCB particles, implying the graphitization degree of h-CCB particles is higher than l-CCB particles. These results suggested that h-CCB particles with high graphitization level serve as conducting fillers in CPCs at low filler content, which has more competitive value in industry applications, owing to its outstanding electrical conductivity.

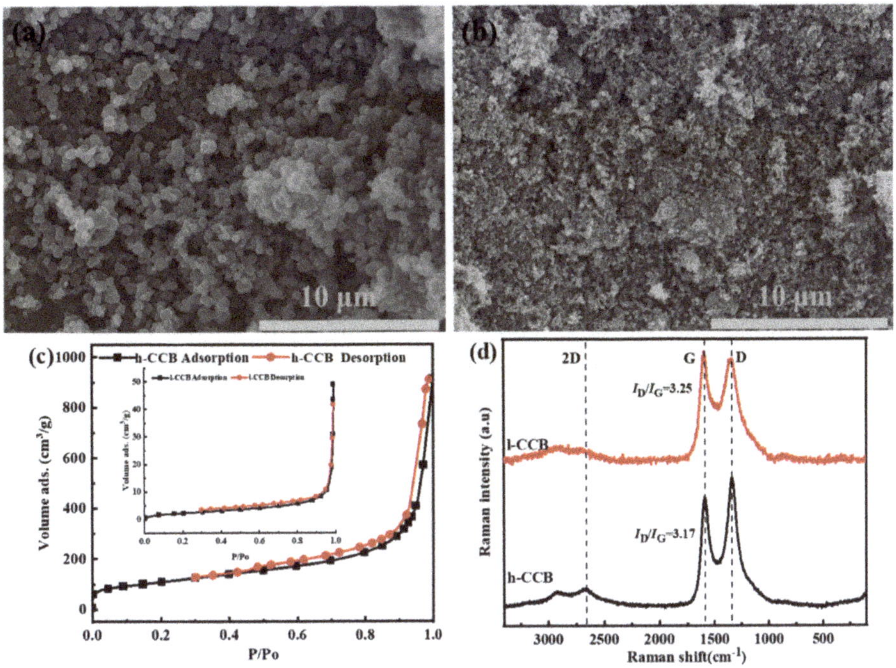

Figure 2. SEM images of CCB particles: (**a**) l-CCB particle; (**b**) h-CCB particles; (**c**) N_2 adsorption/desorption isotherms, (**d**) Raman spectroscopy of nanoscale h-CCB and l-CCB particles.

3.2. Microstructure of UHMWPE/CCB Composites

To vitally confirm the various dispersion and distribution characteristics of CCB particles, we perform optical microscopy (OM) imaging analysis. Figure 3 shows the cross-sectional optical microscopy of UHMWPE/l-CCB composites and UHMWPE/h-CCB composites. The obvious bright parts (UHMWPE region) and dark region (h-CCB or l-CCB rich region) that are connected can be observed clearly, indicating the formation of a segregated h-CCB or l-CCB network. In Figure 3a–c, with the increasing l-CCB content, some conductive networks are gradually constructed. Likewise, this phenomenon also appears in h-CCB filled UHMWPE matrix, as shown in Figure 3e,f. Compared to the loosely stacked l-CCB network, while the strong Van der Waals forces of h-CCB particle are easy to aggregate each other, and gradually form a denser and denser developed conductive network with increasing h-CCB concentrations. This may be attributed to the strong mechanical interlocked interactions of h-CCB particles and UHMWPE matrix, and the Brownian motion of CCB particles in the UHMWPE melt creates the distinctly interfacial transition layer and strongly interfacial adhesion, forming the densely stacked conductive network [35]. To sum up, the optical microscopy observations intuitively confirm the formation of high-quality segregated conductive networks. An interfacial transition layer with the small interface defect is achieved in UHMWPE/h-CCB composites due to the strong interfacial effects. Moreover, we observed that a compacted and oriented conductive pathway near the UHMWPE granules interfacial regions is formed at the optical microscopy of UHMWPE/h-CCB composites (Figure 3e). This result is likely due to the high viscosity of the UHMWPE matrix, which promotes the migration of h-CCB particles during the hot-compression process. With the increase of h-CCB particles dosage, the migration of h-CCB particles attached to the UHMWPE granules interfacial regions become dense and thick, which further reveals good interfacial adhesion between h-CCB and UHMWPE.

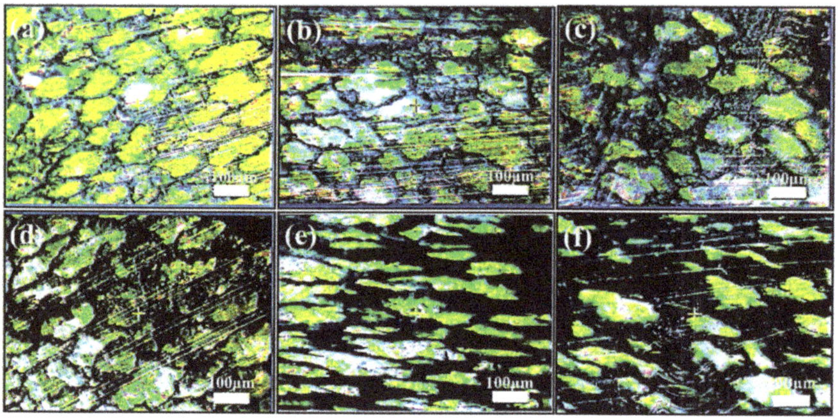

Figure 3. Optical microscopy images of the cross-sectional surface of UHMWPE/CCB composites: (**a**) UHMWPE/1-CCB$_{0.5}$, (**b**)UHMWPE/1-CCB$_1$, (**c**)UHMWPE/l-CCB$_3$, (**d**)UHMWPE/h-CCB$_{0.5}$, (**e**)UHMWPE/h-CCB$_1$ and (**f**)UHMWPE/h-CCB$_3$.

Using a reasonable approach of revealing perfect interfacial bonding in between polymer domains is very important. Figure 4 displays the interfacial microstructure of UHMWPE/CCB composites. The cryo-fractured surface of UHMWPE/CCB composites appears obviously the segregated h-CCB or l-CCB particles located at the interface between adjacent UHMWPE granules. Moreover, the cryo-fractured surface morphology confirms that there are no obvious agglomerates and clusters across the whole cryo-fractured surfaces of two UHMWPE/CCB composites with higher CCB loading. In Figure 4a–c, the micrographs revealed that the l-CCB dispersion and distribution morphology in UHMWPE/l-CCB composites with different l-CCB content is a relatively sparse morphology, the fractured surfaces become uneven, and some interfacial defects appear in the composites with the higher l-CCB loading. In Figure 4a, the red circle indicated the recognized CCB particles, and the phenomena of the selective dispersion and distribution at the interface can be detected by a higher-magnification SEM image. With the increase of the addition of l-CCB, from Figure 4c it can observed that l-CCB conductive pathways between UHMWPE particles begin a loosely conductive networks at 3 wt% 1-CCB concentration. With the CCB content increasing continually, many distinct interfacial voids appear in the cryo-fractured surfaces of UHMWPE composites with high l-CCB concentration. We can obviously observe some cracks of the composites with 5, 7, and 10 wt% l-CCB concentrations at higher magnifications, and two-phase interface of such composite is clear, which suggests the weak interfacial adhesion and compatibilization (Figure S2). This is attributed to the low specific surface area, poor polarity and physical adsorption ability of l-CCB particles. By contrast, UHMWPE/h-CCB composite shows a unique dense segregated structure that h-CCB are orderly distributed at the interfaces of UHMWPE granules. Moreover, the nanosized h-CCB particles are easily decorated onto UHMWPE granules due to their large specific surface area during the mechanical mixing, and then the rough UHMWPE granules experience the plastic deformation under hot-pressing stress, and constrain UHMWPE molecular chains diffusion in between granules. This synergistic interface effect forms a continuous and dense segregated structure in the process of compression molding. In Figure 4d–f, as the amount of h-CCB increases from 0.5 to 3.0 wt%, we observed that both h-CCB and l-CCB particles were orderly dispersed at the interfaces of UHMWPE granules, this result is attributed to the strong volume exclusion effect of the UHMWPE matrix with high viscosity. Moreover, the SEM morphology images of UHMWPE/l-CCB and UHMWPE/h-CCB with different filler content are very similar, and no voids and interfacial defects are found. These results suggested that UHMWPE/l-CCB and UHMWPE/h-CCB composites have excellent mechanical properties at a low loading content. Meanwhile, beyond 3.0 wt% h-CCB content, a unique segregated h-CCB conductive network composed of the mixture of denser stacked h-CCB particles bridging UHMWPE matrix is observed, as seen in Figure S3 at higher magnifications. The evolution of various microstructures two kinds of UHMWPE/CCB composites is mainly attributed to the fabrication process of materials, and interface interactions of CCB-CCB and CCB-polymer matrix. In particular, different filler-matrix interfacial behavior was observed in the composites with CCB particles. In the composite with l-CCB particles, obvious gaps can be found, suggesting that the interactions between phases are relatively poor. By contrast, whereas the incorporation of h-CCB weight fraction into UHMWPE beyond 3 wt%, a compact segregated conductive network between the UHMWPE boundary regions is constructed. Furthermore, the low ratios of h-CCB filled UHMWPE matrix not only greatly guarantees the mechanical properties of the composites, but also realizes the construction of the segregated structure at low conductive percolation threshold values of 0.5~1 wt%.

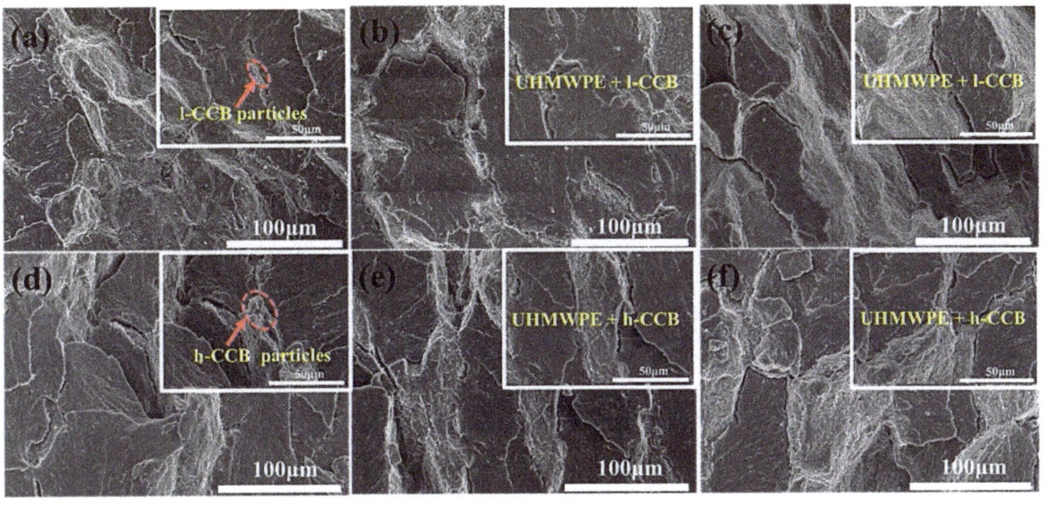

Figure 4. FE-SEM morphology of the cryo-fractured surface of UHMWPE/CCB composites: (**a**) UHMWPE/1-$CCB_{0.5}$; (**b**) UHMWPE/1-CCB_1; (**c**) UHMWPE/1-CCB_3; (**d**) UHMWPE/h-$CCB_{0.5}$; (**e**) UHMWPE/h-CCB_1; and (**f**) UHMWPE/h-CCB_3.

Raman spectroscopy is a fast and useful tool for the characterization of interface structure for carbon-filled polymer composites. To explore the orderly dispersion and distribution of carbon fillers in a polymer matrix, we analyze the interface transition layer structure of h-CCB and l-CCB filled UHMWPE composite via using Raman mapping measurements. Figure 5 shows the polymer-filler interface structure of Raman mapping surface morphology of different UHMWPE/CCB composites. As the CCB content increases from 1.0 wt% to 3.0 wt%, we observed that the distribution of h-CCB or l-CCB significantly affects the interfacial layer microstructure of composite, owing to the formation of the different CCB networks. Moreover, the change of images color from red to blue represents of Raman signal of the UHMWPE matrix phase and CCB particles, respectively. The middle green color transition region is the UHMWPE-CCB co-existed phase, which becomes more and more distinct with the increase of h-CCB content, and in which a distinctly interfacial bonding layer is found. We also found that the segregated h-CCB lamella structures appear a compact segregated structure and interfacial bonding layer with different thicknesses, its h-CCB lamella thickness mainly depends on the content of h-CCB, as observed on the cross-sectional surface of UHMWPE/h-CCB composites as shown in Figure 5c,d. In Figure 5c,d, as the h-CCB content increases from 1 to 3 wt%, the thickness of the compact segregated h-CCB lamella increase from 4.44 μm to 22.65 μm, while the interfacial bonding layer is slightly enhanced, but the increasing trend is no obvious. This implied the strong interfacial interactions of h-CCB and UHMWPE are very helpful for the interfacial structure and properties of the composites. To achieve deeper insights into the different interfacial effects of CCB-filled composites, we analyzed that the presence of different colors represents the different position of Raman spectrum in detail in Raman mapping images of UHMWPE/l-$CCB_{0.5}$, and UHMWPE/l-$CCB_{0.5}$ composites, as is shown in Figure S4a–c. Figure S4b–d displays the corresponding Raman spectrum of Raman mapping images of UHMWPE/l-$CCB_{0.5}$, and UHMWPE/l-$CCB_{0.5}$ composites at the position of UHMWPE matrix materials (A curve), UHMWPE-CCB interfaces (B curve) and CCB particles (C curve). We found that the h-CCB-UHMWPE interfacial structure significantly constrains UHMWPE molecular chains micromovement compared to l-CCB particles, thereby leading to the conformational change of UHMWPE molecular chains in UHMWPE/h-CCB composites. This is related to the existence of the different interface bridging effects of l-CCB-UHMWPE, h-CCB-UHMWPE systems with the incorporation of CCB particles, which is consistent with SEM results. More importantly, the Raman spectrum of UHMWPE/h-CCB composite significantly presents an overlapped characteristic peak at 1348 cm^{-1} and 1587 cm^{-1}. In Figure S4b–d, from this characteristic peak shift change of Raman spectrum of composites, the crystallinity of UHMWPE/h-CCB composite decreases with increasing h-CCB content, we speculated that the addition of h-CCB particles is likely to impact their conformational and crystalline structure of UHMWPE matrices. This further suggested that plentiful h-CCB nanoparticles as an enhanced transition layer is embedded into the UHMWPE matrices, which is contributed to the retention of the mechanical properties of high-loading polymeric-based composites.

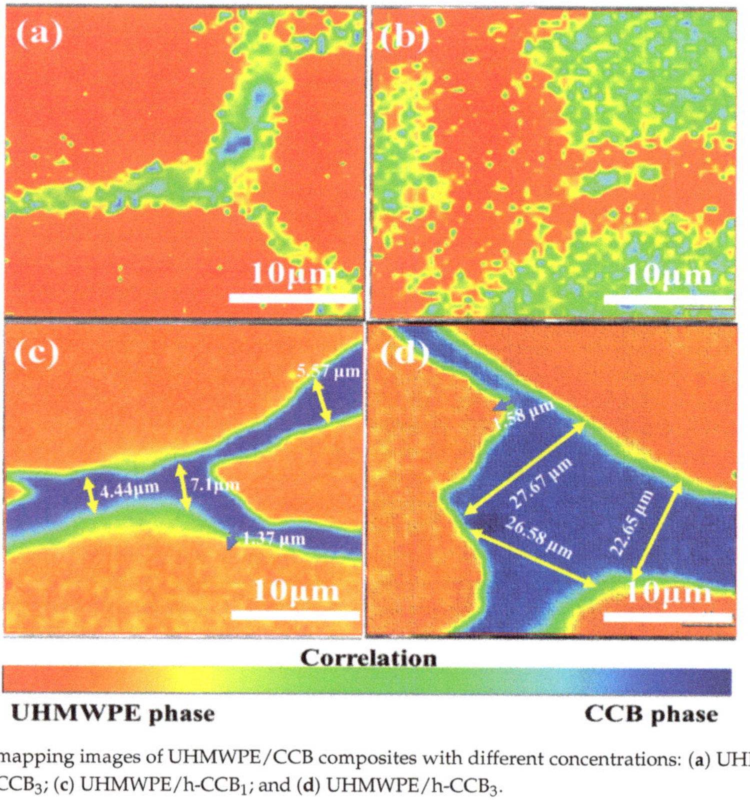

Figure 5. Raman mapping images of UHMWPE/CCB composites with different concentrations: (**a**) UHMWPE/1-CCB_1; (**b**) UHMWPE/1-CCB_3; (**c**) UHMWPE/h-CCB_1; and (**d**) UHMWPE/h-CCB_3.

DSC is one of the most effective methods for characterizing the effect of inorganic CCB in polymer matrices on the crystallization performance and the interaction of filler/polymer under the process of the crystallization of the crystal growth and nucleation. Therefore, further discussion on the thermal property parameters regarding the melting point (T_m), enthalpy of fusion (ΔH_m) and crystallinity (χ_c) are necessary. Figure S5 shows the DSC curves of pure UHMWPE, UHMWPE/l-CCB and UHMWPE/h-CCB composites. The detailed information of these DSC curves is listed in Table S2. In Table S2, the crystallinity (χ_c) of UHMWPE gradually decreases with an increase of CCB content, and the reason for the decrease is mainly ascribed to the extremely high viscosity of the UHMWPE matrix; rigid CCB particles are segregated and evenly dispersed in UHMWPE boundary regions due to the volume exclusion effect of the matrix, limiting the alignment of UHMWPE chains and decreasing the melting points (T_m) of the obtained composites, which is consistent with the reported results of the crystallization properties of high specific surface area MWCNT/UHMWPE [36], large aspect ratio graphene nanoplatelets/UHMWPE [39] and multi-scale mesoporous structural bamboo charcoal/UHMWPE composites [40]. This result further confirms that the crystal structure of UHMWPE is influenced by the h-CCB content—that is, physically interlocking interactions of high polar CCB particles impede the movement of UHMWPE molecular chains, which is also consistent with Raman mapping results. In Figure S5a, the melting curves of DSC find only one melting peak corresponding to that of UHMWPE, which indicates the incorporation of CCB does not affect the crystalline structure of UHMWPE. Besides, the obtained results from the cooling curves Figure S5b show the initial crystallization temperature ($T_{c,\ onset}$) value of the crystallization temperature of UHMWPE is 121.16 °C, while is no obvious change compared to UHMWPE/CCB composites. For instance, when CCB content is 5 wt%, the Tc, onset of h-CCB or the l-CCB modified UHMWPE composites is 120.85 and 120.46 °C, respectively, which is close to the initial crystallization temperature of pure UHMWPE (121.16 °C).

To obtain a better understanding the impact of the interfacial interactions of the addition of various CCB into UHMWPE matrix on the thermal stabilities of the composites, we further evaluate the incorporation of h-CCB and l-CCB content on the thermal stabilities of the composites via using TGA analysis, separately. Figure S6 depicts pure UHMWPE, UHMWPE/h-CCB and UHMWPE/l-CCB composites of TGA curves and DTG curves. The relevant thermal data, including the temperature of 5, 30, 50% weight loss (T_5, T_{30}, and T_{50}) of the composites, the temperature at maximum weight loss rate (T_{max}), heat resistance index (HRI) and charred residue quality are summarized in Table S3. In Table S3, the initial decomposition temperature of UHMWPE/h-CCB and UHMWPE/l-CCB composites with 0.5 wt% CCB concentration has been improved compared to pure UHMWPE, which is due to the interfacial bonding between polar CCB particles and UHMWPE granules, effectively hindering the transformation of heat in UHMWPE molecular chains, obviously enhancing their thermal stabilities. This result is consistent with the microstructure of the composite characterized by SEM. Moreover, the initial decomposition temperature of UHMWPE/h-CCB composites significantly increases as the h-CCB content increases compared to pure UHMWPE and UHMWPE/l-CCB, the T_5, T_{30} and T_{50} of UHMWPE/h-CCB_{10} composites are increased by 9.67, 11.5 and 13.19 °C, respectively. This is attributed to high-quality CCB network structures that play a thermal barrier role, restrict the heat transfer in the UHMWPE matrix and simultaneously delay the thermal decomposition progress of composites as well. Furthermore, the T_5, T_{30}, T_{50} and heat resistance index (HRI) of UHMWPE/h-CCB composites is greater than UHMWPE/l-CCB composites, and these results further illustrate that h-CCB particles are conducive to enhance the interface effect and thermal stability properties of composites.

3.3. Mechanical Properties

The mechanical properties of most polymer composites rely on various factors, including polymer/filler type, method of preparation, dispersion or distribution of fillers [41] and polymer-filler interfacial interaction [42]. In addition, the excellent mechanical properties play an important role in the industrial practical applications of shielding materials as well. It is well-known that the mechanical performance is predominately determined by the interfacial adhesion of the adjacent polymer domains in CPCs materials [43]. Figure 6 shows the mechanical properties of pure UHMWPE, UHMWPE/h-CCB and UHMWPE/l-CCB composites. In Figure 6, when CCB content is 1 wt%, it can be seen that UHMWPE/CCB composites show higher mechanical properties compared to pure UHMWPE, due to the incorporation of good-dispersion CCB particles into UHMWPE matrices as nucleating agents [24], resulting in the stable conformation of the UHMWPE macular chains. Moreover, high loading UHMWPE/h-CCB composites greatly maintain mechanical performance compared to the same addition of UHMWPE/l-CCB composites, the elongation at break value of UHMWPE/h-CCB composites is also significantly higher than that of UHMWPE/l-CCB composites, further suggesting the interfacial adhesion of h-CCB nanoparticles is superior to the interaction of l-CCB, Furthermore, when the increase of h-CCB or l-CCB content to 10 wt%, the tensile strength and elongation at break value of UHMWPE/h-CCB composites are approximately two and six times that of UHMWPE/l-CCB composites, separately, as shown in Figure 6. The mechanical properties improvement in the strength and fracture of high-loading UHMWPE/h-CCB composite is thus attributed to the strong mechanical interlocking in between adjacent UHMWPE granules and h-CCB nanoparticles, good wetting interactions of h-CCB nanoparticles, leading to forming the dense and rigid h-CCB particle networks that can greatly transfer stress and dissipate energy [44]. Numerous matrix fibrils are formed on the tensile fractured surface of UHMWPE/h-CCB composite, and a typical ductile fracture is found. Figure S7 further demonstrated that endowing the good interfacial adhesion between UHMWPE and h-CCB nanoparticles, making the UHMWPE/h-CCB composite more competitive as segregated conductive polymer materials, which can resist high stress in some specific fields. These results are consistent with the above-obtained SEM, OM, Raman mapping imaging and TGA characterization results.

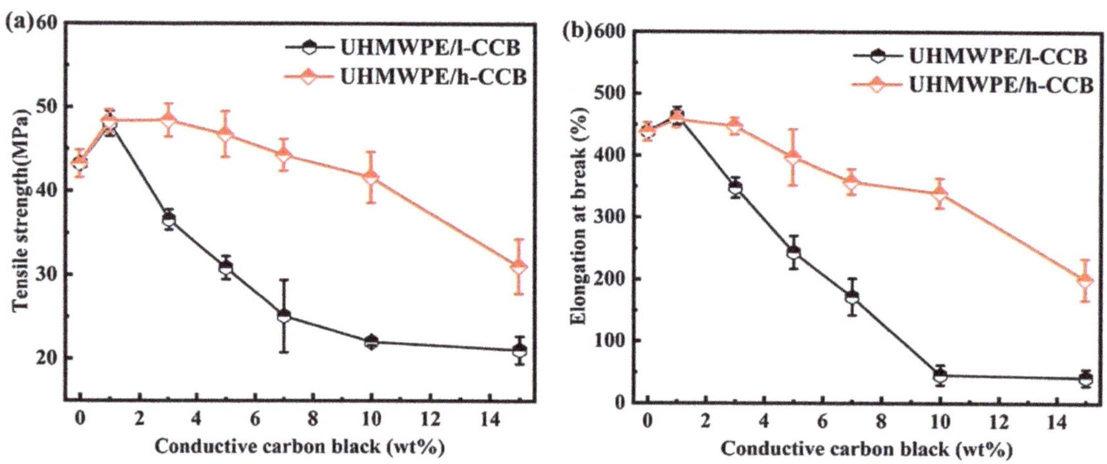

Figure 6. (**a**) tensile strength and (**b**) elongation at break of UHMWPE/CCB composites.

To further elucidate fracture and enhancement mechanism, the tensile fractured surfaces of UHMWPE/h-CCB, and UHMWPE/l-CCB composites, Figure 7 presents the broken tensile surfaces of l-CCB particles filled UHMWPE-based composite compared to a unique UHMWPE/h-CCB composite. The surface on the composite surface is rough because of the presence of h-CCB nanoparticles. No obvious h-CCB particles agglomeration can be observed in the composites, indicating that the dry-blending technique is an effective method to uniformly disperse the h-CCB particles in the polymer matrix. Figure 7a–c show that the surface of the composite material had some small voids, and some l-CCB particles are not well dispersed in the matrix, which was due to the poor interfacial mechanically interlocked in UHMWPE matrix, resulting in the decrease of mechanical properties of composites. Meanwhile, Figure 7d–f show that good interface quality is produced —that is, perfect contact is observed (no voids).

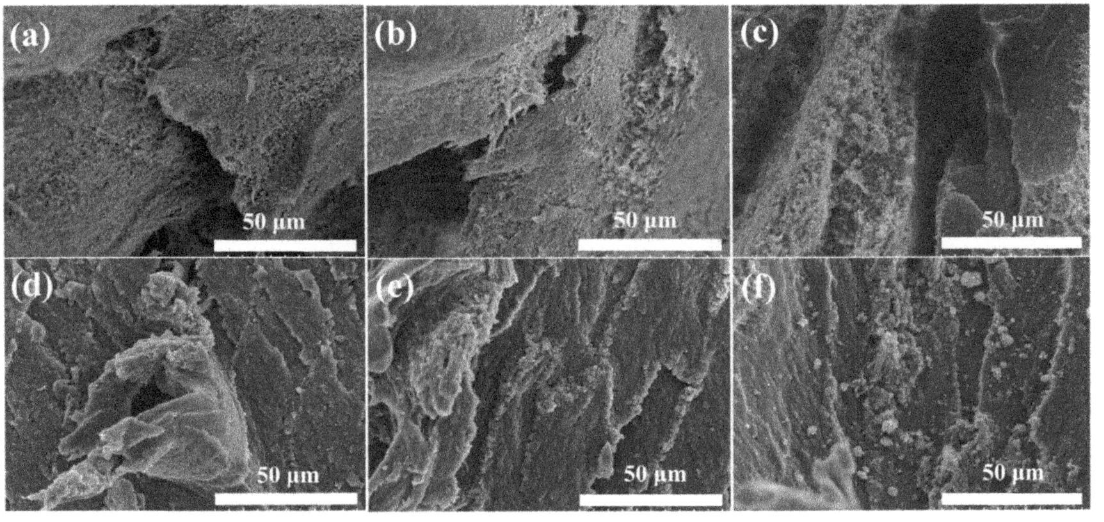

Figure 7. SEM images of the tensile fractured surfaces of UHMWPE/l-CCB composites: (**a**) UHMWPE/l-CCB_5; (**b**) UHMWPE/l-CCB_{10}; and (**c**) UHMWPE/l-CCB_{15}, and UHMWPE/h-CCB composites: (**d**) UHMWPE/h-CCB_5; (**e**) UHMWPE/h-CCB_{10}; and (**f**) UHMWPE/h-CCB_{15}.

3.4. Electrical Conductivity

As one of the key parameters of the electromagnetic interference (EMI) shielding material, high electrical conductivity (σ) plays a vital role in EMI shielding application materials. Figure 8 shows the σ of UHMWPE/h-CCB and UHMWPE/l-CCB composites. In Figure 8a,b, the σ increases gradually with the increase of h-CCB or l-CCB loading weight ratios. Moreover, UHMWPE/h-CCB composites possess a higher σ than that of UHMWPE/l-CCB composites with the same concentration, which should be attributed to the improved conductive network of h-CCB particles. Besides, according to theoretical percolation threshold [45,46]: $\sigma = \sigma_0 (\varphi - \varphi_c)^t$, for $\varphi > \varphi_c$, where φ is the volume fraction of the fillers, φ_c is the volume percolation concentration, σ and σ_0 represents the electrical conductivity of the composites and the conducting fillers, respectively. t represents a critical exponent reflecting the dimensionality of the system. To obtain the φ_c value, the linear fits of log σ vs. log $(\varphi - \varphi_c)$ is performed for each estimated "φ_c", as presented in the interior illustration of Figure 8b, and it is found that φ_c value of UHMWPE/h-CCB is far less than UHMWPE/l-CCB composite, and the percolation threshold of UHMWPE/h-CCB composite is 0.49 wt%. In comparison, we also found that the σ values of UHMWPE/l-CCB composites as a function of CCB content display a linear increasing trend, the estimated percolation threshold value of UHMWPE/l-CCB composite is more than 5 wt%. The φ_c value UHMWPE/h-CCB composite is far lower than that of UHMWPE/l-CCB composite. Furthermore, UHMWPE/h-CCB_{10} already gains a desirable electrical conductivity, and it is exciting that an excellent σ of UHMWPE/h-CCB composites (0.12 S/cm) with low h-CCB loading of 10 wt% far exceeds the target value (1.31×10^{-2} S/cm) for commercial use as EMI shielding materials, which can fulfill the commercial EMI shielding application. The achievement of outstanding σ values in UHMWPE/h-CCB_{10} can be correlated with the different CCB networks formed in the UHMWPE matrix. For UHMWPE/h-CCB_{10} composites, the formation of the interconnecting and dense network reveals that a large number of h-CCB are concentrated at the interfaces of UHMWPE domains and closely overlapped, which is more conducive to the transmission of electrons and interfacial polarity. Moreover, for UHMWPE/l-CCB_{10} composites, the σ values of UHMWPE/l-CCB_{10} composites mainly rely on l-CCB loading to realize higher σ owing to inherent l-CCB's low electrical conductivity and poor interfacial energy characteristics. To further demonstrate the potential application of conductive UHMWPE/CCB composites, in Figure 8c,d a basic electronic setup based on a light-emitting diode (LED), together with various conductive UHMWPE/CCB composites, was assembled. The circuit measurement was carried out in UHMWPE/h-CCB_3 composite and UHMWPE/l-CCB_7 composite, and as expected, the LED bulbs can be glowed by the circuit, demonstrating the percolation behavior of UHMWPE/h-CCB segregated composite at a low percolation threshold, and have desirable σ value. Overall, utilizing an appropriate amount of CCB can not only realize high performance simultaneously, but also achieve the EMI materials with excellent EMI SE.

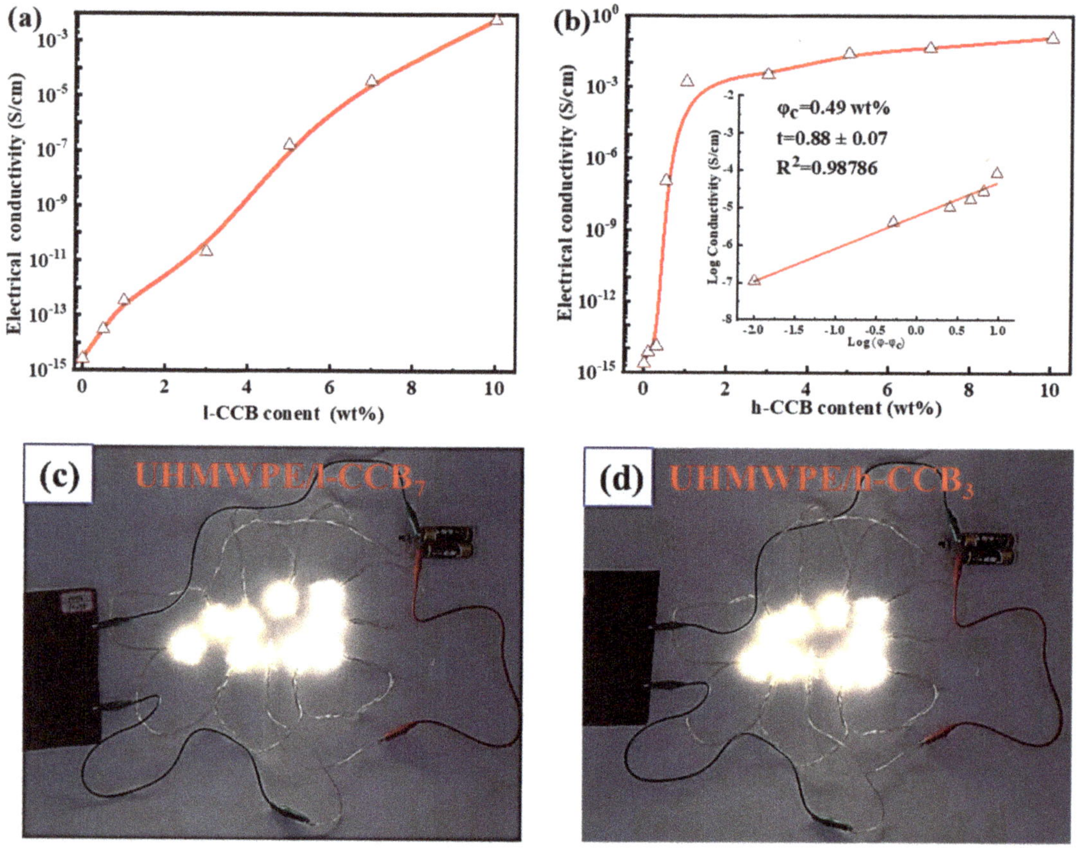

Figure 8. Comparation of σ of UHMWPE/CCB composites with the different CCB content: (**a**) UHMWPE/h-CCB; and (**b**) UHMWPE/l-CCB; the actual conductive circuit of the glowing LED bulbs energized by (**c**) UHMWPE/l-CCB_7; and (**d**) UHMWPE/h-CCB_3 composite.

3.5. Electromagnetic Interference (EMI) Shielding Performance

Figure 9a–f displays the EMI shielding performance and EMI shielding mechanism diagram of segregated UHMWPE/l-CCB and UHMWPE/h-CCB composites with different CCB content. In Figure 9a, it is clear that the EMI SE are continuously improved with increasing CCB filler loading due to the improvement in well-connected conductive network and electrical conductivity in the UHMWPE matrix. For instance, at 10 wt% h-CCB content, the EMI SE of the UHMWPE/h-CCB_{10} composite is found to be a satisfactory EMI SE of 22.3 dB at X band (8.2–12.4 GHz), and which already exceeded the requirement for commercial EMI SE (20 dB)—this signifies 99.9% blocking of the EM wave. Moreover, in Figure 9a, we found that the extent of increase in EMI SE corresponds to the extent of h-CCB incorporated in the UHMWPE matrix. The incorporation of h-CCB in UHMWPE matrix is the most significant contributor to the EMI SE of the composite compared to UHMWPE/l-CCB composite. To comprehend the EMI shielding mechanism for the UHMWPE/h-CCB composite composites, the contribution of absorption SE (SE_A)and reflection SE (SE_R) is disassembled from the total SE (SE_T), and is shown in Figure 9b–d. The detailed SE_A, SE_R and SE_T Values at 8.2 GHz are summarized in Table S4. The relationship between power coefficients [i.e., reflection (R), absorption (A) and Transmission (T)] and frequency of the segregated UHMWPE/l-CCB_{15} and UHMWPE/h-CCB_{15} composites is illustrated in Figure S8. For UHMWPE/h-CCB_{15} composite, the R value raises to 0.78, and the T value keeps at quite a low level. A gradually decreasing trend is observed for A values caused by the huge increase of R. Meanwhile, the UHMWPE/l-CCB_{15} shows a low R value of 0.49 at 8.2 GHz. Moreover, from Figure 9, it could be significantly observed that SE_A contributes overwhelmingly to SE_T, and this result suggests that absorption is the dominant shielding mechanism. Furthermore, we use the mechanism diagram to vividly show the entire process of electromagnetic waves passing through the composites, as shown in Figure 9e,f. The main way for UHMWPE/h-CCB composites to shield EM waves is absorption. When the electromagnetic wave reaches the surface of the composites, part of the electromagnetic wave is reflected, owing to the impedance mismatch. According to the impedance mismatch theory, the EMI SE of the segregated CPCs crucially depends on its high electrical loss and interfacial polarization loss. The desirable EMI shielding performance of CPCs often occurs synchronously with high electrical conductivity. The strong interfacial interaction and compatibility between h-CCB–h-CCB, h-CCB–UHMWPE is usually beneficial to forming more continuous conductive paths and subtle interface polarization, leading to the low percolation threshold and high EMI SE. Moreover, the obtained plentiful interconnecting conductive paths can impact the electrical conductivity of composite and decrease the separation distance between conductive fillers to a great extent. Overall, the high EMI SE of the segregated UHMWPE/CCB composites is attributed to such a dense conductive network with the fine interfacial bonding interactions—such interactions generate electrical loss and polarization loss, leading to the strong absorption and multiple scattering of incident EM waves.

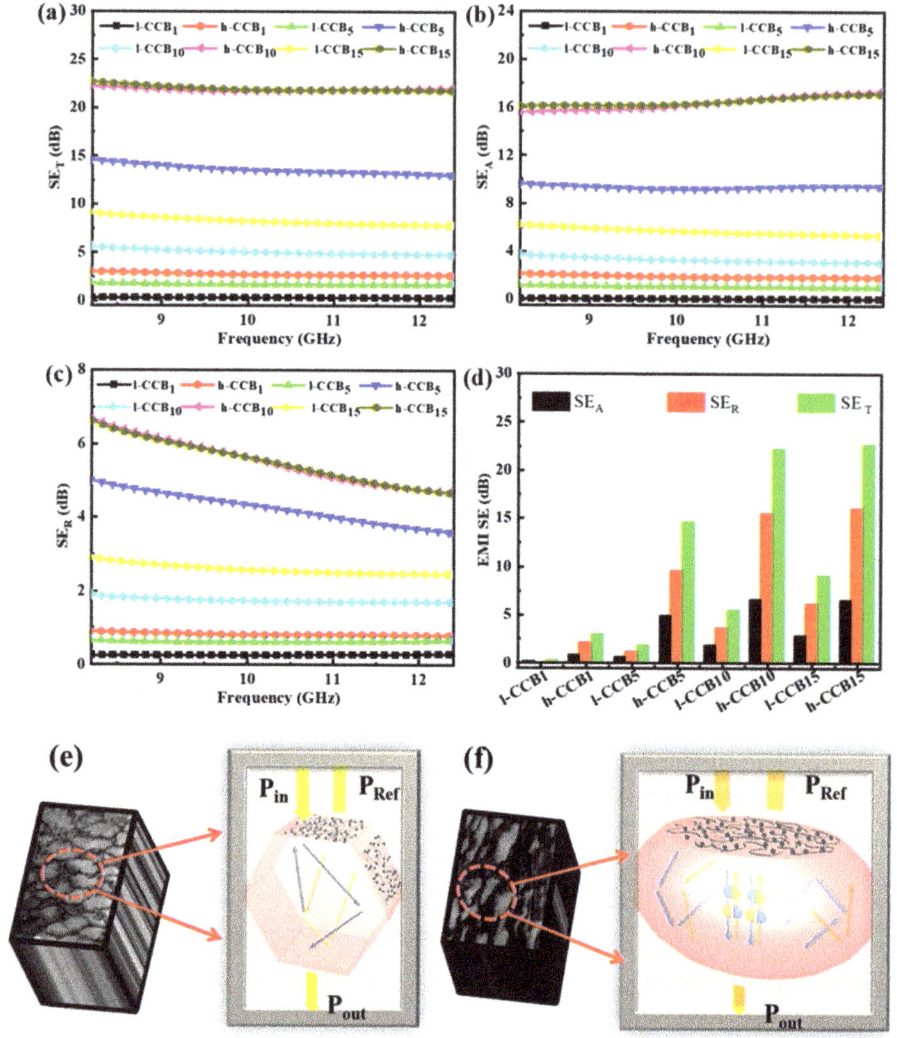

Figure 9. (**a**) SE_T; (**b**) SE_A; and (**c**) SE_R of UHMWPE/h-CCB and UHMWPE/l-CCB composites with different CCB contents; as well as (**d**) Comparison of EMI SE of UHMWPE/h-CCB and UHMWPE/l-CCB composites with different CCB weight ratios at the thickness of approximately 1.5 mm; (**e**) EMI shielding mechanism diagram of UHMWPE/l-CCB; and (**f**) UHMWPE/h-CCB composites.

4. Conclusions

UHMWPE/h-CCB and UHMWPE/l-CCB composites were prepared using simple and effective high-speed mechanical mixing combined with the compression molding method. The mechanical properties of high-loading UHMWPE/h-CCB composite were maintained greatly, and much higher than that of UHMWPE/l-CCB composite, revealing the strong interfacial interactions of UHMWPE matrix and h-CCB particles. The electrical threshold mechanism reveals that a low percolation threshold (0.49 wt%) is obtained in the UHMWPE/h-CCB composite due to the formation of a densely compacted h-CCB network, which is far lower than that of the percolation threshold of UHMWPE/l-CCB. UHMWPE/h-CCB exhibits an evident improvement of electromagnetic interference shielding effectiveness (EMI SE) when the dosage of h-CCB exceeds its electrical percolation threshold, the UHMWPE/h-CCB composite exhibits a satisfactory EMI SE value of 22.3 dB at 10 wt% h-CCB content, which is as four times that of UHMWPE/l-CCB composite with same l-CCB dosage (5.6 dB). Moreover, the tensile strength and elongation at break value of UHMWPE/h-CCB_{10} composites are approximately two and six times that of UHMWPE/l-CCB_{10} composites. In a word, the addition of h-CCB particles not only effectively increases the thermal stability of segregated conducting UHMWPE-based composite, but also ensures the excellent mechanical properties and acceptable EMI SE in practical applications. This low-cost and high-performance segregated conductive polymer composite is a promising candidate for EMI shielding and antistatic fields, which will expect to open the door for next-generation cost-effective EMI shielding materials to cater academic and industrial applications in the future.

Supplementary Materials: The following are available online at https://www.mdpi.com/article/10.3390/nano11082074/s1. Figure S1. Pore size distributions of h-CCB and l-CCB nanoparticles, Figure S2. SEM morphologies of quenched section of the segregated UHMWPE composites with the different l-CCB content: (a) UHMWPE/l-CCB_5, (b) UHMPWPE/l-CCB_7, (c) UHMWPE/l-CCB_{10}, Figure S3. SEM morphologies of quenched section of the segregated UHMWPE composites with the different h-CCB content: (a) UHMWPE/h-CCB_5, (b) UHMWPE/h-CCB_7, (c) UHMWPE/h-CCB_{10}, Figure S4. (a) Raman mapping images of the UHMWPE/l-$CCB_{0.5}$ composites, and (b) the corresponding Raman spectrum of comparation of UHMWPE/l-$CCB_{0.5}$ composite at the different position of Raman mapping images, (c) Raman mapping images of UHMWPE/h-$CCB_{0.5}$ composite, and (d) the corresponding Raman spectrum of comparation of UHMWPE/h-$CCB_{0.5}$ composite at the different position of Raman mapping images. Figure S5. DSC curves of pure UHMWPE, UHMWPE/l-CCB and UHMWPE/h-CCB composites: (a) second heating curved; (b) cooling curved, Figure S6. (a) TGA and (b) DTA curved of UHMWPE, UHMWPE/h-CCB and UHMWPE/l-CCB composites under nitrogen atmosphere, Figure S7. SEM micrographs of the tensile fractured surface of UHMWPE/l-CCB, and UHMWPE/h-CCB composites: (a) UHMWPE/l-CCB_{10}, (b) UHMWPE/l-CCB_{15}, (c) UHMWPE/h-CCB_{10}, (d) UHMWPE/h-CCB_{15}, Figure S8. Relationship between power coefficients and frequency of the segregated (a) UHMWPE/h-CCB_{15} and (b) UHMWPE/l-CCB_{15} composites, Table S1. Physical properties of two types of conductive carbon black (CCB) nanoparticles, Table S2. DSC values of pure UHMWPE, UHMWPE/l-CCB and UHMWPE/h-CCB composites, Table S3. Thermal stabilities of pure UHMWPE, UHMWPE/h-CCB and UHMWPE/l-CCB composites obtained from TGA curves, Table S4. Comparison of SE_T, SE_R, and SE_A at the frequency of 8.2 GHZ for UHMWPE/l-CCB and UHMWPE/h-CCB composites.

Author Contributions: Conceptualization, C.C. and Q.Q.; methodology, H.C.; validation, B.H., X.S. and L.X.; formal analysis, C.C.; investigation, H.C.; resources, Q.C.; data curation, H.C.; writing—original draft preparation, H.C.; writing—review and editing, C.C.; supervision, Q.Q. and X.S.; project administration, Q.C.; funding acquisition, Q.Q. All authors have read and agreed to the published version of the manuscript.

Funding: The study was funded by the grant from the National Key Research and Development Program of China (No. 2016YFB0302300).

Data Availability Statement: Data is contained within the article and Supplementary Materials.

Conflicts of Interest: The authors declare no conflict of interest.

References

1. Kumar, P.; Narayan Maiti, U.; Sikdar, A.; Kumar Das, T.; Kumar, A.; Sudarsan, V. Recent advances in polymer and polymer composites for electromagnetic interference shielding: Review and future prospects. *Polym. Rev.* **2019**, *59*, 687–738. [CrossRef]
2. Jiang, D.; Murugadoss, V.; Wang, Y.; Lin, J.; Ding, T.; Wang, Z.; Shao, Q.; Wang, C.; Liu, H.; Lu, N.; et al. Electromagnetic interference shielding polymers and nanocomposites—A review. *Polym. Rev.* **2019**, *59*, 280–337. [CrossRef]
3. Wanasinghe, D.; Aslani, F. A review on recent advancement of electromagnetic interference shielding novel metallic materials and processes. *Compos. Part B* **2019**, *176*, 107207. [CrossRef]
4. Zhao, B.; Hamidinejad, M.; Wang, S.; Bai, P.; Che, R.; Zhang, R.; Park, C.B. Advances in electromagnetic shielding properties of composite foams. *J. Mater. Chem. A* **2021**, *9*, 8896–8949. [CrossRef]

5. Zhang, W.; Wei, L.; Ma, Z.; Fan, Q.; Ma, J. Advances in waterborne polymer/carbon material composites for electromagnetic interference shielding. *Carbon* **2021**, *177*, 412–426. [CrossRef]
6. Zhang, C.; Lv, Q.; Liu, Y.; Wang, C.; Wang, Q.; Wei, H.; Liu, L.; Li, J.; Dong, H. Rational design and fabrication of lightweight porous polyimide composites containing polyaniline modified graphene oxide and multiwalled carbon nanotube hybrid fillers for heat-resistant electromagnetic interference shielding. *Polymer* **2021**, *224*, 123742. [CrossRef]
7. Yao, Y.; Jin, S.; Zou, H.; Li, L.; Ma, X.; Lv, G.; Gao, F.; Lv, X.; Shu, Q. Polymer-based lightweight materials for electromagnetic interference shielding: A review. *J. Mater. Sci.* **2021**, *56*, 6549–6580. [CrossRef]
8. Xu, H.; Li, Y.; Han, X.; Cai, H.; Gao, F. Carbon black enhanced wood-plastic composites for high-performance electromagnetic interference shielding. *Mater. Lett.* **2021**, *285*, 129077. [CrossRef]
9. Wu, N.; Hu, Q.; Wei, R.; Mai, X.; Naik, N.; Pan, D.; Guo, Z.; Shi, Z. Review on the electromagnetic interference shielding properties of carbon based materials and their novel composites: Recent progress, challenges and prospects. *Carbon* **2021**, *176*, 88–105. [CrossRef]
10. Liang, J.; Chen, J.; Shen, H.; Hu, K.; Zhao, B.; Kong, J. Hollow porous bowl-like nitrogen-doped cobalt/carbon nanocomposites with enhanced electromagnetic wave absorption. *Chem. Mater.* **2021**, *33*, 1789–1798. [CrossRef]
11. Xiao, W.; Han, X.; Niu, X.; Lin, J.; Han, X.; He, A.; Jiang, Q.; Nie, H. Hierarchical porous carbon nanotube skeleton supported polydimethylsiloxane composite with electrical continuity for high-performance electromagnetic shielding. *Adv. Mater. Technol.* **2021**, 2100013. [CrossRef]
12. Li, X.; Wang, G.; Yang, C.; Zhao, J.; Zhang, A. Mechanical and EMI shielding properties of solid and microcellular TPU/nanographite composite membranes. *Polym. Test.* **2021**, *93*, 106891. [CrossRef]
13. Deng, H.; Lin, L.; Ji, M.; Zhang, S.; Yang, M.; Fu, Q. Progress on the morphological control of conductive network in conductive polymer composites and the use as electroactive multifunctional materials. *Prog. Polym. Sci.* **2014**, *39*, 627–655. [CrossRef]
14. Pang, H.; Xu, L.; Yan, D.-X.; Li, Z.-M. Conductive polymer composites with segregated structures. *Prog. Polym. Sci.* **2014**, *39*, 1908–1933. [CrossRef]
15. Song, P.; Liu, L.; Fu, S.; Yu, Y.; Jin, C.; Wu, Q.; Zhang, Y.; Li, Q. Striking multiple synergies created by combining reduced graphene oxides and carbon nanotubes for polymer nanocomposites. *Nanotechnology* **2013**, *24*, 125704. [CrossRef]
16. Song, P.; Cao, Z.; Cai, Y.; Zhao, L.; Fang, Z.; Fu, S. Fabrication of exfoliated graphene-based polypropylene nanocomposites with enhanced mechanical and thermal properties. *Polymer* **2011**, *52*, 4001–4010. [CrossRef]
17. Liu, S.; Qin, S.; Jiang, Y.; Song, P.; Wang, H. Lightweight high-performance carbon-polymer nanocomposites for electromagnetic interference shielding. *Compos. Part A Appl. Sci. Manuf.* **2021**, *145*. [CrossRef]
18. Yu, W.C.; Zhang, G.Q.; Liu, Y.H.; Xu, L.; Yan, D.X.; Huang, H.D.; Tang, J.H.; Xu, J.Z.; Li, Z.M. Selective electromagnetic interference shielding performance and superior mechanical strength of conductive polymer composites with oriented segregated conductive networks. *Chem. Eng. J.* **2019**, *373*, 556–564. [CrossRef]
19. Jia, L.-C.; Yan, D.-X.; Cui, C.-H.; Ji, X.; Li, Z.-M. A Unique double percolated polymer composite for highly efficient electromagnetic interference shielding. *Macromol. Mater. Eng.* **2016**, *301*, 1232–1241. [CrossRef]
20. Zhang, Q.; Wang, J.; Guo, B.-H.; Guo, Z.-X.; Yu, J. Electrical conductivity of carbon nanotube-filled miscible poly(phenylene oxide)/polystyrene blends prepared by melt compounding. *Compos. Part B Eng.* **2019**, *176*, 107213. [CrossRef]
21. Feng, D.; Wang, Q.; Xu, D.; Liu, P. Microwave assisted sinter molding of polyetherimide/carbon nanotubes composites with segregated structure for high-performance EMI shielding applications. *Compos. Sci. Technol.* **2019**, *182*, 107753. [CrossRef]
22. Jia, L.C.; Yan, D.X.; Jiang, X.; Pang, H.; Gao, J.F.; Ren, P.G.; Li, Z.M. Synergistic effect of graphite and carbon nanotubes on improved electromagnetic interference shielding performance in segregated composites. *Ind. Eng. Chem. Res.* **2018**, *57*, 11929–11938. [CrossRef]
23. Cui, C.-H.; Yan, D.-X.; Pang, H.; Jia, L.-C.; Bao, Y.; Jiang, X.; Li, Z.-M. Towards efficient electromagnetic interference shielding performance for polyethylene composites by structuring segregated carbon black/graphite networks. *Chin. J. Polym. Sci.* **2016**, *34*, 1490–1499. [CrossRef]
24. Cheng, H.; Cao, C.; Zhang, Q.; Wang, Y.; Liu, Y.; Huang, B.; Sun, X.-L.; Guo, Y.; Xiao, L.; Chen, Q.; et al. Enhancement of electromagnetic interference shielding performance and wear resistance of the UHMWPE/PP blend by constructing a segregated hybrid conductive carbon black-polymer network. *ACS Omega* **2021**, *6*, 15078–15088. [CrossRef] [PubMed]
25. Wang, Y.; Yang, J.; Zhou, S.; Zhang, W.; Chuan, R. Electrical properties of graphene nanoplatelets/ultra-high molecular weight polyethylene composites. *J. Mater. Sci. Mater. Electron.* **2017**, *29*, 91–96. [CrossRef]
26. Dyachkova, T.; Gutnik, I.; Nagdaev, V.; Maksimkin, A.; Burakova, E.; Galunin, E.; Memetov, N.; Khan, Y.; Dayyoub, T. Studying the surface of UHMWPE films modified with graphene nanoplatelets using a Raman mapping method. *Fuller. Nanotub. Carbon Nanostruct.* **2020**, *28*, 561–564. [CrossRef]
27. Sun, Z.-F.; Ren, P.-G.; Zhang, Z.-W.; Ren, F. Synergistic effects of conductive carbon nanofillers based on the ultrahigh-molecular-weight polyethylene with uniform and segregated structures. *J. Appl. Polym. Sci.* **2019**, *136*, 47317. [CrossRef]
28. Pang, H.; Yan, D.X.; Bao, Y.; Chen, J.B.; Chen, C.; Li, Z.M. Super-tough conducting carbon nanotube/ultrahigh-molecular-weight polyethylene composites with segregated and double-percolated structure. *J. Mater. Chem.* **2012**, *22*, 23568–23575. [CrossRef]
29. Mondal, S.; Ravindren, R.; Shin, B.; Kim, S.; Lee, H.; Ganguly, S.; Das, N.C.; Nah, C. Electrical conductivity and electromagnetic interference shielding effectiveness of nano-structured carbon assisted poly(methyl methacrylate) nanocomposites. *Polym. Eng. Sci.* **2020**, *60*, 2414–2427. [CrossRef]

30. Huang, C.J.; Cheng, Q.F. Learning from nacre: Constructing polymer nanocomposites. *Compos. Sci. Technol.* **2017**, *150*, 141–166. [CrossRef]
31. Zeng, H.; Wu, J.; Pei, H.; Zhang, Y.; Ye, Y.; Liao, Y.; Xie, X. Highly thermally conductive yet mechanically robust composites with nacre-mimetic structure prepared by evaporation-induced self-assembly approach. *Chem. Eng. J.* **2021**, *405*, 126865. [CrossRef]
32. Zhang, T.; Sun, J.; Ren, L.; Yao, Y.; Wang, M.; Zeng, X.; Sun, R.; Xu, J.-B.; Wong, C.-P. Nacre-inspired polymer composites with high thermal conductivity and enhanced mechanical strength. *Compos. Part A Appl. Sci. Manuf.* **2019**, *121*, 92–99. [CrossRef]
33. Zhu, S.; Shi, R.; Qu, M.; Zhou, J.; Ye, C.; Zhang, L.; Cao, H.; Ge, D.; Chen, Q. Simultaneously improved mechanical and electromagnetic interference shielding properties of carbon fiber fabrics/epoxy composites via interface engineering. *Compos. Sci. Technol.* **2021**, *207*, 108696. [CrossRef]
34. Wang, K.; Shen, L.; Song, C.; Zhang, Y.; Chen, P. The electrical performance and conductive network of reduced graphene oxide-coated ultra-high-molecular-weight polyethylene fibers through electrostatic interaction and covalent bonding. *J. Appl. Polym. Sci.* **2020**, *137*, 48946. [CrossRef]
35. Zhao, S.; Li, G.; Liu, H.; Dai, K.; Zheng, G.; Yan, X.; Liu, C.; Chen, J.; Shen, C.; Guo, Z. Positive Temperature Coefficient (PTC) evolution of segregated structural conductive polypropylene nanocomposites with visually traceable carbon black conductive network. *Adv. Mater. Interfaces* **2017**, *4*, 201700265. [CrossRef]
36. Wang, R.; Zheng, Y.; Chen, L.; Chen, S.; Zhuo, D.; Wu, L. Fabrication of high mechanical performance UHMWPE nanocomposites with high-loading multiwalled carbon nanotubes. *J. Appl. Polym. Sci.* **2019**, *137*, 48667. [CrossRef]
37. Zhou, H.; Xiao, Z.; Wang, Y.; Hao, X.; Xie, Y.; Song, Y.; Wang, F.; Wang, Q. Conductive and fire-retardant wood/polyethylene composites based on a continuous honeycomb-like nanoscale carbon black network. *Constr. Build. Mater.* **2020**, *233*, 117369. [CrossRef]
38. Li, Y.; Huang, X.; Zeng, L.; Li, R.; Tian, H.; Fu, X.; Wang, Y.; Zhong, W.-H. A review of the electrical and mechanical properties of carbon nanofiller-reinforced polymer composites. *J. Mater. Sci.* **2018**, *54*, 1036–1076. [CrossRef]
39. Alam, F.; Choosri, M.; Gupta, T.K.; Varadarajan, K.M.; Choi, D.; Kumar, S. Electrical, mechanical and thermal properties of graphene nanoplatelets reinforced UHMWPE nanocomposites. *Mater. Sci. Eng. B* **2019**, *241*, 82–91. [CrossRef]
40. Li, S.; Huang, A.; Chen, Y.-J.; Li, D.; Turng, L.-S. Highly filled biochar/ultra-high molecular weight polyethylene/linear low density polyethylene composites for high-performance electromagnetic interference shielding. *Compos. Part B Eng.* **2018**, *153*, 277–284. [CrossRef]
41. Kadar, R.; Abbasi, M.; Figuli, R.; Rigdahl, M.; Wilhelm, M. Linear and nonlinear rheology combined with dielectric spectroscopy of hybrid polymer nanocomposites for semiconductive applications. *Nanomaterials* **2017**, *7*, 23. [CrossRef] [PubMed]
42. Ke, K.; Potschke, P.; Gao, S.; Voit, B. An ionic liquid as interface linker for tuning piezoresistive sensitivity and toughness in poly(vinylidene fluoride)/carbon nanotube composites. *ACS Appl. Mater. Interfaces* **2017**, *9*, 5437–5446. [CrossRef] [PubMed]
43. Xie, Y.; Tang, J.; Ye, F.; Liu, P. Microwave-assisted sintering to rapidly construct a segregated structure in low-melt-viscosity poly(lactic acid) for electromagnetic interference shielding. *ACS Omega* **2020**, *5*, 26116–26124. [CrossRef] [PubMed]
44. Song, Y.; Zheng, Q. Concepts and conflicts in nanoparticles reinforcement to polymers beyond hydrodynamics. *Prog. Mater. Sci.* **2016**, *84*, 1–58. [CrossRef]
45. Wang, S.; Huang, Y.; Zhao, C.; Chang, E.; Ameli, A.; Naguib, H.E.; Park, C.B. Theoretical modeling and experimental verification of percolation threshold with MWCNTs' rotation and translation around a growing bubble in conductive polymer composite foams. *Compos. Sci. Technol.* **2020**, *199*, 108345. [CrossRef]
46. Abbasi, H.; Antunes, M.; Velasco, J.I. Recent advances in carbon-based polymer nanocomposites for electromagnetic interference shielding. *Prog. Mater. Sci.* **2019**, *103*, 319–373. [CrossRef]

nanomaterials

MDPI

Article

Assembly of Copolymer and Metal−Organic Framework HKUST-1 to Form $Cu_{2-x}S$/CNFs Intertwining Network for Efficient Electrocatalytic Hydrogen Evolution

Yuanjuan Bai [1,2,*], Yanran Li [2], Gonggang Liu [1] and Jinbo Hu [1,*]

1 Hunan Province Key Laboratory of Materials Surface & Interface Science and Technology, College of Materials Science and Engineering, Central South University of Forestry and Technology, Changsha 410004, China; liugonggang@csuft.edu.cn
2 State Key Laboratory of Molecular Engineering of Polymers, Department of Macromolecular Science, Fudan University, Shanghai 200438, China; 17110440007@fudan.edu.cn
* Correspondence: byj@cqu.edu.cn or yuanjuanbai@126.com (Y.B.); hjb1999@hotmail.com (J.H.)

Abstract: The construction of complex intertwined networks that provide fast transport pathways for ions/electrons is very important for electrochemical systems such as water splitting, but a challenge. Herein, a three dimensional (3-D) intertwined network of $Cu_{2-x}S$/CNFs (x = 0 or 0.04) has been synthesized through the morphology-preserved thermal transformation of the intertwined PEG-*b*-P4VP/ HKUST-1 hybrid networks. The strong interaction between PEG chains and Cu^{2+} is the key to the successful assembly of PEG-*b*-P4VP nanofibers and HKUST-1, which inhibits the HKUST-1 to form individual crystalline particles. The obtained $Cu_{2-x}S$/CNFs composites possess several merits, such as highly exposed active sites, high-speed electronic transmission pathways, open pore structure, etc. Therefore, the 3-D intertwined hierarchical network of $Cu_{2-x}S$/CNFs displays an excellent electrocatalytic activity for HER, with a low overpotential (η) of 276 mV to reach current densities of 10 mA cm^{-2}, and a smaller Tafel slope of 59 mV dec^{-1} in alkaline solution.

Keywords: assembly; metal-organic frameworks; hydrogen evolution reaction; $Cu_{2-x}S$

Citation: Bai, Y.; Li, Y.; Liu, G.; Hu, J. Assembly of Copolymer and Metal−Organic Framework HKUST-1 to Form $Cu_{2-x}S$/CNFs Intertwining Network for Efficient Electrocatalytic Hydrogen Evolution. *Nanomaterials* **2021**, *11*, 1505. https://doi.org/10.3390/nano11061505

Academic Editors: Mohammad Malikan, Shahriar Dastjerdi and Danil N. Dybtsev

Received: 13 May 2021
Accepted: 3 June 2021
Published: 7 June 2021

Publisher's Note: MDPI stays neutral with regard to jurisdictional claims in published maps and institutional affiliations.

1. Introduction

Electrochemical water splitting is a critical energy conversion process for producing clean and sustainable hydrogen, which is composed of two half-cell reactions: oxygen evolution reaction (OER) and hydrogen evolution reaction (HER) [1–3] Electrolysis is a process that consumes electricity, therefore a catalyst is needed to reduce the potential. Pt-based electrocatalysts exhibit the best performance for H_2 evolution in strongly acidic electrolytes, however their HER activities are substantially diminished under alkaline conditions [4,5] Consequently, considerable attempts have been devoted to developing sustainable, highly efficient, and non-precious electrocatalysts to meet a target of Pt-based catalysts replacement. In recent years, transition metal sulfides (TMSs) have been widely investigated and have demonstrated their potential as HER catalysts due to their high catalytic activity and chemical stability [6] Metal-organic frameworks (MOFs) are an intriguing class of porous crystalline materials constructed by the coordination of metal ions or clusters with organic linkers [7–11] For example, HKUST-1({$[Cu_3(C_9H_3O_6)_2(H_2O)_3]$}$_n$, aka Cu-BTC) is one of the very first permanently porous MOFs, which has been widely studied for multiple applications [12–14] The microporosity and tunable functionality of MOFs make them ideal template precursors to fabricate various transition metal-based carbon composites, including TMSs, by means of the pyrolysis process under a specific atmosphere [6,15,16] However, MOFs may undergo structural collapse during pyrolysis, resulting in the dramatic decrease in surface area, the wreck of well-defined MOF pore/channel structures, and the uneven distribution of active components, which significantly reduces the electrochemical performances of the MOFs derived materials. Therefore, the performance of the existing MOF derivatives' electrocatalysts is still struggles to match precious-metal-based materials.

One of the best ways to solve the above issues is the assembly of crystalline MOF nanoparticles into well-aligned one, two, or three dimensional (1-, 2- or 3-D) superstructures, used as an appropriate template as both scaffold and directing agent for the epitaxial growth of MOFs [17–19] After a pyrolysis process, the MOFs superstructures would be converted into transition-metal-based materials with ordered stacking and porous nanostructure. The choice of the template is vital to the success of the assembly. Commonly used templates include metal-based materials (e.g., Te nanowire [20]), carbon materials (carbon nanotubes (CNTs), carbon nanofiber (CNFs) and so on) and polymers [21–23] Among of them, 1-D polymer nanowires or nanofiber templates with unique high surface-to-volume ratio, adjustable size, and surface modifiability attract wide research interests [24,25] In particular, 1-D polyacrylonitrile (PAN) nanowires substrate prepared by electrospinning is the most widely used in many research studies [23,26] For instance, Centrone et al. reported a microwave irradiation approach to grow MIL-47 on 1-D PAN nanowires [27] Han et al. proposed that the uniform and stable growth of MOFs on PAN nanowires can be enhanced through two types of chemical modification methods [23] These studies have shown that the surface-exposed functional groups play a key role in the MOFs' growth on the PAN substrate, which act as binding sites for metal species of the target MOFs. Nevertheless, due to the difficulties in controlling the number of functional groups, and the compatibility between functional groups and MOFs, MOFs may fall off and be unevenly distributed on the surface of PAN.

In 2012, Chen and coworkers reported 1-D poly (ethylene glycol)-*b*-poly(4-vinylpyridine) (PEG-*b*-P4VP) nanowires composed of a PEG shell and slightly crosslinked P4VP core [28] When the aspect ratio of nanowires is high enough, the 3-D intertwining network can be formed. The follow-up research confirmed that the PEG-*b*-P4VP intertwined nanowires are excellent templates for the nucleation and growth of small MOF crystals due to their high aspect ratio (an average diameter of 30 nm and length up to several microns) and original abundant nucleation sites for metal ions [29] In the electrochemical application, the construction of complex intertwined conductive networks is desired, which would provide fast transport pathways for mass and charge, and then enhance the performance of catalysts [30,31] It is likely that a TMSs/carbon composite intertwining network could be prepared by the vulcanization of the as-prepared PEG-*b*-P4VP@MOFs network; but this hypothesis remains unexplored.

Herein, we have successfully constructed a novel intertwined $Cu_{2-x}S$/CNFs ($x = 0$ or 0.04; CNFs = carbon nanofibers) network. The self-assembling strategy is used to control the growth of HKUST-1 nanocrystals on the 1-D super long PEG-*b*-P4VP nanofibers to form PEG-*b*-P4VP@HKUST-1 composites under room temperature. The PEG shell of the PEG-*b*-P4VP nanofibers provide numerous nucleation sites for HKUST-1, while the P4VP core with a positive charge repels the metal ions (precursor of the HKUST-1). As a result, hybridization of the PEG-*b*-P4VP nanofibers by the HKUST-1 occurs selectively in the PEG shell. Afterwards, the $Cu_{2-x}S$/CNFs composite materials with an intertwined network structure can be prepared with un-changed morphology by using in-situ thermal calcination PEG-*b*-P4VP@ HKUST-1 precursors. Compared with $Cu_{2-x}S$/C composites derived from individual HKUST-1 crystal, $Cu_{2-x}S$/CNFs significantly enhance the HER performance, which is strongly related to the novel network superstructure.

2. Results and Discussion

Figure 1 shows the forming process of the 3-D network of $Cu_{2-x}S$/CNFs nanocomposites. First, the core-crosslinked polymeric linear-like micelles with a PEG shell and a slightly crosslinked P4VP core, designated as PEG-*b*-P4VP nanofibers (NFs), is synthesized in a water/methanol mixed solvent according to a reported method [28]. Second, as the assembly substrate, PEG-*b*-P4VP NFs are dispersed in mixed solvent of water and N, N-Dimethylformamide (DMF) containing $Cu(CH_3COO)_2 \cdot H_2O$ ($Cu(OAC)_2$). During this process, Cu^{2+} ions were favorably absorbed onto the PEG chain surface by the electrostatic interaction. Along to the previous step, trimesic acid (BTC) ligands solutions are added to assemble with Cu^{2+} ions via the coordination interactions [12–14]. Then, a uniform PEG/HKUST-1 hybrid shell is generated in-situ on the surface of the P4VP core, resulting in a PEG-*b*-P4VP@HKUST-1 composites with core-shell structure. Ultimately, the PEG-*b*-P4VP@HKUST-1 composites are converted into a 3-D hierarchical network of $Cu_{2-x}S$/CNFs after the sulfurization reaction between PEG-*b*-P4VP@HKUST-1 and thiourea in an argon flow. The $Cu_{2-x}S$/CNFs hierarchical network have many advantages for HER in alkaline solution including: (1) the continuous and conductive network of the CNFs with a hierarchically porous structure can enable fast charge and mass transfer; and (2) the high active specific surface areas of the $Cu_{2-x}S$/CNFs offer abundant electrocatalytic active sites that are easily accessible in HER.

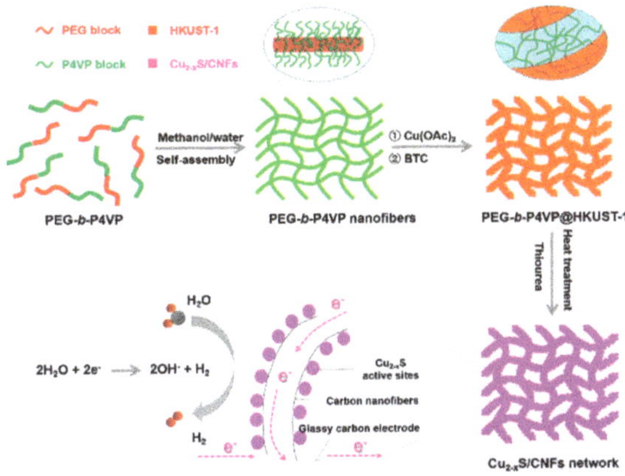

Figure 1. Schematic diagram of the synthesis of 3-D $Cu_{2-x}S$/CNFs hierarchical network electrode.

Field emission scanning electron microscope (FESEM) and transmission electron microscopy (TEM) in Figure 2a–c reveal that the PEG-*b*-P4VP nanofibers have a 3-D feature assembled by a series of stacked 1-D entangled ultrafine nanofibers with uniform diameters of about 30 nm and extra-long lengths above 2 μm. From the typical FESEM and TEM images in Figure 2d–f, we can see the structure of PEG-*b*-P4VP@HKUST-1 composites comprise a tightly crowded nanosized HKUST-1 crystal layer encapsulating the PEG-*b*-P4VP nanofibers with uniform size and well-defined shape. Notably, the generated composites inherit the intertwining network structure of their precursors. No unassembled irregular HKUST-1 crystals are observed, suggesting that the nucleation and growth of MOF crystals are localized on the surface of the PEG-*b*-P4VP templates. In addition, it is found that the polymer chains penetrate the crystalline structure of HKUST-1, and the particle size was less than 100 nm. Consequently, the interaction is PEG-*b*-P4VP nanofibers and HKUST-1 crystals, as it is extended from the core surface to the deep crystalline structure, which is important for the stability of HKUST-1 nanoparticles. The calculation result shows that the HKUST-1 content in PEG-b-P4VP/HKUST-1 composites is around 91 wt % (experimental section, Supporting Information). Because the individual composite nanofibers were dispersible in the suspension, the resultant architecture of the 3-D network shows a loose morphology (Figure S1). Figure 3 shows the X-ray diffraction (XRD) patterns of the PEG-b-P4VP/HKUST-1 composites. Based on the HKUST-1 structure data [12–14], the observed pattern and the simulated pattern show high similarity, confirming the formation of pure crystalline HKUST-1. It is found that the packing density of the HKUST-1 crystals be tuned by simply adjusting precursor concentration (Figure S2). In other words, the number of nucleation sites are controllable. As the addition of Cu^{2+} and organic ligands increased, the number of HKUST-1 nanoparticles on the surface of PEG-*b*-P4VP nanofibers gradually increased. In the end, the 0.24 g of $Cu(OAC)_2$ and 0.21 g of BTC ligands were selected to apply for defined PEG-*b*-P4VP@HKUST-1.

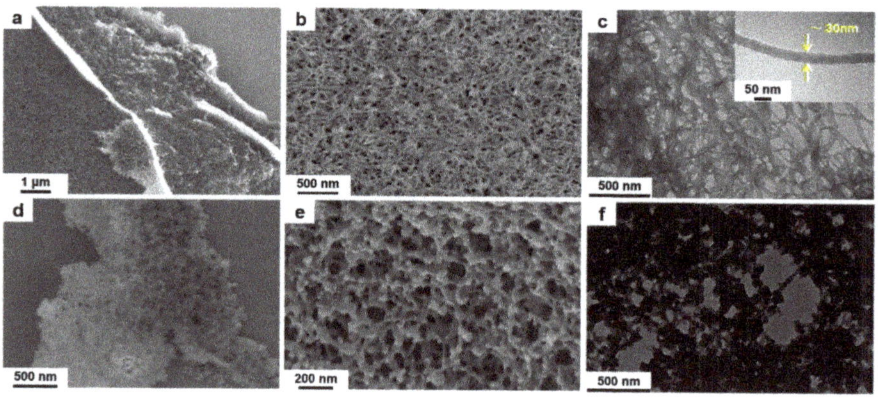

Figure 2. Morphology of the as-prepared (**a–c**) PEG-*b*-P4VP nanofibers and (**d–f**) PEG-*b*-P4VP@HKUST-1 hybrids as observed by SEM and TEM with different magnifications. The inset TEM image in Figure 2c shows the diameter of PEG-*b*-P4VP nanowire is approximately 30 nm.

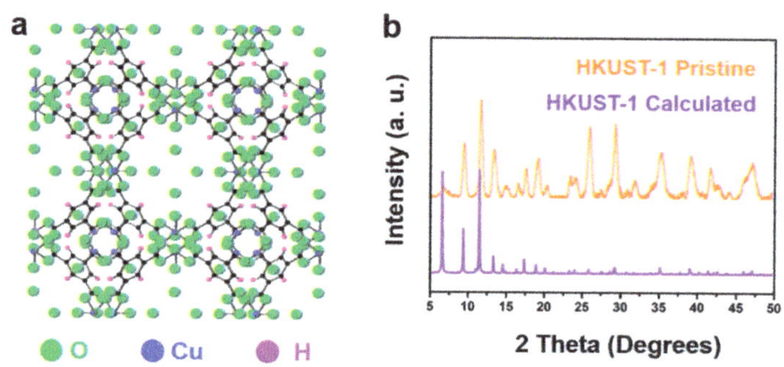

Figure 3. (**a**) The structure diagram of HKUST-1. (**b**) XRD pattern of the PEG-*b*-P4VP@HKUST-1 composites.

Characterization using XRD confirmed the structural identity and phase purity of the as-synthesized samples. As shown in Figure 4a, the XRD peaks of the final product after thermal treatment are well matched to those of the typical crystalline structures of Cu_2S (JCPDS card no 09-0328) and $Cu_{1.96}S$ phase (JCPDS card no 29-0578). Therefore, the value of x can be determined to be 0 or 0.04. SEM (Figure 4b) and TEM (Figure 4c) images show that the 3-D intertwining network structure can be well-maintained for the $Cu_{2-x}S$ /CNFs even after calcination with the presence of thiourea under 400 °C. Closer observation of the $Cu_{2-x}S$ /CNFs (Figure 4d) reveals that $Cu_{2-x}S$ nanoparticle with coarse surfaces are uniformly packed along the ultra-long CNFs. No scattered $Cu_{2-x}S$ nanoparticles are observed, suggesting that the nanoparticles are sturdily attached to the CNFs. High-resolution TEM (HRTEM) (Figure 4e) observes clearly resolved and well-defined lattice fringes, revealing the high crystallinity in agreement with XRD results. The distance between the adjacent lattice planes is 0.339 nm, corresponding to standard spacing of (302) plane of Cu_2S. By contrast, for individual HKUST-1 crystals prepared from the assembly of $Cu(OAc)_2$ and BTC ligands without the PEG-*b*-P4VP template, the collapse of the porous structure and random aggregation of the nanoparticles occurred after sulfurization (Figure S3). The product of the individual HKUST-1 crystals after sulfurization treatment is denoted as $Cu_{2-x}S/C$. The X-ray spectrometry (EDS) spectrum of the $Cu_{2-x}S$/CNFs (Figure S4a) shows Cu and S signals attributable to $Cu_{2-x}S$ and C signals to CNFs. The corresponding EDS element mapping images (Figure S4b–e) demonstrate the homogeneous distribution of Cu, S, and C elements in the hierarchical nanostructure.

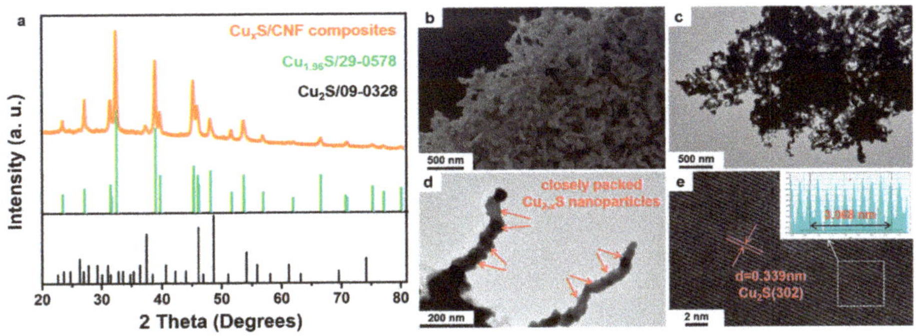

Figure 4. (**a**) XRD pattern, (**b**) SEM, (**c**,**d**) TEM and (**e**) HRTEM images of $Cu_{2-x}S$/CNFs composites (inset image is the intensity plot of d-spacing for the (302) plane of $Cu_{2-x}S$ in e).

Electrochemical water splitting is a surface chemical process. The surface of materials plays a major role in determining catalysts behaviors. Consequently, the elemental compositions and valence states of $Cu_{2-x}S$/CNFs network are studied in detail by X-ray photoelectron spectroscopies (XPS). Figure 5a displays clear a survey spectrum, which confirms the successful synthesis of $Cu_{2-x}S$/CNFs. Regarding the Cu(2p) XPS spectrum as exhibited in Figure 5b, two strong peaks for Cu 2p orbit can be assigned to the Cu $2p_{3/2}$ and Cu $2p_{1/2}$ [32,33]. The binding energy of Cu $2p_{3/2}$ can be fitted into two peaks simulated at 932.7 and 934.6 eV [32,33]. The shakeup satellite peaks appearing at 943.7 eV might result from the partial surface oxidation of the samples when they made contact with air. In the S 2p region of the high resolution XPS spectra in Figure 5c, the binding energies (BEs) at 162.8 and 161.8 eV are ascribed to S $2p_{3/2}$ and S $2p_{1/2}$, respectively [34]. The two doublet peaks simulated at 163.7 and 164.9 eV correspond to the BEs of S-S bond. This result provided further evidence of the surface oxidation of $Cu_{2-x}S$ nanoparticles caused by air exposure. Raman spectroscopy in Figure S5 is used to investigate the nature of the carbon of $Cu_{2-x}S$/CNFs. The two peaks located at 1351 and 1582 cm^{-1} match well with the D (disordered carbon) and G (graphitized carbon) bands of carbon, respectively. The D-band to G-band intensity ratio (I_D/I_G) of the samples is calculated to be around 1.07, implying the limited graphitization degree of $Cu_{2-x}S$/CNFs.

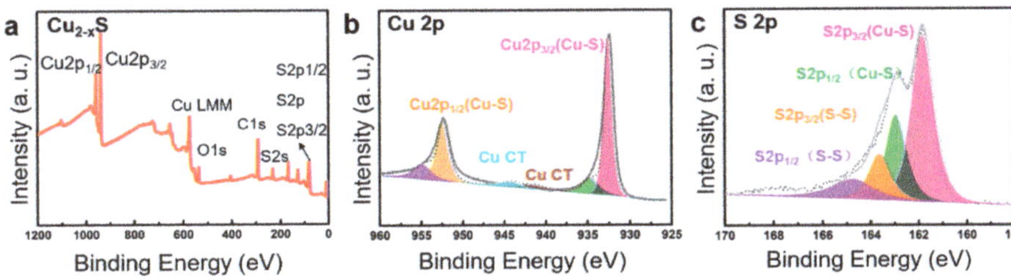

Figure 5. XPS survey spectra (**a**) of $Cu_{2-x}S$/CNFs. High-resolution (**b**) Cu 2p and (**c**) S 2p XPS spectra for $Cu_{2-x}S$/CNFs.

The electrocatalytic HER performance of $Cu_{2-x}S$/CNFs is tested in 1.0 M KOH in a typical three-electrode setup with a scan rate of 5 mV s^{-1}. Pt/C, bare glassy carbon (GC) electrode, and $Cu_{2-x}S/C$ composites are also examined for comparison. Figure 6a shows the linear sweep voltammetry (LSV) curves on the reversible hydrogen electrode

(RHE) scale after iR correction. The bare GC electrode shows nearly no catalytic activity, while Pt/C exhibits excellent activity for HER. $Cu_{2-x}S$/C composites show the limited HER performance with an overpotential of 432 mV at the current density of 10 mA cm^{-2}. However, the as-prepared $Cu_{2-x}S$/CNFs exhibit much superior HER activity compared to $Cu_{2-x}S$/C composites, and only demand overpotentials of 276 and 337 mV to achieve current densities of 10, and 50 mA cm^{-2}, respectively, fully demonstrating the importance of hierarchical intertwining network structures in the catalyst. The Tafel slope is an important indicator of electron-transfer kinetics and the rate-determining step in the HER process. The linear portions of the Tafel plots are fitted to the Tafel equation: $\eta = b \log j + a$, where j is the current density and b is the Tafel slope. As shown in Figure 6b, Pt/C, $Cu_{2-x}S$/C composites, and $Cu_{2-x}S$/CNFs has Tafel slopes of 30, 102, and 59 mV dec^{-1}, respectively. Based on the kinetic mechanism for alkaline HER, the HER over $Cu_{2-x}S$/CNFs follows a Volmer—Heyrovsky mechanism, indicating that the Heyrovsky process (electrochemical desorption of hydrogen atoms) is the rate determining step [4,5,35] Electrochemical stability is another vital criterion to evaluate the new catalysts. As observed in Figure 6c, the $Cu_{2-x}S$/CNFs retained 94% of its initial HER activity after a 12 h test, while the overpotential increase of the $Cu_{2-x}S$/C electrode under the same condition was as high as 24% after only 8 h electrocatalysis. Meanwhile, the cyclic voltammetry (CV) durability tests of the $Cu_{2-x}S$/CNFs electrode for HER in alkaline media is carried out at a scan rate of 100 mV s^{-1} (Figure 6d). After 2000 CV sweeps, the polarization curve shows a negligible difference compared with the initial curve, suggesting superior stability of $Cu_{2-x}S$/CNFs in the long-term electrochemical process.

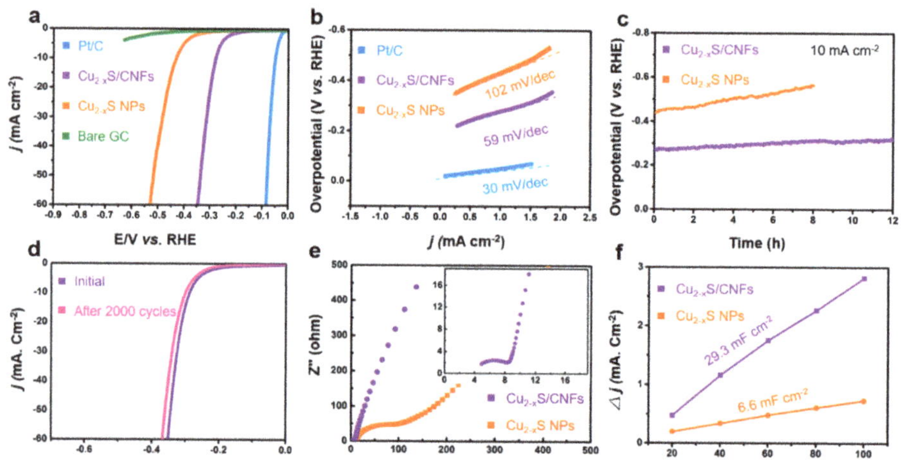

Figure 6. (**a**) LSV polarization curves and (**b**) corresponding Tafel plots of the $Cu_{2-x}S$/CNFs, $Cu_{2-x}S$/CNFs, 20% Pt/C, and a bare glassy carbon electrode derived from the polarization curves. (**c**) Chronopotentiometric curves at 10 mA cm^{-2} for $Cu_{2-x}S$/CNFs and $Cu_{2-x}S$/C composites over 10 h. (**d**) Accelerated HER polarization curves of $Cu_{2-x}S$/CNFs. (**e**) Nyquist plots of $Cu_{2-x}S$/CNFs and $Cu_{2-x}S$/C composites at 300 mV overpotential in 1 M KOH (inset image is the magnified Nyquist plot of $Cu_{2-x}S$/CNFs). (**f**) Capacitive current densities at 0.2 V vs. RHE as a function of scan rate of $Cu_{2-x}S$/CNFs and $Cu_{2-x}S$/C composites.

To further study the influence factors of catalytic kinetics, electrochemical impedance spectroscopy (EIS) and cyclic voltammetry (CV) are conducted. EIS analysis is performed to investigate the electrode/electrolyte interface properties of two catalysts with different structures. Figure 6e shows the Nyquist plots of the $Cu_{2-x}S$/CNFs and $Cu_{2-x}S$/C electrode at an overpotential of 300 mV (vs. RHE) in the frequency range of 100 kHz to 0.1 Hz. It can be observed that the charge-transfer resistance (R_{ct}) is only 6.4 Ω for $Cu_{2-x}S$/CNFs, but 58 Ω for the $Cu_{2-x}S/C$ electrode, in accordance with the HER results. This reflects the highly faradaic efficiency and fast electron transfer of the catalysts during the reaction. The smaller value of R_{ct} of $Cu_{2-x}S$/CNFs electrocatalysts may be ascribed to their 3-D hierarchical intertwining network structure, which increased the contact of the active sites with the electrolyte, leading to a significant acceleration of the interfacial electrocatalytic reactions. We further tested the electrochemical double layer capacitances (C_{dl}) (Figure 6f) of the catalysts by a simple cyclic voltammetry method, to relate the catalytic activity with the electrochemical surface area (ECSA) (Figure S6). The values are measured to be 29.3 and 6.6 mF cm^{-2}, for $Cu_{2-x}S$/CNFs and $Cu_{2-x}S/C$, respectively, revealing the higher active surface area of $Cu_{2-x}S$/CNFs. The large ECSA indicates that $Cu_{2-x}S$/CNFs exposes higher accessible active sites, which is one of the possible reasons for the excellent HER performance.

3. Conclusions

In conclusion, $Cu_{2-x}S$/CNFs with 3-D hierarchical intertwining network structures have successfully been designed and fabricated. They are derived from a 3-D network of PEG-*b*-P4VP@HKUST-1. The key to the successful assembly of the PEG-*b*-P4VP and HKUST-1 is the strong interaction between Cu^{2+} ions and PEG chains. Benefiting from the unique hierarchical structure and uniformly distributed active sites, the as-prepared $Cu_{2-x}S$/CNFs $Cu_{2-x}S$/CNFs exhibit high intrinsic HER activity in alkaline medium. The overpotential is 276 and 337 mV at the current of 10 and 50 mA cm^{-2}, respectively. The Tafel slope is calculated to be 59 mV/decade, and the high activity can be maintained for more than 12 h. The assembly strategy of HKUST-1 and PEG-*b*-P4VP herein can be extended to the hybrid of other MOF-based materials and copolymer, which shows great potential, not only for catalysts, but also for gas sensors, energy storage, and environmental science.

Supplementary Materials: The following are available online at https://www.mdpi.com/article/10.3390/nano11061505/s1, Figure S1: SEM and optical images of PEG-*b*-P4VP/HKUST-1; Figure S2: TEM images of PEG-*b*-P4VP@HKUST-1 composites at different weight of $Cu(OAc)_2$ and BTC; Figure S3: SEM images of individual HKUST-1 crystals and $Cu_{2-x}S/C$ composites; Figure S4: EDS spectrum of the $Cu_{2-x}S$/CNFs and the corresponding element mapping image; Figure S5: Raman spectrum of $Cu_{2-x}S$/CNFs; Figure S6: CV curves of the $Cu_{2-x}S$/CNFs and $Cu_{2-x}S/C$ composites with different scan rates.

Author Contributions: Conceptualization, methodology, writing—original draft preparation, data curation, writing—review and editing were done by Y.B. and J.H.; software, formal analysis, investigation, visualization were done by G.L. and Y.L.; validation, supervision, funding acquisition and project administration were done by Y.B. and Y.L. All authors have read and agreed to the published version of the manuscript.

Funding: This research was funded by the National Natural Science Foundations of China (no. 22005348, 21908251), China Postdoctoral Science Foundation (no. 2020M670970), Hunan Provincial Natural Science Foundation of China (no. 2020JJ5961), Scientific Research Project of Education Department of Hunan Province (no. 19C1915), and research startup foundation of Central South University of Forestry and Technology (no. 2017YJ003).

Institutional Review Board Statement: Not applicable.

Informed Consent Statement: Not applicable.

Data Availability Statement: Data available in a publicly accessible repository.

Conflicts of Interest: The authors declare no conflict of interest.

References

1. Durovic, M.; Hnat, J.; Bouzek, K. Electrocatalysts for the hydrogen evolution reaction in alkaline and neutral media. A comparative review. *J. Power Sources* **2021**, *493*, 229708. [CrossRef]
2. Peng, L.S.; Wei, Z.D. Design and product engineering of high-performance electrode catalytic materials for water electrolysis. *Prog. Chem.* **2018**, *30*, 14–28.

3. Bai, Y.J.; Fang, L.; Xu, H.T.; Gu, X.; Zhang, H.J.; Wang, Y. Strengthened synergistic effect of metallic M_xP_y(M = Co, Ni, and Cu) and carbon layer via peapod like architecture for both hydrogen and oxygen evolution reactions. *Small* **2017**, *13*, 1603718. [CrossRef] [PubMed]
4. Menezes, P.W.; Indra, A.; Das, C.; Walter, C.; Göbel, C.; Gutkin, V.; Schmeisser, D.; Driess, M. Uncovering the Nature of Active Species of Nickel Phosphide Catalysts in High-Performance Electrochemical Overall Water Splitting. *ACS Catal.* **2017**, *7*, 103–109. [CrossRef]
5. Wen, L.L.; Yu, J.; Xing, C.C.; Liu, D.L.; Lyu, X.J.; Cai, W.P.; Li, X.Y. Flexible vanadium-doped Ni_2P nanosheet arrays grown on carbon cloth for an efficient hydrogen evolution reaction. *Nanoscale* **2019**, *11*, 4198–4203. [CrossRef] [PubMed]
6. Liu, P.Y.; Shi, K.; Chen, W.Z.; Gao, R.; Liu, Z.L.; Hao, H.G.; Wang, Y.Q. Enhanced electrocatalytic nitrogen reduction reaction performance by interfacial engineering of MOF-based sulfides $FeNi_2S_4$/NiS hetero-interface. *Appl. Catal. B Environ.* **2021**, *287*, 119956. [CrossRef]
7. Batten, S.R.; Neville, S.M.; Turner, D.R. *Coordination Polymers: Design, Analysis and Application*; Royal Society of Chemistry: London, UK, 2009.
8. MacGillivray, L.R.; Lukehart, C.M. *Metal-Organic Framework Materials*; John Wiley & Sons: Chichester, UK, 2014.
9. Tuttle, R.R.; Folkman, S.J.; Rubin, H.N.; Finke, R.G.; Reynolds, M.M. Copper Metal–Organic Framework Surface Catalysis: Catalyst Poisoning, IR Spectroscopic, and Kinetic Evidence Addressing the Nature and Number of the Catalytically Active Sites En Route to Improved Applications. *ACS Appl. Mater. Interfaces* **2020**, *12*, 39043. [CrossRef] [PubMed]
10. Nasani, R.; Saha, M.; Mobin, S.M.; Martins, L.M.D.R.S.; Pombeiro, A.J.L.; Kirillov, A.M.; Mukhopadhyay, S. Copper–organic frameworks assembled from in situ generated 5-(4-pyridyl) tetrazole building blocks: Synthesis, structural features, topological analysis and catalytic oxidation of alcohols. *Dalton Trans.* **2014**, *43*, 9944–9954. [CrossRef] [PubMed]
11. Gu, J.Z.; Wen, M.; Cai, Y.; Shi, Z.F.; Arol, A.S.; Kirillova, M.V.; Kirillov, A.M. Metal–Organic Architectures Assembled from Multifunctional Polycarboxylates: Hydrothermal Self-Assembly, Structures, and Catalytic Activity in Alkane Oxidation. *Inorg. Chem.* **2019**, *58*, 2403–2412. [CrossRef] [PubMed]
12. Chui, S.S.-Y.; Lo, M.-F.; Charmant, J.P.; Orpen, A.G.; Williams, I.D. A Chemically Functionalizable Nanoporous Material $[Cu_3(TMA)_2(H_2O)_3]_n$. *Science* **1999**, *283*, 1148–1150. [CrossRef]
13. Ge, L.; Wang, L.; Rudoiph, V.; Zhu, Z.H. Hierarchically structured metal–organic framework/ vertically-aligned carbon nanotubes hybrids for CO_2 capture. *RSC Adv.* **2013**, *3*, 25360–25366. [CrossRef]
14. Zhan, G.W.; Fan, L.L.; Zhou, S.F.; Yang, X. Fabrication of Integrated Cu_2O@HKUST-1@Au Nanocatalysts via Galvanic Replacements toward Alcohols Oxidation Application. *ACS Appl. Mater. Interfaces* **2018**, *10*, 35234–35243. [CrossRef] [PubMed]
15. Zhang, B.Y.; Guan, L.; Yu, L.; Lou, X.W. Formation of double-shelled zinc–cobalt sulfide dodecahedral cages from bimetallic zeolitic imidazolate frameworks for hybrid supercapacitors. *Angew. Chem. Int. Ed.* **2017**, *56*, 7141–7145. [CrossRef] [PubMed]
16. Tao, K.Y.; Gong, Y.; Lin, J.H. Low-temperature synthesis of NiS/MoS_2/C nanowires/nanoflakes as electrocatalyst for hydrogen evolution reaction in alkaline medium via calcining/sulfurizing metal-organic frameworks. *Electrochim. Acta* **2018**, *274*, 74–83. [CrossRef]
17. Hu, Q.; Wang, Z.Y.; Huang, X.W.; Qin, Y.J.; Yang, H.P.; Ren, X.Z.; Zhang, Q.L.; Liu, J.H.; He, C.X. A unique space confined strategy to construct defective metal oxides within porous nanofibers for electrocatalysis. *Energy Environ. Sci.* **2020**, *13*, 5097–5103. [CrossRef]
18. Wang, Z.Y.; Gui, M.X.; Asif, M.; Yu, Y.; Dong, S.; Wang, H.T.; Wang, W.; Wang, F.; Xiao, F.; Liu, H.F. A facile modular approach to the 2D oriented assembly MOF electrode for non-enzymatic sweat biosensors. *Nanoscale* **2018**, *10*, 6629–6638. [CrossRef] [PubMed]
19. Zheng, Y.; Qiao, S.Z. Direct Growth of Well-Aligned MOF Arrays onto Various Substrates. *Chem* **2017**, *2*, 751–752. [CrossRef]
20. Zhang, W.; Wu, Z.Y.; Jiang, H.L.; Yu, S.H. Nanowire-Directed Templating Synthesis of Metal−Organic Framework Nanofibers and Their Derived Porous Doped Carbon Nanofibers for Enhanced Electrocatalysis. *J. Am. Chem. Soc.* **2014**, *136*, 14385–14388. [CrossRef]
21. Li, C.; Wang, Y.Y.; Li, H.Y.; Liu, J.; Song, J.P.; Fusaro, L.; Hu, Z.Y.; Chen, Y.X.; Li, Y.; Su, B.L. Weaving 3-D highly conductive hierarchically interconnected nanoporous web by threading MOF crystals onto multi walled carbon nanotubes for high performance Li–Se batter. *J. Energy Chem.* **2021**, *59*, 396–404. [CrossRef]
22. Li, J.; Jiao, C.M.; Zhu, J.H.; Zhong, L.B.; Kang, T.; Aslam, S.; Wang, J.Y.; Zhao, S.F.; Qiu, Y.J. Hybrid co-based MOF nanoboxes/CNFs interlayer as microreactors for polysulfides-trapping in lithium-sulfur batteries. *J. Energy Chem.* **2021**, *57*, 469–476. [CrossRef]
23. Han, W.G.; Dong, F.; Han, W.L.; Tang, Z.C. A strategy to construct uniform MOFs/PAN nanowire derived bead-like Co_3O_4 for VOC catalytic combustion. *Chem. Commun.* **2020**, *56*, 14307–14310. [CrossRef]
24. Zhou, S.Y.; Apostolopoulou-Kalkavoura, V.; da Costa, M.V.T.; Bergstrom, L.; Stromme, M.; Xu, C. Elastic Aerogels of Cellulose Nanofibers@Metal–Organic Frameworks for Thermal Insulation and Fire Retardancy. *Nano Micro Lett.* **2020**, *12*, 9. [CrossRef]
25. Liu, P.Y.; Zhao, J.J.; Dong, Z.P.; Liu, Z.L.; Wang, Y.Q. Interwoving polyaniline and a metal-organic framework grown in situ for enhanced supercapacitor behavior. *J. Alloys Compd.* **2021**, *854*, 157181. [CrossRef]
26. Wang, L.; Feng, X.; Ren, L.T.; Piao, Q.H.; Zhong, J.Q.; Wang, Y.B.; Li, H.W.; Chen, Y.F.; Wang, B. Flexible solid-state supercapacitor based on a metal-organic framework interwoven by electrochemically-deposited PANI. *J. Am. Chem. Soc.* **2015**, *137*, 4920–4923. [CrossRef] [PubMed]

27. Centrone, A.; Yang, Y.; Speakman, S.; Bromberg, L.; Rutledge, G.C.; Hatton, T.A. Growth of metal-organic frameworks on polymer surfaces. *J. Am. Chem. Soc.* **2010**, *132*, 15687–15691. [CrossRef] [PubMed]
28. Zhang, K.K.; Jiang, M.; Chen, D.Y. DNA/Polymeric micelle self-assembly mimicking chromatin compaction. *Angew. Chem. Int. Ed.* **2012**, *51*, 8744–8747. [CrossRef] [PubMed]
29. Yi, J.Q.; Li, H.D.; Jiang, L.; Zhang, K.K.; Chen, D.Y. Solution-based fabrication of a highly catalytically active 3-D network constructed from 1-D metal-organic framework-coated polymeric worm-like micelles. *Chem. Commun.* **2015**, *51*, 10162–10165. [CrossRef]
30. Cheng, Q.H.; Yang, C.; Tao, K.; Han, L. Inlaying ZIF-derived Co_3S_4 hollow nanocages on intertwined polypyrrole tubes conductive networks for high-performance supercapacitors. *Electrochim. Acta* **2020**, *341*, 136042. [CrossRef]
31. Yu, Q.; Lv, J.S.; Liu, Z.H.; Xu, M.; Yang, W.; Owusu, K.A.; Mai, L.Q.; Zhao, D.Y.; Zhou, L. Macroscopic synthesis of ultrafine N-doped carbon nanofibers for superior capacitive energy storage. *Sci. Bull.* **2019**, *64*, 1617–1624. [CrossRef]
32. Liu, I.P.; Teng, H.; Lee, Y.L. Highly electrocatalytic carbon black/copper sulfide composite counter electrodes fabricated by a facile method for quantum-dot-sensitized solar cells. *J. Mater. Chem. A* **2017**, *5*, 23146–23157. [CrossRef]
33. Yilmaz, G.; Yam, K.M.; Zhang, C.; Fan, H.J.; Ho, G.W. In Situ Transformation of MOFs into Layered Double Hydroxide Embedded Metal Sulfides for Improved Electrocatalytic and Supercapacitive Performance. *Adv. Mater.* **2017**, *29*, 1606814. [CrossRef] [PubMed]
34. Kang, C.; Lee, Y.; Kim, I.; Hyun, S.; Lee, T.H.; Yun, S.; Yoon, W.S.; Moon, Y.; Lee, J.; Kim, S.; et al. Highly efficient nanocarbon coating layer on the nanostructured copper sulfide-metal organic framework derived carbon for advanced sodium-ion battery anode. *Materials* **2019**, *12*, 1324. [CrossRef] [PubMed]
35. Bai, Y.J.; Zhang, H.J.; Xiao, L.; Liu, L.; Xu, H.T.; Qiu, H.J.; Wang, Y. Novel peapod-like Ni_2P nanoparticles with improved electrochemical properties for hydrogen evolution and lithium storage. *Nanoscale* **2015**, *7*, 1446–1453. [CrossRef] [PubMed]

nanomaterials

MDPI

Article

Nonlinear Free and Forced Vibrations of a Hyperelastic Micro/Nanobeam Considering Strain Stiffening Effect

Amin Alibakhshi [1], Shahriar Dastjerdi [2], Mohammad Malikan [3,*] and Victor A. Eremeyev [3,4]

1 Department of Mechanical Engineering, Science and Research Branch, Islamic Azad University, Tehran 1477893855, Iran; alibakhshiamin@yahoo.com
2 Civil Engineering Department, Division of Mechanics, Akdeniz University, Antalya 07058, Turkey; dastjerdi_shahriar@yahoo.com
3 Department of Mechanics of Materials and Structures, Faculty of Civil and Environmental Engineering, Gdansk University of Technology, 80-233 Gdansk, Poland; victor.eremeev@pg.edu.pl
4 Department of Civil and Environmental Engineering and Architecture (DICAAR), Università degli Studi di Cagliari, Via Marengo 2, 09123 Cagliari, Italy
* Correspondence: mohammad.malikan@pg.edu.pl

Abstract: In recent years, the static and dynamic response of micro/nanobeams made of hyperelasticity materials received great attention. In the majority of studies in this area, the strain-stiffing effect that plays a major role in many hyperelastic materials has not been investigated deeply. Moreover, the influence of the size effect and large rotation for such a beam that is important for the large deformation was not addressed. This paper attempts to explore the free and forced vibrations of a micro/nanobeam made of a hyperelastic material incorporating strain-stiffening, size effect, and moderate rotation. The beam is modelled based on the Euler–Bernoulli beam theory, and strains are obtained via an extended von Kármán theory. Boundary conditions and governing equations are derived by way of Hamilton's principle. The multiple scales method is applied to obtain the frequency response equation, and Hamilton's technique is utilized to obtain the free undamped nonlinear frequency. The influence of important system parameters such as the stiffening parameter, damping coefficient, length of the beam, length-scale parameter, and forcing amplitude on the frequency response, force response, and nonlinear frequency is analyzed. Results show that the hyperelastic microbeam shows a nonlinear hardening behavior, which this type of nonlinearity gets stronger by increasing the strain-stiffening effect. Conversely, as the strain-stiffening effect is decreased, the nonlinear frequency is decreased accordingly. The evidence from this study suggests that incorporating strain-stiffening in hyperelastic beams could improve their vibrational performance. The model proposed in this paper is mathematically simple and can be utilized for other kinds of micro/nanobeams with different boundary conditions.

Keywords: hyperelastic micro/nanobeam; extended modified couple stress theory; strain-stiffening effect; nonlinear frequency response

Citation: Alibakhshi, A.; Dastjerdi, S.; Malikan, M.; Eremeyev, V.A. Nonlinear Free and Forced Vibrations of a Hyperelastic Micro/Nanobeam Considering Strain Stiffening Effect. *Nanomaterials* **2021**, *11*, 3066. https://doi.org/10.3390/nano11113066

Academic Editor: Kosei Ueno

Received: 24 October 2021
Accepted: 12 November 2021
Published: 14 November 2021

1. Introduction

For many decades, vibration analysis of mechanical structures was a major topic among scientists [1–10]. Over recent decades, a surge of interest in studying hyperelastic materials was shown. The main characteristic of hyperelastic materials is that their strain-stress diagram is nonlinear and may undergo large deformations [11–13]. Hyperelastic materials play a vital role in soft systems and structures, e.g., soft robotics [14], human organs [15,16], soft actuators [17,18], soft sensors [19,20], and soft energy harvesters [19–22]. Data from previous studies show that various mechanical structures such as beams, plates, membranes, and shells were made of hyperelastic materials [23–35]. It was reported that hyperelastic beams are an appropriate candidate to fabricate systems with high performance. For this reason, this paper focuses on beams made of hyperelastic materials. In light of the applications and properties of hyperelastic beams, it is becoming extremely difficult to ignore their investigation in different situations.

There are numerous published works that investigate dependent and time-independent responses of hyperelastic beams. For example, the nonlinear postbuckling of a hyperelastic beam-like structure was investigated by Lubbers et al. [36]. They employed the neo-Hookean hyperelastic model in conjunction with empirical tests and the finite element technique in their study. Wang and coworkers [37] studied nonlinear vibration of hyperelastic beams utilizing time history diagrams and frequency responses, who employed a compressible neo-Hookean constitutive law. He and coworkers [38] developed the Euler–Bernoulli beam model in a new finite strain framework to model a neo-Hookean hyperelastic beam. Xu and Liu [39] proposed an improved method to dynamically explore the response of a beam-like hyperelastic structure, where a Yeoh model was utilized to capture the material nonlinearity. Nonlinear dynamic characteristics of a soft hyperelastic beam were investigated by Wang et al. [40], employing a compressible neo-Hookean model and variational approach. Wang and Zhu [41] studied the nonlinear oscillation of a harmonically excited hyperelastic beam. They utilized the frequency-amplitude response, time histories, and a compressible neo-Hookean model in their investigation. The finite bending of a beam made of hyperelastic materials was analyzed by Bacciocchi and Tarantino [42]. They utilized a compressible Mooney-Rivlin hyperelastic material model to physical nonlinearity of the beam. Dadgar–Rad and Sahraee [43], by considering the incompressibility condition, investigated the large deformation response of a beam made of hyperelastic materials, where a neo-Hookean model was employed as the hyperelastic constitutive model. Bacciocchi and Tarantino [44] conducted a finite anticlastic bending analysis of hyperelastic beams using two hyperelastic models, namely Mooney–Rivlin and Saint Venant–Kirchhoff. Lanzoni and coworker [45] studied the nonuniform bending of a beam made of the hyperelastic beam, taking the Mooney–Rivlin into account. The large deformation response of hyperelastic beams was explored by Dadgar–Rad and Firouzi [46]. They incorporated Fung's quasilinear viscoelasticity theory and Mooney-Rivlin model.

Results from earlier studies demonstrate that few researchers addressed the modelling of hyperelastic beams with the strain-stiffening effect. Furthermore, previous studies have notably investigated a beam-like hyperelastic structure on a large scale and have not considered the hyperelastic beams in micro/nanoscales. However, fabrication of such beams in smaller scales was feasible, and hence analyzing hyperelastic micro/nanobeam and proposing more sophisticated theories should be developed for such structures. A challenging problem that arises in this domain is accurate modelling for hyperelasticity in micro/nanoscales. More specifically, in nanoscale, it is necessary to capture the size effect. Because hyperelastic materials may undergo large deformation and large rotation, these conditions should be considered on micro/nanoscale. One of the problems that it investigates in hyperelasticity is the strain-stiffening effect. This effect may improve or limit the performance of hyperelastic micro/nanobeams. Therefore, incorporating strain-stiffening with simple mathematical modelling in micro/nanoscale is essential. Specifically, to our knowledge, no study has considered large deformation, strain-stiffening, and moderate rotation for hyperelastic micro/nanobeams.

This paper aims to propose a sophisticated model for a micro/nanobeam made of hyperelastic materials that incorporate the small-scale and strain-stiffening effects of nonlinear elasticity. The nonlinear equations of motion are derived via Hamilton's principle and an extended von-Kármán theory. The frequency-amplitude plot and nonlinear resonance plot are presented by considering different system parameters. The results are discussed in detail, and influential parameters on free and forced vibrations of the hyperelastic micro/nanobeam are identified.

2. Governing Equations

The schematic view of the hyperelastic micro/nanobeam is illustrated in Figure 1, where the length, width, and height of the beam are denoted by L, b, and d, respectively.

A clamped-clamped boundary condition is assumed to the beam, and a harmonic transverse mechanical load is applied to it. It is considered that the length of the beam is much greater than the depth. In addition, the shear deformation and rotary inertia are neglected. Thus, we use the Euler–Bernoulli (E-B) beam theory to define the displacement field.

The displacement field for the beam is established according to the Euler–Bernoulli beam equation, namely,

$$\begin{aligned} u_x &= -z\frac{\partial w(x,t)}{\partial x} \\ u_y &= 0 \\ u_z &= w(x,t) \end{aligned} \tag{1}$$

where $w(x,t)$ stands for the transverse displacement of any point on the neutral axis. The strain-displacement relations originated for the Euler–Bernoulli beam theory are modelled based on an extended von Kármán equation, in which large deformation, moderate rotation, and transverse strain are included, namely [47,48]

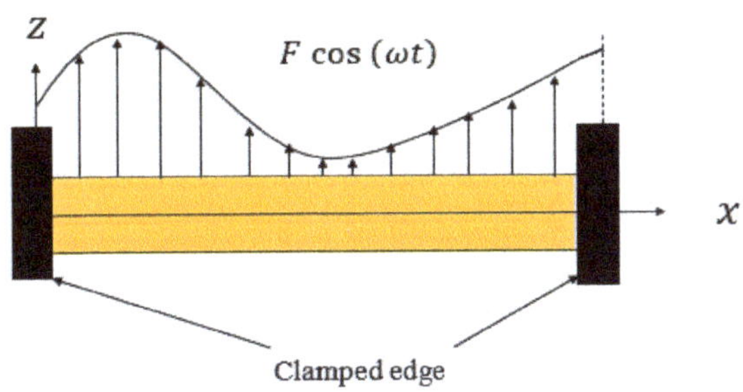

Figure 1. Schematic representation of a clamped-clamped hyperelastic beam.

$$\begin{aligned} \varepsilon_1 &= \frac{1}{2}\left(\frac{\partial w}{\partial x}\right)^2 - z\frac{\partial^2 w}{\partial x^2} \\ \varepsilon_3 &= \frac{1}{2}\left(\frac{\partial w}{\partial x}\right)^2 \end{aligned} \tag{2}$$

Other components of the extended von Kármán equation are equal to zero.

The strain energy of the hyperelastic micro/nanobeam is decomposed into two parts, i.e., the potential due to the hyperelasticity and the potential due to small-scale effects.

For hyperelastic materials, a strain energy function is used to obtain the strain energy of the system. Numerous hyperelastic strain energy functions can capture the strain stiffening, for instance, the standard Gent, the Arruda–Boyce, and modified versions of the Standard Gent model [49–51]. In this work, for simplicity, a standard Gent model is considered, in which the strain-stiffening effect is incorporated, namely [52,53]

$$\Psi_1 = \frac{\mu}{2}\left[(I_1-3) + \frac{1}{2\,J_m}(I_1-3)^2 + \ldots + \frac{1}{(n+1)J_m^n}(I_1-3)^{n+1}\right] \tag{3}$$

where μ is the shear modulus; I_1 denotes the first invariant of the deformation tensor; J_m is a dimensionless parameter that is called the stiffening parameter.

For simplicity, the second-order expansion of the standard Gent model is utilized, such that

$$\Psi_1 = \frac{\mu}{2}\left[(I_1-3) + \frac{1}{2\,J_m}(I_1-3)^2\right] \tag{4}$$

The first principal invariant of the right Cauchy–Green deformation tensor in terms of the extended von Kármán strains is formulated as [54]

$$I_1 = 2(\varepsilon_1+\varepsilon_2+\varepsilon_3)+3 \tag{5}$$

Substituting Equation (2) into Equation (5), the first principal invariant is reformulated as

$$I_1 = 2\left[\left(\frac{\partial w}{\partial x}\right)^2 - z\frac{\partial^2 w}{\partial x^2}\right] + 3 \tag{6}$$

Substituting Equation (6) into Equation (4), the Gent strain energy function as a function of transverse displacement is obtained below

$$\Psi_1 = \int_0^L \left[\mu A\left(\frac{\partial w}{\partial x}\right)^2 + \frac{\mu A}{J_m}\left(\frac{\partial w}{\partial x}\right)^4 + \frac{\mu}{J_m} I\left(\frac{\partial^2 w}{\partial x^2}\right)^2\right] \mathrm{d}x \tag{7}$$

It is mentioned that Equation (7) was obtained by considering the following relations

$$\begin{aligned} I &= \textstyle\int_A z^2\, \mathrm{d}y\mathrm{d}z = b\frac{d^3}{12} \\ A &= \textstyle\int_A \mathrm{d}y\mathrm{d}z = bd \\ 0 &= \textstyle\int_A z\, \mathrm{d}y\mathrm{d}z \end{aligned} \tag{8}$$

In Equation (8), A is the cross-section area, and I represents the second moment of the cross-section.

The potential of the small-scale effect is considered through the use of an extended modified couple stress theory, such that [47]

$$\Psi_2 = \frac{1}{2}\left(2\mu A\ell^2\right) \int_0^L \left(\frac{\partial^2 w}{\partial x^2}\right)^2 \mathrm{d}x \tag{9}$$

where ℓ is a length-scale parameter.

Comparing Equation (9) with previous studies, for the moderate rotation, a coefficient 2 appears in the equation in comparison to the small rotation [55].

Finally, the total strain energy of the hyperelastic micro/nanobeam is written as

$$U_s = \Psi_1 + \Psi_2 \tag{10}$$

The moving beam generates the kinetic energy in the system, which is formulated as

$$U_k = \frac{1}{2}\rho A \int_0^L \left(\frac{\partial w}{\partial t}\right)^2 \mathrm{d}t \tag{11}$$

where ρ stands for the mass-density of the hyperelastic beam.

The transverse applied periodic loading does the work of the following form

$$W_F = \int_0^L F\cos(\omega t) w\, \mathrm{d}x \tag{12}$$

in which F is the amplitude and ω indicates the excitation frequency.

The work generated from the viscous damping is expressed as

$$W_D = -c_D \int_0^L \frac{\partial w}{\partial t} w\, \mathrm{d}x \tag{13}$$

where c_D is the viscous damping coefficient.

To derive boundary conditions and governing equation, Hamilton's principle is utilized, namely

$$\delta \int_{t_1}^{t_2} [U_k - U_S]\mathrm{d}t + \delta \int_{t_1}^{t_2} [\delta W_F + \delta W_D]\mathrm{d}t = 0 \tag{14}$$

Substituting Equations (10)–(13) into Equation (14), we obtain the following equations

$$\rho A\frac{\partial^2 w}{\partial t^2} + C_D\frac{\partial w}{\partial t} + \frac{2\mu I}{J_m}\frac{\partial^4 w}{\partial x^4} + 2\mu A\ell^2\frac{\partial^4 w}{\partial x^4} - 2\mu A\frac{\partial^2 w}{\partial x^2} - \frac{12\mu A}{J_m}\left(\frac{\partial w}{\partial x}\right)^2\frac{\partial^2 w}{\partial x^2} = F\cos(\omega t) \quad (15)$$

and boundary conditions for the double-clamped micro/nanobeam

$$w(0) = 0,\ w(L) = 0,\ \frac{dw(0)}{dx} = 0,\ \frac{dw(L)}{dx} = 0 \quad (16)$$

The above equations are made dimensionless to simplify and generalize the vibration analysis of the micro/nanobeam. The following nondimensional quantities are introduced, such that

$$\begin{gathered}\hat{x} = \frac{x}{L},\ \hat{w} = \frac{w}{L},\ \hat{t} = t\sqrt{\frac{\mu I}{\rho A L^4}},\ \hat{c} = \frac{cL^4}{\mu I}\sqrt{\frac{\mu I}{\rho A L^4}},\ \Omega = \omega\sqrt{\frac{\rho A L^4}{\mu I}} \\ \eta_1 = \frac{2\mu A\ell^2}{\mu I},\ \eta_2 = -\frac{2\mu A L^2}{\mu I},\ \beta = -\frac{12\mu A L^2}{\mu I J_m},\ \hat{F} = \frac{FL^3}{\mu I}\end{gathered} \quad (17)$$

Utilizing the above equations, the dimensionless partial differential equation governing the transverse vibration of the beams is obtained as (the hat notation is omitted for convenience).

$$\frac{\partial^2 w}{\partial t^2} + c\frac{\partial w}{\partial t} + \frac{1}{J_m}\frac{\partial^4 w}{\partial x^4} + \eta_1\frac{\partial^4 w}{\partial x^4} + \eta_2\frac{\partial^2 w}{\partial x^2} + \beta\left(\frac{\partial w}{\partial x}\right)^2\frac{\partial^2 w}{\partial x^2} = F\cos(\Omega t) \quad (18)$$

Subsequently, the boundary conditions become

$$w(0) = 0,\ w(1) = 0,\ \frac{dw(0)}{dx} = 0,\ \frac{dw(1)}{dx} = 0 \quad (19)$$

The system is continuous, and therefore there are infinite modes of vibration. In this paper, the first mode is considered only, with the aid of the separation of variable technique and the Galerkin method. Based on the separation of variable technique, we assume the transverse response is approximated as

$$w(x,t) = W(x)q(t) \quad (20)$$

in which $q(t)$ is the time-dependent coordinate of vibration; $W(x)$ stands for the mode shape of a double-clamped beam that is given below [56]

$$W(x) = \sqrt{\frac{2}{3}}\,[1 - \cos(2\,\pi\,x)] \quad (21)$$

The function expressed in Equation (21) satisfies conditions in Equation (19).

According to the Galerkin method, Equation (20) is substituted in Equation (18), and the resulting equation is multiplied by Equation (21), and integration over [0 1] is taken, which results in

$$\ddot{q} + c\dot{q} + \omega_0^2 q + \alpha q^3 = f\cos(\Omega t) \quad (22)$$

In which

$$\begin{gathered}\omega_0 = \left(\frac{\int_0^1\left\{\eta_1 W'''' W + \frac{1}{J_m}W'''' W + \eta_2 W'' W\right\}dx}{\int_0^1 W^2 dx}\right)^{\frac{1}{2}} \\ \alpha = \frac{\int_0^1(\beta W' W' W'' W)dx}{\int_0^1 W^2 dx} \\ f = \frac{\int_0^1(FW)dx}{\int_0^1 W^2 dx}\end{gathered} \quad (23)$$

In Equation (23), ω_0 indicates dimensionless linear natural frequency.

3. Solution Method

This section is divided into two parts. In the first one, the forced vibration is solved using the Multiple Scales Method (MSM) [57], and in the second one, the free vibration is solved using Hamilton Approach (HA).

3.1. Forced Vibration Solution

To implement the MSM, the forced vibration equation, Equation (22), is converted to a perturbated form by introducing the following parameters

$$c = 2\varepsilon\, c_d,\ \alpha = \varepsilon\alpha_1,\ f = \varepsilon\, f_1 \tag{24}$$

where ε is a dimensionless quantity that measures the strength of the nonlinearity of the beam and is called the gauge parameter.

Substituting Equation (24) into Equation (22), we obtain

$$\ddot{q} + 2\varepsilon\, c_d\dot{q} + \omega_0^2 q + \varepsilon\alpha_1 q^3 = \varepsilon\, f_1 \cos(\Omega t) \tag{25}$$

In line with the MSM, the original time is replaced with new time scales as $T_n = \varepsilon^n\, t; n = 1, 2, \ldots$ and therefore, the ODE is converted to a PDE.

New differential operators based on new time scales are $D_n = \partial/\partial T_n$, and original time first and second derivatives in terms of these operators are expressed as

$$\begin{aligned} \tfrac{\mathrm{d}}{\mathrm{d}t} &= D_0 + \varepsilon\, D_1 + \varepsilon^2\, D_2 + \ldots \\ \tfrac{\mathrm{d}^2}{\mathrm{d}t^2} &= {D_0}^2 + 2\,\varepsilon\, D_0 D_1 + \varepsilon^2 (D_1^2 + 2\, D_0 D_2\,) + \ldots \end{aligned} \tag{26}$$

The governing equation includes a nonlinear cubic term. Therefore, a first-order perturbation approximation is accurate enough, such that

$$q = q_0 + \varepsilon\, q_1 \tag{27}$$

$q_n,\ n = 0, 1$ are independent of the gauge parameter ε. For this reason, we can equal the coefficient of each power of ε to zero.

Combining Equations (25)–(27), and equating coefficients of ε^0 and ε^1 to zero, the following PDEs are attained

Coefficients of ε^0

$$D_0^2 q_0 + \omega_0^2 q_0 = 0 \tag{28}$$

Coefficients of ε^1

$$D_0^2 q_1 + \omega_0^2 q_1 = -2D_0 D_1 q_0 - 2D_0 q_0 - \alpha_1 {q_0}^3 + f_1 \cos(\Omega t) \tag{29}$$

The solution of Equation (28) takes the following form

$$q_0 = A(T_1) e^{i\,\omega_0 T_0} + \overline{A}(T_1) e^{-i\,\omega_0 T_0} \tag{30}$$

in which $A(T_1)$ is a complex-valued function and $\overline{A}(T_1)$ is its complex conjugate.

Substituting Equation (30) into Equation (29), the following equation is obtained as

$$D_0^2 q_1 + \omega_0^2 q_1 = \left[-3\alpha_1 A^2\overline{A} - 2ic_d A\omega_0 - 2i\omega_0 \frac{dA}{dT_1}\right] e^{i\,\omega_0 T_0} + f_1 \cos(\Omega t) + \mathrm{CC} + \mathrm{NST} \tag{31}$$

In the above equation, the terms inside the box bracket shows secular terms, CC stands for complex conjugates of previous terms, and NST is an abbreviation for terms with higher degrees of $e^{i\,\omega_0 T_0}$ (nonsecular terms).

By equating secular terms to zero, the frequency-amplitude relation can be obtained. However, the external loading can also give rise to secure terms. This fact is considered in two states, i.e., the primary resonance and the secondary resonance. In this paper, the primary resonance is analyzed, which states that

$$\Omega = \omega_0 + \varepsilon\,\sigma \tag{32}$$

Writing the trigonometric function in Equation (31) and using Equation (32), we obtain

$$3\alpha_1 A^2\overline{A} + 2ic_d A\omega_0 + 2i\omega_0\frac{dA}{dT_1} - \frac{1}{2}f_1 e^{i\sigma T_1} = 0 \tag{33}$$

The complex-valued function A is written as

$$A = \frac{1}{2}ae^{i\theta},\ \overline{A} = \frac{1}{2}ae^{-i\theta} \tag{34}$$

in which a and θ are the amplitude and phase, which are functions of T_1.

Substituting Equation (34) into Equation (33) and then separating the resulting equation into real and imaginary parts yields

Imaginary parts:

$$\frac{\mathrm{da}}{\mathrm{dT_1}} = -ac_d + \frac{1}{2\omega_0}f_1\sin(\sigma T_1 - \theta) \tag{35}$$

Real parts:

$$a\frac{\mathrm{d}\theta}{\mathrm{d}T_1} = \frac{3}{8\omega_0}\alpha_1 a^3 - \frac{1}{2\omega_0}f_1\cos(\sigma T_1 - \theta) \tag{36}$$

Equations (35) and (36) are converted to an autonomous equation by introducing $\gamma = (\sigma T_1 - \theta)$, which results in

$$\frac{\mathrm{da}}{\mathrm{dT_1}} = -ac_d + \frac{1}{2\omega_0}f_1\sin(\gamma) \tag{37}$$

$$a\frac{\mathrm{d}\gamma}{\mathrm{d}T_1} = \sigma a - \frac{3}{8\omega_0}\alpha_1 a^3 + \frac{1}{2\omega_0}f_1\cos(\gamma) \tag{38}$$

A bounded response is acquired while $\frac{\mathrm{da}}{\mathrm{dT_1}} = a\frac{\mathrm{d}\gamma}{\mathrm{d}T_1} = 0$, whereby one can obtain

$$ac_d = \frac{1}{2\omega_0}f_1\sin(\gamma) \tag{39}$$

$$\sigma a - \frac{3}{8\omega_0}\alpha_1 a^3 = -\frac{1}{2\omega_0}f_1\cos(\gamma) \tag{40}$$

After some mathematical manipulation and using the fact $\sin^2(\gamma) + \cos^2(\gamma) = 1$, we obtain the frequency-amplitude response as

$$[c_d a]^2 + \left[\sigma a - \frac{3}{8\omega_0}\alpha_1 a^3\right]^2 = \left[\frac{1}{2\omega_0}f_1\right]^2 \tag{41}$$

3.2. Free Vibration Solution

In this subsection, the nonlinear frequency of the micro/nanobeam with neglecting the external force and damping is obtained via Hamilton's approach. The initial conditions for the vibration of the hyperelastic beam are expressed as

$$q(0) = a_0,\ \dot{q}(0) = 0 \tag{42}$$

where a_0 stands for the maximum amplitude of the vibration. Based on Hamilton's principle, the nonlinear frequency is derived as [58]

$$\omega_{nl} = \sqrt{\omega_0^2 + \frac{49}{70}\alpha a_0^2} \tag{43}$$

4. Result and Discussion

The effects of several parameters such as the stiffening parameter, the length scale parameter, and forcing amplitude and damping on the frequency response and nonlinear frequency of the system are analyzed. The material and geometrical parameters of the hyperelastic microbeam are given in Table 1.

Table 1. Material and geometrical parameters.

Parameters	Value
Length	$L = 30$ μm
Width	$b = 10$ μm
Height	$d = 0.65$ μm
Young's modulus	$E = 3$ GPa
Shear modulus	$\mu = E/3 = 1$ GPa

4.1. Frequency Response

Figure 2 depicts the influence of the gauge parameter ε on the frequency response under the following parameter $f_1 = 0.5$, $\ell = 0$, $c_d = 0.004$, and $J_m = 100$. As the gauge parameter ε is decreased, the nonlinearity of the system increases. Mathematically speaking, with the decrease of ε, the value of nonlinear terms in the equation of motion becomes higher in comparison to the value of linear terms. Depending on the accuracy, an arbitrary value for ε can be adopted, which in this paper it is chosen as $\varepsilon = 1$.

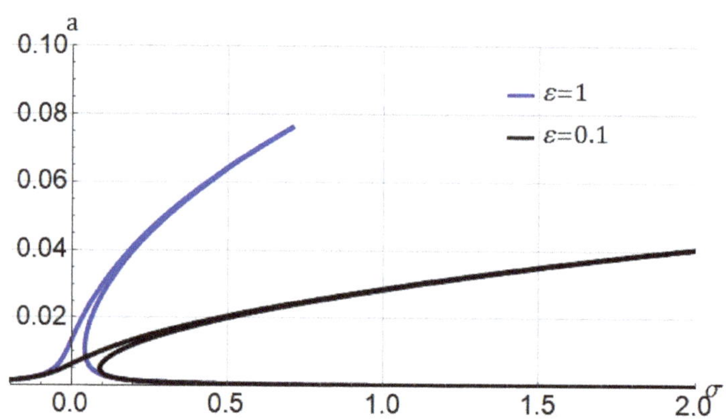

Figure 2. Influence of gauge parameter (ε) on frequency response of system. Systems parameters: $\ell = 0$; $c_d = 0.004$; $J_m = 100$; $f_1 = 0.5$.

Illustrated in Figure 3 is the influence of the damping coefficient c_d on the frequency response of the system while considering the following parameters $\ell = 0$, and $J_m = 100$. From the figure, it is concluded that increasing the damping coefficient decreases the response amplitude of the hyperelastic micro/nanobeam. The damping in hyperelastic materials mainly originates from the viscosity of matter. In the remaining part of the numerical simulation, as a test case, the damping coefficient is taken as $c_d = 0.004$.

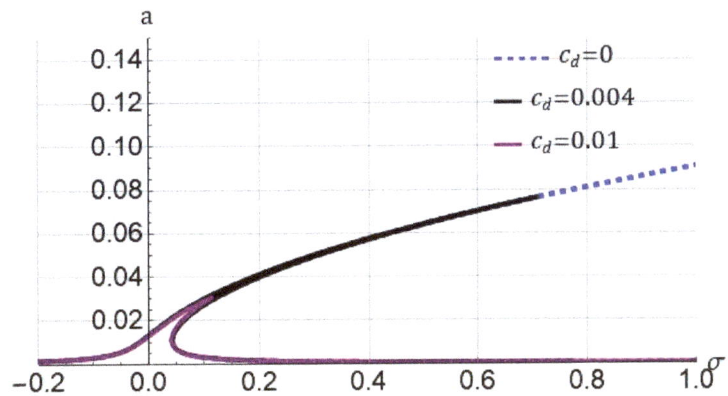

Figure 3. Influence of damping coefficient (c_d) on frequency response of system. Systems parameters: $\ell = 0$; $J_m = 100$; $f_1 = 0.5$.

Figure 4 represents the impact of the length scale parameter ℓ on the nonlinear resonant vibration of the hyperelastic beam. As seen, increasing the size effect, the response amplitude decreases, and the hardening nonlinearity becomes weaker. This result is in agreement with that shown in the literature for linear materials. Obtaining the accurate small-length scall parameter in the experimental test is a crucial task for engineers. Finding an exact value for the length scale parameter for the hyperelastic beam in the experimental test should be carried out to improve the performance of hyperelastic microbeams.

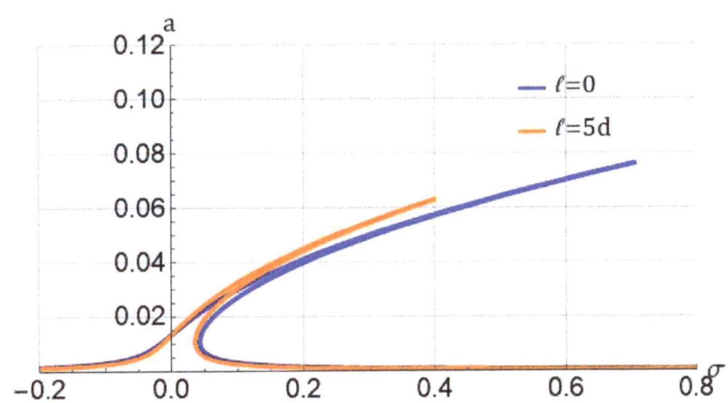

Figure 4. Influence of length scale parameter (ℓ) on frequency response of system. Systems parameters: $J_m = 100$; $f_1 = 0.5$; $c_d = 0.004$.

The influence of the stiffening parameter J_m on the frequency-amplitude plot is shown in Figure 5. It is concluded that as the stiffening parameter is decreased, the hardening nonlinearity gets stronger. When the stiffening parameter is equal to $J_m = \infty$, i.e., the conversion of the Gent model to the neo-Hookean model, the system's response is linear. It is noted that if the stiffening parameter is smaller, the strain-stiffening effect is stronger. As reported by Amabili, a stiffening parameter in a range $J_m = 30 - 100$ stands for rubber materials, and values less than them stand for biological tissues [12].

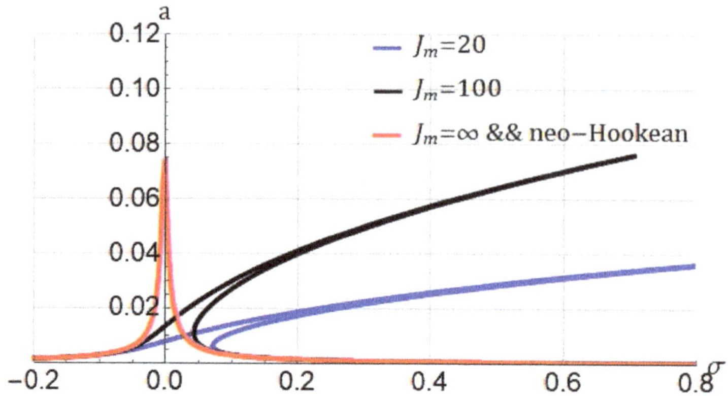

Figure 5. Influence of stiffening parameter (J_m) on frequency response of system. Systems parameters: $\ell = 0$; $f_1 = 0.5$; $c_d = 0.004$.

The influence of the amplitude of the external loading f_1 on the resonant characteristics of the hyperelastic micro/nanobeam is analyzed in Figure 6. Increasing f_1 the response amplitude increases, and the frequency response becomes wider. Moreover, the forcing amplitude cannot alter the nonlinear nature of the system and only quantitatively alter the resonant behaviour.

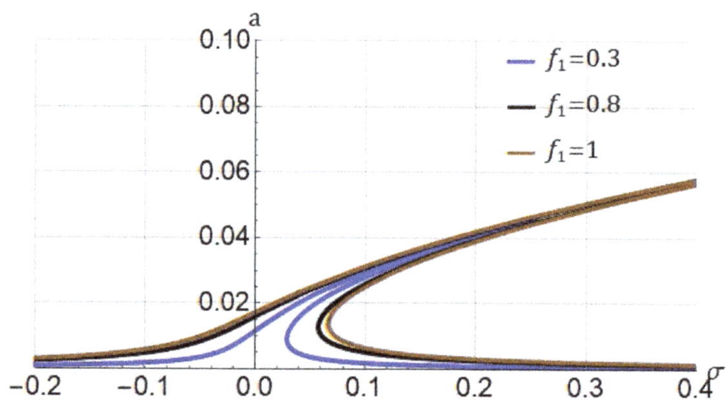

Figure 6. Influence of forcing amplitude (f_1) on frequency response of system. Systems parameters: $\ell = 0$; $J_m = 100$; $c_d = 0.004$.

We analyze the influence of the strain-stiffening parameter on the force-response in Figure 7. The system parameters are $\ell = 0$; $\sigma = 0.05$; $c_d = 0.004$. We can see that by increasing the value of the strain-stiffening parameter, a higher value of forcing amplitude is required to cause the jump phenomenon. Moreover, by increasing the strain-stiffening parameter, the system becomes stable and for the neo-Hookean model.

We show the influence of the length-scale parameter on the force-response in Figure 8. With the inclusion of the effect of size, the jump phenomenon arises for higher values of forcing amplitude.

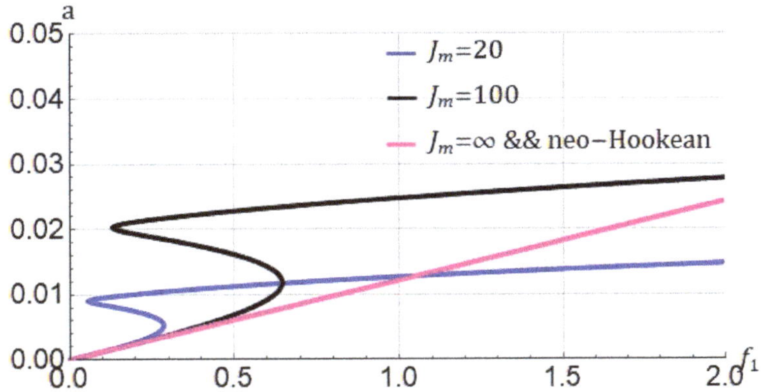

Figure 7. Influence of stiffening parameter (J_m) on force response of system. Systems parameters: $\ell = 0$; $\sigma = 0.05$; $c_d = 0.004$.

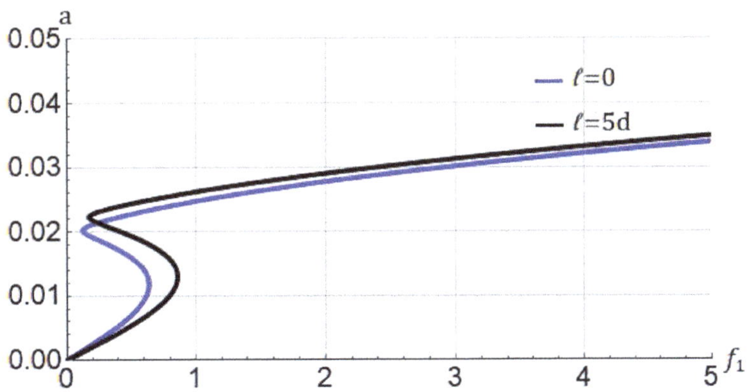

Figure 8. Influence of stiffening parameter (ℓ) on force response of system. Systems parameters: $J_m = 100$; $\sigma = 0.05$; $c_d = 0.004$.

4.2. Nonlinear Frequency

The previous section demonstrates the results for the forced vibration of the Gent hyperelastic beam. Herein, the nonlinear frequency of the system given in Equation (42) is evaluated.

Illustrated in Figure 9 is the nonlinear frequency versus the maximum amplitude when $\ell = 0$, and $J_m = 100$. It is found that by increasing the maximum amplitude a_0 the nonlinear frequency increases.

As depicted in Figure 10, the nonlinear frequency for variations of the length of the beam is presented. As the length is increased, the dimensionless nonlinear frequency increases accordingly.

The nonlinear frequency versus the stiffening parameter J_m is presented in Figure 11. Increasing J_m, the nonlinear frequency decreases.

As depicted in Figure 12, the nonlinear frequency versus the length scale parameter is presented. As the size effect is increased, the nonlinear frequency increases accordingly.

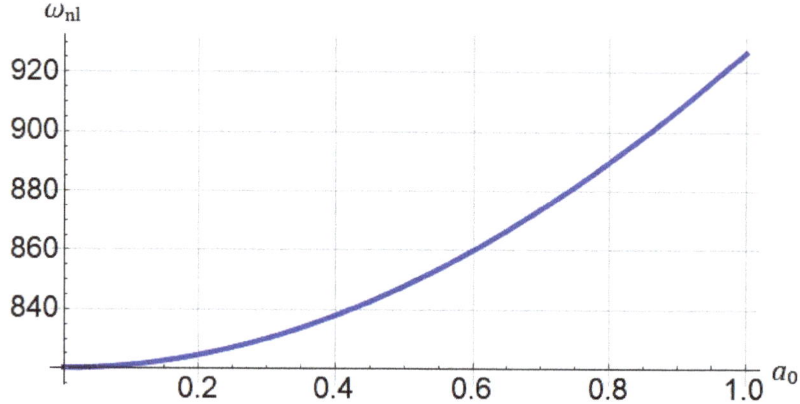

Figure 9. Influence of maximum amplitude (a_0) on nonlinear frequency of system. Systems parameters: $\ell = 0$; $J_m = 100$, $L = 30$ μm.

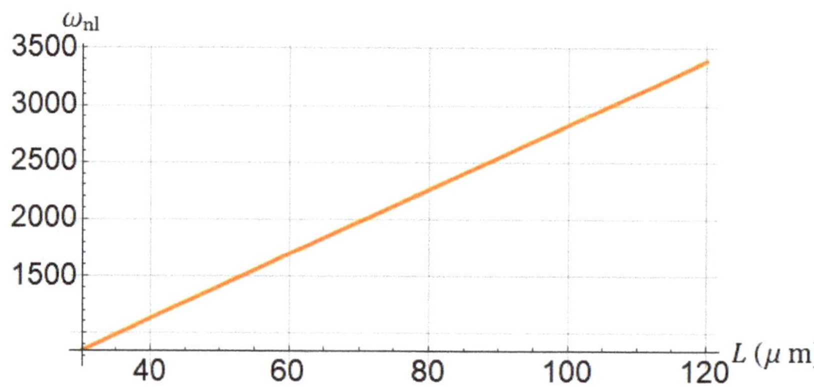

Figure 10. Influence of length of micro/nanobeam (L) on nonlinear frequency of system. Systems parameters: $\ell = 0$; $J_m = 100$; $a_0 = 0.5$.

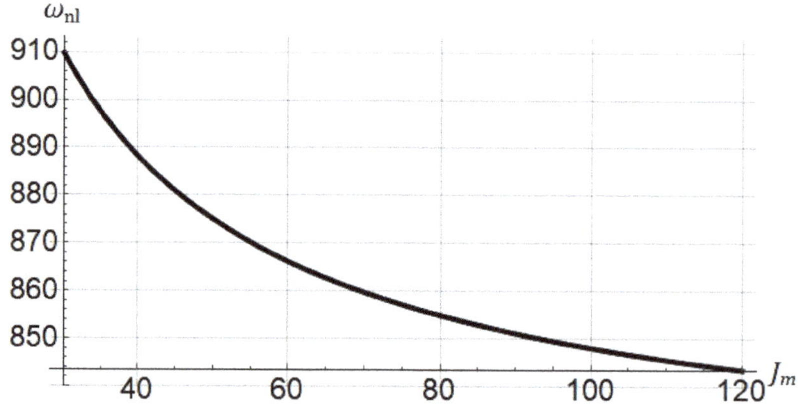

Figure 11. Influence of stiffening parameter (J_m) on nonlinear frequency of system. Systems parameters: $L = 30$ μm; $\ell = 0$; $a_0 = 0.5$.

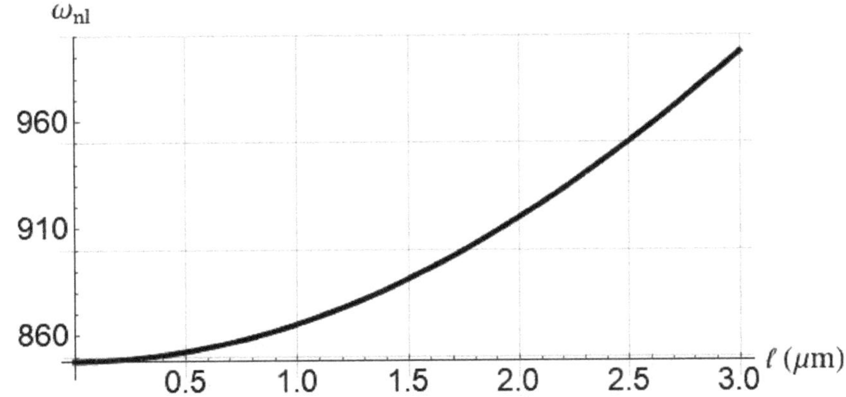

Figure 12. Influence of length scale parameter ℓ on nonlinear frequency of system. Systems parameters: $L = 30$ μm; $J_m = 100$; $a_0 = 0.5$.

5. Discussion on the Strain-Stiffening

Rubber-like materials can be deformed by stretching. In the beginning, we can stretch rubbers easily, but if the stretch is large enough, the stretching process becomes difficult. This is due to the strain-stiffening effect in rubber-like materials. The strain-stiffening is a nonlinear behavior that is seen even in soft biological materials such as liver and brain tissue [59]. We can use this property in hyperelastic materials so as to evade damage. The strain-stiffening can also be connected to the molecular-statistical point of view in nonlinear elasticity. The stiffening parameters J_m in the Gent model relates to the number of rigid links in a single chain N using $J_m = 3(N-1)$. N is also called the classical number of Kuhn segments [60]. The results of Figures 5, 7 and 11, can also be interpreted based on molecular-statistical point of view. We see that altering J_m, the number of segments changes accordingly. Therefore, this change affects the frequency/force response of the hyperelastic microbeam. Taken together, the results of this paper can help researchers who would like to analyze the hyperelastic microbeam via molecular-statistical hyperelastic models such as generalized neo-Hookean model.

6. Conclusions

In this paper, nonlinear, free, and forced oscillations of a hyperelastic micro/nanobeam were investigated with the inclusion of the small-scale effect, strain-stiffening effect, and moderate rotation. A developed Euler–Bernoulli beam theory was utilized to model the beam, and the energies and works that appeared in the system were formulated. The equation of motion was derived using Hamilton's principle and the Galerkin decomposition method. Frequency-amplitude curves and the nonlinear natural frequency diagrams were illustrated by analytically solving the equation of motion. This paper concludes that:

- Increasing the strain-stiffening effect leads to increasing hardening nonlinearity.
- For the neo-Hookean model with $J_m = \infty$, the nonlinearity vanishes, and the response is transformed into a linear type.
- As the stiffening parameter J_m is increased, the nonlinear natural frequency decreases.
- The length of the micro/nanobeam, the damping, and size effects were identified as influential parameters in the system.

Author Contributions: Conceptualization, A.A. and S.D.; methodology, A.A.; software, A.A.; validation, A.A. and S.D.; formal analysis, A.A.; investigation, M.M. and V.A.E.; resources, A.A. and M.M.; data curation, A.A.; writing—original draft preparation, A.A. and S.D.; writing—review and editing, A.A., S.D., M.M. and V.A.E.; visualization, A.A., S.D.; supervision, V.A.E.; project administration, M.M. and V.A.E.; funding acquisition, S.D. and M.M. All authors have read and agreed to the published version of the manuscript.

Funding: V.A. Eremeyev acknowledges the financial support of the Ministry of Education and Science of the Russian Federation as part of the program of the Mathematical Center for Fundamental and Applied Mathematics under the agreement №075-15-2019-1621.

Data Availability Statement: Data sharing is not applicable.

Conflicts of Interest: The authors declare no conflict of interest.

References

1. Avey, M.; Tornabene, F.; Dimitri, R.; Kuruoglu, N. Free vibration of thin-walled composite shell structures reinforced with uniform and linear carbon nanotubes: Effect of the elastic foundation and nonlinearity. *Nanomaterials* **2021**, *11*, 2090. [CrossRef]
2. Yi, X.; Li, B.; Wang, Z. Vibration analysis of fluid conveying carbon nanotubes based on nonlocal timoshenko beam theory by spectral element method. *Nanomaterials* **2019**, *9*, 1780. [CrossRef] [PubMed]
3. Wang, Y.; Xie, K.; Fu, T.; Shi, C. Bending and elastic vibration of a novel functionally graded polymer nanocomposite beam reinforced by graphene nanoplatelets. *Nanomaterials* **2019**, *9*, 1690. [CrossRef] [PubMed]
4. Liu, Y.F.; Wang, Y.Q. Thermo-electro-mechanical vibrations of porous functionally graded piezoelectric nanoshells. *Nanomaterials* **2019**, *9*, 301. [CrossRef]
5. Zhang, Y.; Zhang, F. Vibration and buckling of shear deformable functionally graded nanoporous metal foam nanoshells. *Nanomaterials* **2019**, *9*, 271. [CrossRef]
6. Dastjerdi, S.; Akgöz, B.; Civalek, Ö. On the effect of viscoelasticity on behavior of gyroscopes. *Int. J. Eng. Sci.* **2020**, *149*, 103236. [CrossRef]
7. Liu, D. Free vibration of functionally graded graphene platelets reinforced magnetic nanocomposite beams resting on elastic foundation. *Nanomaterials* **2020**, *10*, 2193. [CrossRef] [PubMed]
8. Selim, M.M.; Musa, A. Nonlinear vibration of a pre-stressed water-filled single-walled carbon nanotube using shell model. *Nanomaterials* **2020**, *10*, 974. [CrossRef]
9. Kiani, K.; Żur, K.K. Dynamic behavior of magnetically affected rod-like nanostructures with multiple defects via nonlocal-integral/differential-based models. *Nanomaterials* **2020**, *10*, 2306. [CrossRef] [PubMed]
10. Civalek, Ö.; Akbaş, Ş.D.; Akgöz, B.; Dastjerdi, S. Forced vibration analysis of composite beams reinforced by carbon nanotubes. *Nanomaterials* **2021**, *11*, 571. [CrossRef] [PubMed]
11. Hackett, R.M. Hyperelasticity Primer. In *Hyperelasticity Primer*; Springer: Cham, Switzerland, 2015; pp. 1–170. [CrossRef]
12. Amabili, M. *Nonlinear Mechanics of Shells and Plates in Composite, Soft and Biological Materials*; Cambridge University Press: Cambridge, UK, 2018.
13. Chaves, E.W.V. *Notes on Continuum Mechanics*, 1st ed.; Oñate, E., Ed.; Lecture Notes on Numerical Methods in Engineering and Sciences; Springer: Amsterdam, The Netherlands, 2013; p. 673. Available online: https://www.springer.com/gp/book/9789400759855 (accessed on 21 October 2021).
14. Maloisel, G.; Knoop, E.; Schumacher, C.; Bacher, M. Automated Routing of Muscle Fibers for Soft Robots. *IEEE Trans. Robot.* **2021**, *37*, 996–1008. [CrossRef]
15. Gasser, T.C.; Ogden, R.W.; Holzapfel, G.A. Hyperelastic modelling of arterial layers with distributed collagen fibre orientations. *J. R. Soc. Interface* **2006**, *3*, 15–35. [CrossRef] [PubMed]
16. Holzapfel, G.A.; Ogden, R.W. Constitutive modelling of arteries. *Proc. R. Soc. A Math. Phys. Eng. Sci.* **2010**, *466*, 1551–1597. [CrossRef]
17. Moseley, P.; Florez, J.M.; Sonar, H.A.; Agarwal, G.; Curtin, W.; Paik, J. Modeling, Design, and Development of Soft Pneumatic Actuators with Finite Element Method. *Adv. Eng. Mater.* **2016**, *18*, 978–988. [CrossRef]
18. Zhang, C.; Wei, W.; Sun, H.; Zhu, Q. Study on the properties of different dielectric elastomers applying to actuators. *Sensors Actuators A Phys.* **2021**, *329*, 112806. [CrossRef]
19. Zhang, H.; Wang, M.Y. Multi-Axis Soft Sensors Based on Dielectric Elastomer. *Soft Robot.* **2016**, *3*, 3–12. [CrossRef]
20. Chen, D.; Pei, Q. Electronic Muscles and Skins: A Review of Soft Sensors and Actuators. *Chem. Rev.* **2017**, *117*, 11239–11268. [CrossRef]
21. Dong, L.; Grissom, M.D.; Prasad, M.G.; Fisher, F.T. Application of mechanical stretch to tune the resonance frequency of hyperelastic membrane-based energy harvesters. *Sens. Actuators A Phys.* **2016**, *252*, 165–173. [CrossRef]
22. Chen, L.; Yang, S. Enhancing the Electromechanical Coupling in Soft Energy Harvesters by Using Graded Dielectric Elastomers. *Micromachines* **2021**, *12*, 1187. [CrossRef]
23. Heidari, H.; Alibakhshi, A.; Azarboni, H.R. Chaotic Motion of a Parametrically Excited Dielectric Elastomer. *Int. J. Appl. Mech.* **2020**, *12*, 2050033. [CrossRef]
24. Breslavsky, I.D.; Amabili, M.; Legrand, M. Nonlinear vibrations of thin hyperelastic plates. *J. Sound Vib.* **2014**, *333*, 4668–4681. [CrossRef]

25. Alibakhshi, A.; Heidari, H. Analytical approximation solutions of a dielectric elastomer balloon using the multiple scales method. *Eur. J. Mech. A/Solids* **2019**, *74*, 485–496. [CrossRef]
26. Alibakhshi, A.; Heidari, H. Nonlinear resonance analysis of dielectric elastomer actuators under thermal and isothermal conditions. *Int. J. Appl. Mech.* **2020**, *12*, 2050100. [CrossRef]
27. Alibakhshi, A.; Heidari, H. Nonlinear dynamic responses of electrically actuated dielectric elastomer-based microbeam resonators. *J. Intell. Mater. Syst. Struct.* **2021**. [CrossRef]
28. Du, P.; Dai, H.H.; Wang, J.; Wang, Q. Analytical study on growth-induced bending deformations of multi-layered hyperelastic plates. *Int. J. Non-Linear Mech.* **2020**, *119*, 103370. [CrossRef]
29. Tripathi, A.; Bajaj, A.K. Topology optimization and internal resonances in transverse vibrations of hyperelastic plates. *Int. J. Solids Struct.* **2016**, *81*, 311–328. [CrossRef]
30. Amabili, M.; Breslavsky, I.D.; Reddy, J.N. Nonlinear higher-order shell theory for incompressible biological hyperelastic materials. *Comput. Methods Appl. Mech. Eng.* **2019**, *346*, 841–861. [CrossRef]
31. Breslavsky, I.D.; Amabili, M.; Legrand, M. Static and Dynamic Behavior of Circular Cylindrical Shell Made of Hyperelastic Arterial Material. *J. Appl. Mech. Trans. ASME* **2016**, *83*, 051002. [CrossRef]
32. Soares, R.M.; Amaral, P.F.T.; Silva, F.M.A.; Gonçalves, P.B. Nonlinear breathing motions and instabilities of a pressure-loaded spherical hyperelastic membrane. *Nonlinear Dyn.* **2020**, *99*, 351–372. [CrossRef]
33. Vedeneev, V. Nonlinear steady states of hyperelastic membrane tubes conveying a viscous non-Newtonian fluid. *J. Fluids Struct.* **2020**, *98*, 103113. [CrossRef]
34. Soares, R.M.; Gonçalves, P.B. Nonlinear vibrations of a rectangular hyperelastic membrane resting on a nonlinear elastic foundation. *Meccanica* **2018**, *53*, 937–955. [CrossRef]
35. Soares, R.M.; Gonalves, P.B. Nonlinear vibrations and instabilities of a stretched hyperelastic annular membrane. *Int. J. Solids Struct.* **2012**, *49*, 514–526. [CrossRef]
36. Lubbers, L.A.; van Hecke, M.; Coulais, C. A nonlinear beam model to describe the postbuckling of wide neo-Hookean beams. *J. Mech. Phys. Solids* **2017**, *106*, 191–206. [CrossRef]
37. Wang, Y.; Ding, H.; Chen, L.Q. Nonlinear vibration of axially accelerating hyperelastic beams. *Int. J. Non-Linear Mech.* **2018**, *99*, 302–310. [CrossRef]
38. He, L.; Lou, J.; Dong, Y.; Kitipornchai, S.; Yang, J. Variational modeling of plane-strain hyperelastic thin beams with thickness-stretching effect. *Acta Mech.* **2018**, *229*, 4845–4861. [CrossRef]
39. Xu, Q.; Liu, J. An improved dynamic model for a silicone material beam with large deformation. *Acta Mech. Sin. Xuebao* **2018**, *34*, 744–753. [CrossRef]
40. Wang, Y.; Ding, H.; Chen, L.Q. Vibration of axially moving hyperelastic beam with finite deformation. *Appl. Math. Model.* **2019**, *71*, 269–285. [CrossRef]
41. Wang, Y.; Zhu, W. Nonlinear transverse vibration of a hyperelastic beam under harmonic axial loading in the subcritical buckling regime. *Appl. Math. Model.* **2021**, *94*, 597–618. [CrossRef]
42. Bacciocchi, M.; Tarantino, A.M. Bending of hyperelastic beams made of transversely isotropic material in finite elasticity. *Appl. Math. Model.* **2021**, *100*, 55–76. [CrossRef]
43. Dadgar-Rad, F.; Sahraee, S. Large deformation analysis of fully incompressible hyperelastic curved beams. *Appl. Math. Model.* **2021**, *93*, 89–100. [CrossRef]
44. Bacciocchi, M.; Tarantino, A.M. Finite anticlastic bending of hyperelastic laminated beams with a rubberlike core. *Mech. Adv. Mater. Struct.* **2021**. [CrossRef]
45. Lanzoni, L.; Tarantino, A.M. Nonuniform bending theory of hyperelastic beams in finite elasticity. *Int. J. Non-Linear Mech.* **2021**, *135*, 103765. [CrossRef]
46. Dadgar-Rad, F.; Firouzi, N. Large deformation analysis of two-dimensional visco-hyperelastic beams and frames. *Arch. Appl. Mech.* **2021**, *91*, 4279–4301. [CrossRef]
47. Reddy, J.N.; Srinivasa, A.R. Non-linear theories of beams and plates accounting for moderate rotations and material length scales. *Int. J. Non-Linear Mech.* **2014**, *66*, 43–53. [CrossRef]
48. Srinivasa, A.R.; Reddy, J.N. A model for a constrained, finitely deforming, elastic solid with rotation gradient dependent strain energy, and its specialization to von Kármán plates and beams. *J. Mech. Phys. Solids* **2013**, *61*, 873–885. [CrossRef]
49. Alibakhshi, A.; Imam, A.; Haghighi, S.E. Effect of the second invariant of the Cauchy–Green deformation tensor on the local dynamics of dielectric elastomers. *Int. J. Non-Linear Mech.* **2021**, *137*, 103807. [CrossRef]
50. Alibakhshi, A.; Heidari, H. Nonlinear dynamics of dielectric elastomer balloons based on the Gent-Gent hyperelastic model. *Eur. J. Mech. A/Solids* **2020**, *82*, 103986. [CrossRef]
51. Mangan, R.; Destrade, M. Gent models for the inflation of spherical balloons. *Int. J. Non-Linear Mech.* **2015**, *68*, 52–58. [CrossRef]
52. Liu, L.; Liu, Y.; Luo, X.; Li, B.; Leng, J. Electromechanical instability and snap-through instability of dielectric elastomers undergoing polarization saturation. *Mech. Mater.* **2012**, *55*, 60–72. [CrossRef]
53. Gent, A.N. A new constitutive relation for rubber. *Rubber Chem. Technol.* **1996**, *69*, 59–61. [CrossRef]
54. Breslavsky, I.D.; Amabili, M.; Legrand, M. Physically and geometrically non-linear vibrations of thin rectangular plates. *Int. J. Non-Linear Mech.* **2014**, *58*, 30–40. [CrossRef]

55. Ghayesh, M.H.; Farokhi, H.; Amabili, M. Nonlinear dynamics of a microscale beam based on the modified couple stress theory. *Compos. Part. B Eng.* **2013**, *50*, 318–324. [CrossRef]
56. Azarboni, H.R.; Rahimzadeh, M.; Heidari, H.; Keshavarzpour, H.; Edalatpanah, S.A. Chaotic dynamics and primary resonance analysis of a curved carbon nanotube considering influence of thermal and magnetic fields. *J. Braz. Soc. Mech. Sci. Eng.* **2019**, *41*, 294. [CrossRef]
57. Nayfeh, A.H.; Mook, D.T. *Nonlinear Oscillations*; Wiley-VCH: Weinheim, Germany, 1995. [CrossRef]
58. Choulaie, M.; Bagheri, A.; Khademifar, A. Nonlinear Vibration and Stability Analysis of Beam on the Variable Viscoelastic Foundation. *J. Comput. Appl. Mech.* **2017**, *48*, 99–110. [CrossRef]
59. Raayai-Ardakani, S.; Cohen, T. Capturing strain stiffening using Volume Controlled Cavity Expansion. *Extrem. Mech. Lett.* **2019**, *31*, 100536. [CrossRef]
60. Horgan, C.O.; Saccomandi, G. A molecular-statistical basis for the Gent constitutive model of rubber elasticity. *J. Elast.* **2002**, *68*, 167–176. [CrossRef]

nanomaterials

MDPI

Article

Hyperelastic Microcantilever AFM: Efficient Detection Mechanism Based on Principal Parametric Resonance

Amin Alibakhshi [1], Sasan Rahmanian [2], Shahriar Dastjerdi [3], Mohammad Malikan [4,*], Behrouz Karami [5], Bekir Akgöz [3] and Ömer Civalek [3,6]

1 Department of Mechanical Engineering, Science and Research Branch, Islamic Azad University, Tehran 1477893855, Iran; alibakhshiamin@yahoo.com
2 Department of Systems Design Engineering, University of Waterloo, Waterloo, ON N2L 3G1, Canada; s223rahm@uwaterloo.ca
3 Division of Mechanics, Civil Engineering Department, Akdeniz University, Antalya 07058, Turkey; dastjerdi_shahriar@yahoo.com (S.D.); bekirakgoz@akdeniz.edu.tr (B.A.); civalek@yahoo.com (Ö.C.)
4 Department of Mechanics of Materials and Structures, Faculty of Civil and Environmental Engineering, Gdańsk University of Technology, Gabriela Narutowicza 11/12, 80-233 Gdańsk, Poland
5 Department of Mechanical Engineering, Marvdasht Branch, Islamic Azad University, Marvdasht 73711-13119, Iran; behrouz.karami@miau.ac.ir
6 Department of Medical Research, China Medical University Hospital, China Medical University, Taichung 406, Taiwan
* Correspondence: mohammad.malikan@pg.edu.pl

Citation: Alibakhshi, A.; Rahmanian, S.; Dastjerdi, S.; Malikan, M.; Karami, B.; Akgöz, B.; Civalek, Ö. Hyperelastic Microcantilever AFM: Efficient Detection Mechanism Based on Principal Parametric Resonance. *Nanomaterials* **2022**, *12*, 2598. https://doi.org/10.3390/nano12152598

Academic Editor: James Murray Hill

Received: 9 June 2022
Accepted: 26 July 2022
Published: 28 July 2022

Abstract: The impetus of writing this paper is to propose an efficient detection mechanism to scan the surface profile of a micro-sample using cantilever-based atomic force microscopy (AFM), operating in non-contact mode. In order to implement this scheme, the principal parametric resonance characteristics of the resonator are employed, benefiting from the bifurcation-based sensing mechanism. It is assumed that the microcantilever is made from a hyperelastic material, providing large deformation under small excitation amplitude. A nonlinear strain energy function is proposed to capture the elastic energy stored in the flexible component of the device. The tip–sample interaction is modeled based on the van der Waals non-contact force. The nonlinear equation governing the AFM's dynamics is established using the extended Hamilton's principle, obeying the Euler–Bernoulli beam theory. As a result, the vibration behavior of the system is introduced by a nonlinear equation having a time-dependent boundary condition. To capture the steady-state numerical response of the system, a developed Galerkin method is utilized to discretize the partial differential equation to a set of nonlinear ordinary differential equations (ODE) that are solved by the combination of shooting and arc-length continuation method. The output reveals that while the resonator is set to be operating near twice the fundamental natural frequency, the response amplitude undergoes a significant drop to the trivial stable branch as the sample's profile experiences depression in the order of the picometer. According to the performed sensitivity analysis, the proposed working principle based on principal parametric resonance is recommended to design AFMs with ultra-high detection resolution for surface profile scanning.

Keywords: atomic force microscopy; hyperelastic microcantilever; softening resonance; non-contact cantilever; shooting and arc-length continuation method; developed Galerkin method

1. Introduction

Atomic force microscopy (AFM) is a device used to study materials' properties and surface structure in micro- and nanometer dimensions [1–3]. Flexibility, multiple potential signals, and the ability to operate the device in different modes have enabled researchers to study AFMs at different levels and under different environmental conditions. This device makes it possible to examine conductive or insulating surfaces, soft or hard, cohesive or powdery, biological and organic or inorganic. The general architecture of an AFM includes a microcantilever with a sharp tip at the free end contacting with the surface of samples [4]. The AFMs are applicable devices that have been widely used in so many scientific purposes, such as chemistry [5], materials science [6], medical and biophysics [7], and biological and colloidal applications [8].

Such as many structures that are influenced by the time [9–12], the operation of AFMs may be time-dependent and, consequently, under dynamic conditions. Thus, most previous studies in this field have focused on the dynamic performance of AFMs. The chaotic motion and bifurcation of a tapping mode AFM were examined by Yagasaki [13]. Bahrami and Nayfeh [14] explored the nonlinear oscillation of a tapping mode AFM by using a high-dimensional Galerkin discretization technique and a four-step Adams–Moulton method. They modeled the microcantilever based on the Euler–Bernoulli beam theory. The flexural and torsional vibrations of an AFM were investigated by Kahrobaiyan et al. [15], who performed a sensitivity analysis. Arafat and co-investigators [16] analyzed the frequency–amplitude response of a contact mode AFM analytically using the multiple time-scale method. Dastjerdi and Abbasi [17] studied the free vibration of a cracked AFM, incorporating the influence of crack size and its location. They utilized the transfer matrix method to solve the cracked system. Also, they concluded that existing a crack (with a specific size and location) on the AFM cantilever can be beneficial for controlling some unwanted phenomena. Mahmoudi et al. [18] studied the resonance of a non-contact AFM using harmonic balance and multiple time-scale methods. Saeidi et al. [19] explored forced vibrations of an AFM, taking the temperature and contact effects into account. Ahmadi and co-workers [20] studied free and forced oscillations of AFM with rectangular and V-shaped and employed the finite element analysis. Kouchaksaraei and Bahrami [21] proposed a new multifrequency excitation for AFM in a non-contact regime. They analyzed the sensitivity to the Hamaker constant and the initial tip–sample distance to enhance the performance of the AFM. In most previous studies investigating dynamic characteristics of AFMs, the structural materials have been a linear type. This means that the linear elasticity governs their behavior with small strain and deformation.

There has been an increasing interest in soft and hyperelastic materials in recent decades, because they are lightweight, cheap, compatible with many flexible structures, and show ease of fabrication. The relation between the strain and stress for hyperelastic materials is nonlinear. Besides, the strain in hyperelastic materials is moderate or large, and they undergo large deformation. These are the main differences between hyperelastic materials and linear elastic materials. Hyperelastic materials are structural materials of many electromechanical systems such as actuators, sensors, and energy harvesters. More specifically, hyperelastic materials have been widely used for electro-active-based systems [22–25]. Up to now, hyperelastic materials have been designed in different geometries, e.g., beams [26–29], plates [30,31], and shells [32,33]. Reviewing the previous studies shows a growing body of literature on using hyperelastic materials in different systems. Due to these advantages of hyperelastic materials and their broad applications, this paper proposes an AFM made of such materials, unique in simulation and application. Most recently, the dielectric elastomer actuators have been proposed [34] for AFM, and because these kinds of actuators are hyperelastic, exploring hyperelastic-based AFM may be of significant importance.

As is reported in the literature review, it can be seen that there is a lack of nonlinear study on vibration analysis of AFM cantilevers made of hyperelastic material. Consequently, thorough research on nonlinear frequency analysis of AFM cantilevers has been conducted based on the hyperelasticity approach. This paper is structured as follows. Section 2 derives the equation of motion based on a mathematical model containing material and geometrical nonlinearities, size effects, and finite rotations. In Section 3, the governing equation is discretized with a developed Galerkin decomposition method and solved through the shooting method combined with the arc-length continuation method. Finally, in Section 4, numerical results have been provided to study different dynamical behaviors of the AFM cantilever under the assumed conditions.

2. Mathematical Modeling

Figure 1 demonstrates the schematic view of an AFM cantilever simulated in this paper. The system consists of a hyperelastic microcantilever with a sharp tip at the free end. A Cartesian coordinate system (x, y, z) is employed to define the configuration of the AFM. The base vectors in $x, y, z-$directions are e_1, e_2, and e_3, respectively. The microcantilever's length, thickness, and width are denoted via L, d, and b, respectively.

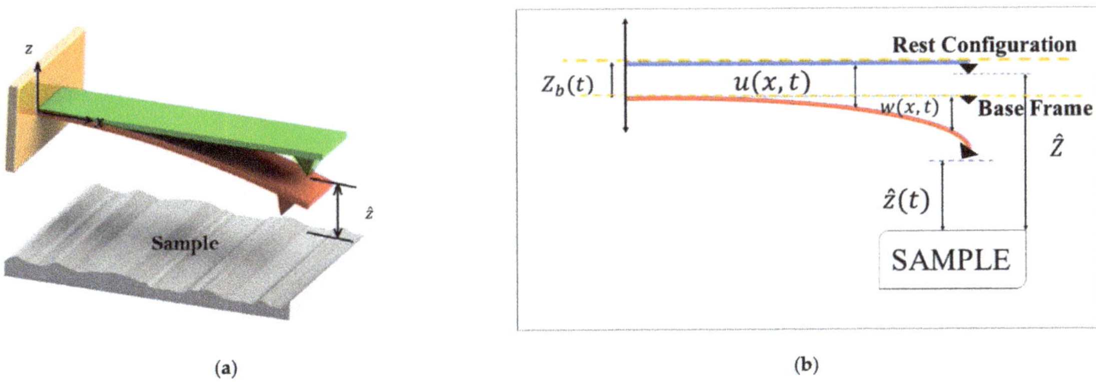

Figure 1. The schematic view of an AFM's cantilever scanning a sample. (**a**) 3D view of the AFM. (**b**) 2D view of the cantilever.

Figure 1a is a 3D schematic and perhaps does not provide more details. On the other hand, Figure 1b gives more details. The AFM consists of a microcantilever mounting on a base with piezoelectric patches as the excitation source (see Figure 1b). As depicted in Figure 1b, $\hat{Z}$ is the distance between the tip and the sample when the microbeam is located at rest, and $w(x, t)$ is the transverse displacement of the microcantilever with respect to the base frame. The transverse base excitation is $Z_b(t) = d_0 \sin(\omega t)$, in which d_0 and ω are the amplitude and frequency, respectively. The total deflection is $u(x, t) = w(x, t) + Z_b(t)$. Regarding these parameters, the instantaneous tip–sample separation is indicated by $\hat{z}(t) = \hat{Z} - w(L, t) - Z_b(t)$.

The governing equations are derived under the following assumptions: (1) the effect of rotary inertia and shear deformation is disregarded, i.e., the microbeam is modeled based on the Euler-Bernoulli beam theory. (2) both geometric and material nonlinearities are considered. (3) the rotation of the microcantilever is considered to be a moderate type, which means that $\partial w/\partial x$ is moderate. In other words, the rotation is of order $\sqrt{\epsilon}$, in which $\epsilon \ll 1$ stands for a small parameter (for more details on deriving moderate rotation see [35]). (4) the size effect is considered using the modified couple stress theory. (5) the axial displacement is neglected.

In reality, a sample subjected to probing may have an inhomogeneous surface, thus leading to surface depression or an increase in surface height. During probing, these increases in surface height may affect the performance of the hyperelastic AFM. As shown in Figure 2, the location of surface depression and increase in surface height assumed in the system are shown. In Figure 2, δ refers to any variation in the height of the sample profile with respect to the baseline, i.e., $\hat{Z} + \delta = \hat{Z} + 0 = \hat{Z}$.

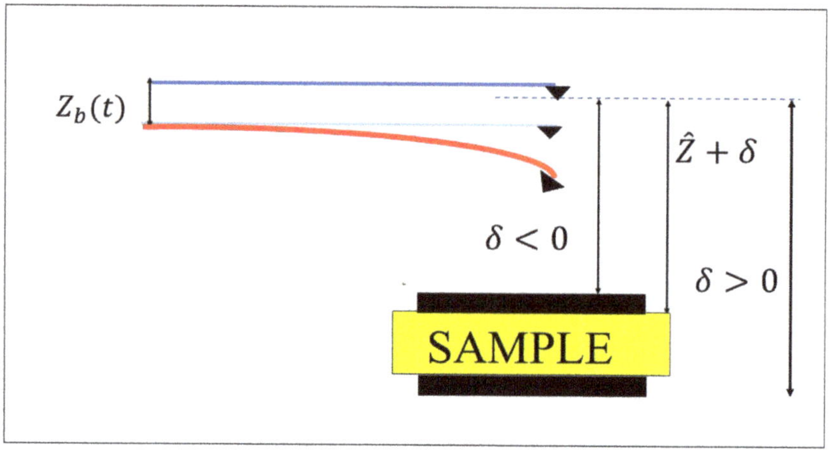

Figure 2. Schematic representation of the increase and decrease in surface profile.

2.1. Beam Theory with Finite Rotation and Deformation

The components of the displacement vector $u = u_1\hat{e}_1 + u_2\hat{e}_2 + u_3\hat{e}_3$ can be given as [36].

$$\begin{cases} u_1 = z\theta_x, \qquad u_2 = 0 \\ u_3 = w(x,t) \\ \theta_x \equiv -\frac{\partial w}{\partial x} \end{cases} \tag{1}$$

In Equation (1), θ_x is a defined parameter, namely, the minus of the slope of the transverse displacement.

The strain–displacement relations for the system are obtained according to the finite deformation. To this end, the von Kármán theory originated from Green's strain theory is used as follows [37]

$$\begin{cases} E_{xx} = \varepsilon_{xx}{}^{(0)} + z\varepsilon_{xx}{}^{(1)} \\ E_{zz} = \varepsilon_{zz}{}^{(0)} \end{cases} \tag{2}$$

in which

$$\begin{cases} \varepsilon_{xx}{}^{(0)} = \frac{1}{2}\left(\frac{\partial w}{\partial x}\right)^2 \\ \varepsilon_{xx}{}^{(1)} = \frac{\partial \theta_x}{\partial x} \\ \varepsilon_{zz}{}^{(0)} = \frac{1}{2}\theta_x{}^2 \end{cases} \tag{3}$$

In Equation (2), E_{xx} is the strain in the x-direction, and E_{zz} stands for the strain in the z-direction.

The potential energy is expressed in terms of the strain energy function for hyperelastic materials. The strain energy can be formulated based on the displacement as mentioned above and strain components incorporating the finite deformation, finite rotation, and size effect as follows [38,39].

$$\Psi = \frac{1}{2}\left(a_1 E_{xx}^2 + a_2 E_{zz}^2 + a_3\left(\theta_x{}'\right)^2 + 2a_4 E_{xx}E_{zz}\right) \tag{4}$$

in which

$$\begin{cases} a_1 = a_2 = 2\mu + \lambda \\ a_4 = \lambda \\ a_3 = 2\mu\ell^2 \end{cases} \tag{5}$$

In Equation (5), ℓ is the internal length-scale parameter capturing the size effect, which is defined as the square root of the ratio of the moduli of curvature to the shear [38,40] and is generally quantified as one-half or one-fourth of the thickness of the structure in theoretical analyses. It is noted that the value of this parameter from experimental evidence for hyperelastic microstructures seems to be rare. For more details on experimental methods of calculating ℓ, you may refer to [41].

Moreover, in Equation (15), $\mu = E/2(1+\nu)$ stands for the shear modulus, and $\lambda = [E\nu]/[(1+\nu)(1-2\nu)]$ is the Lamé's constant (ν is the Poisson's ratio). It is mentioned that Equation (5) has shown its applicability for hyperelastic structures in previous literature.

The hyperelastic potential energy is therefore calculated as:

$$U_{el} = \int_{\overline{V}} \Psi d\overline{V} \tag{6}$$

where $\overline{V}$ is the volume of the microcantilever.

Substituting Equation (2) into Equation (4), then inserting it into Equation (6), the final form of the hyperelastic energy is presented as:

$$\begin{aligned} U_{el} = \int_0^L \Bigg[& \tfrac{1}{8}a_1 A\left(\tfrac{\partial w}{\partial x}\right)^4 + \tfrac{1}{2}a_1 I\left(\tfrac{\partial^2 w}{\partial x^2}\right)^2 + \tfrac{1}{8}a_2 A\left(\tfrac{\partial w}{\partial x}\right)^4 + \tfrac{1}{2}a_3 A\left(\tfrac{\partial^2 w}{\partial x^2}\right)^2 \\ & + \tfrac{1}{4}a_4 A\left(\tfrac{\partial w}{\partial x}\right)^4 \Bigg] dx \end{aligned} \tag{7}$$

With the cross-section area $A = bd$ and the second moment of area $I = bd^3/12$.

2.2. Tip–Sample Interaction

At the free end of the microcantilever, there is the interaction force between the tip and sample. In the present paper, as a test case, a non-contact tip–sample interaction so-called the van der Waals non-contact force is applied, i.e., [14],

$$F_{\mathrm{vdW}} = -\frac{HR}{6\hat{z}^2} \tag{8}$$

in which F_{vdW} stands for the van der Waals non-contact force, H is the Hamaker constant, and R is the radius of the spherical tip apex.

The system's potential energy due to the tip–sample interaction is $V_{\mathrm{tp}} = -\int F_{\mathrm{vdW}}\, d\hat{z}$. By calculating this integration, the final form of the tip–sample interaction potential energy for the non-contact region will be formulated as [14]:

$$U_{\mathrm{vdW}} = -\frac{HR}{6\left[\hat{Z} - w(L,t) - Z_b(t)\right]} \tag{9}$$

2.3. Kinetic Energy

The total kinetic energy of the AFM cantilever can be expressed below (ρ is the mass density of the microcantilever).

$$U_k = \frac{1}{2}\int_0^L \rho A\left[\left(\frac{\partial w}{\partial t}\right) + \left(\frac{\partial Z_b(t)}{\partial t}\right)\right]^2 dx \tag{10}$$

2.4. Surrounding Damping Force

The constitutive material of the AFM is hyperelastic, for example, it is rubbery or elastomeric. For such materials, viscoelasticity plays a major role that should be considered in the analysis. In this paper, a linear damping model is assumed, while other complicated types of damping can also be used.

The amount of work created by the viscous surrounding medium is formulated as (c_d is the viscous damping coefficient).

$$W_D = -c_d \int_0^L \left[\left(\frac{\partial w}{\partial t} \right) + \left(\frac{\partial Z_b(t)}{\partial t} \right) \right] w dx \tag{11}$$

2.5. Hamilton's Principle and Equation of Motion

The governing equation can be easily derived using variational approaches by obtaining the energies and works that appear in the system [42–44]. The partial differential equation governing the motion of the AFM is derived using the following Hamilton's principle as one of the variational approaches:

$$\int_{t_1}^{t_2} [U_k - U_s] dt + \int_{t_1}^{t_2} \delta W_D dt = 0 \tag{12}$$

where U_s shows the total potential energy of the system, which for the non-contact region is: $U_s = U_{el} + U_{\mathrm{vdW}}$.

Eventually, substituting the expression of the kinetic and potential energies and the work of damping force into Hamilton's principle gives the following formulation:

$$\begin{aligned} \rho A \frac{\partial^2 w}{\partial t^2} + \rho A \quad & \frac{\partial^2 Z_b(t)}{\partial t^2} + c_d \frac{\partial w}{\partial t} + c_d \frac{\partial Z_b(t)}{\partial t} - \frac{3}{2} a_1 A \left(\frac{\partial^2 w}{\partial x^2} \right) \left(\frac{\partial w}{\partial x} \right)^2 \\ & + a_1 I \left(\frac{\partial^4 w}{\partial x^4} \right) - \frac{3}{2} a_2 A \left(\frac{\partial^2 w}{\partial x^2} \right) \left(\frac{\partial w}{\partial x} \right)^2 + a_3 A \left(\frac{\partial^4 w}{\partial x^4} \right) \\ & - 3 a_4 A \left(\frac{\partial^2 w}{\partial x^2} \right) \left(\frac{\partial w}{\partial x} \right)^2 = 0 \end{aligned} \tag{13}$$

It is noteworthy that the mathematical definition of the boundary conditions is obtained from Hamilton's principle. Consequently, the boundary condition definition can be formulated as the following equations.

$$\begin{cases} w(0,t) = \frac{\partial w}{\partial x}(0,t) = 0 \\ \qquad \frac{\partial^2 w}{\partial x^2}(L,t) = 0 \\ a_1 I \frac{\partial^3 w}{\partial x^3}(L,t) + a_3 A \frac{\partial^3 w}{\partial x^3}(L,t) = -\frac{HR}{6\left[\hat{Z} - w(L,t) - \hat{y}(t)\right]^2} \end{cases} \tag{14}$$

The boundary conditions for cantilevers state that the deflection and slope at the point $x = 0$ (at the fixed end) are equal to zero, and the bending moment and shear force at $x = L$ (at the free end) are equal to zero. From Equation (14), it is observed that the difference between the boundary conditions of the cantilever and AFM appears in the shear force in such a way that for the AFM it is not equal to zero any longer and is nonhomogeneous and time-dependent due to tip–sample interaction.

2.6. Nondimensionalization

In this section, the equation of motion and boundary conditions are made dimensionless. Hence, the dimensionless parameters can be presented as follows [45]:

$$\begin{aligned} & x^* = \frac{x}{L},\ w^* = \frac{w}{\hat{Z}}, \bar{d}_0 = \frac{d_0}{\hat{Z}},\ t^* = t\sqrt{\frac{EI}{\rho A L^4}}, c = \frac{c_d L^4}{EI}\sqrt{\frac{EI}{\rho A L^4}}, \Omega = \omega\sqrt{\frac{\rho A L^4}{EI}} \\ & \mathrm{d}_1 = \frac{a_1 I}{EI}, \mathrm{d}_2 = \frac{a_3 A}{EI} = \frac{2\mu A \ell^2}{EI}, \mathrm{d}_3 = \frac{3 a_1 A \hat{Z}^2}{2EI}, \mathrm{d}_4 = \frac{3 a_2 A \hat{Z}^2}{2EI}, \mathrm{d}_5 = \frac{3 a_4 A \hat{Z}^2}{EI} \\ & \overline{Z} = \frac{\hat{Z}}{\hat{Z}},\ \mathrm{d}_6 = \frac{HRL^3}{6EI\hat{Z}^3} \end{aligned} \tag{15}$$

The non-dimensional form of the equation of motion and boundary conditions are obtained as (the asterisk notation is disregarded for simplification):

$$\begin{gathered}\frac{\partial^2 w}{\partial t^2}+c\frac{\partial w}{\partial t}-\bar{d}_0\,\Omega^2\sin(\Omega t)+c\bar{d}_0\,\Omega\cos(\Omega t)+\underbrace{(\mathrm{d}_1+\mathrm{d}_2)}_{\alpha}\frac{\partial^4 w}{\partial x^4}\\ -\underbrace{(\mathrm{d}_3+\mathrm{d}_4+\mathrm{d}_5)}_{\beta}\left(\frac{\partial^2 w}{\partial x^2}\right)\left(\frac{\partial w}{\partial x}\right)^2=0\end{gathered}\tag{16}$$

with the dimensionless boundary conditions for the non-contact region

$$\begin{gathered}w(0,t)=\frac{\partial w}{\partial x}(0,t)=\frac{\partial^2 w}{\partial x^2}(1,t)=0\\ \underbrace{(\mathrm{d}_1+\mathrm{d}_2)}_{\alpha}\frac{\partial^3 w}{\partial x^3}(1,t)=-\frac{\mathrm{d}_6}{\left[(\overline{Z})-w(1,t)-\bar{d}_0\sin(\Omega t)\right]^2}\end{gathered}\tag{17}$$

3. Discretization of the Governing Motion's Equation

In this section, the governing partial differential equation is discretized by using the Galerkin method. Because the boundary condition is time-dependent, implementing the Galerkin method directly may be difficult. To solve such non-homogeneous boundary conditions, the following process is implemented.

It is assumed that the deflection has the following form [46]

$$w(x,t)=\mathrm{W}(x,t)+F(t)\,G(x)\tag{18}$$

in which $F(t)$ is obtained from non-homogeneous boundary conditions. $F(t)$ can be expressed as the following formulation.

$$F(t)=-\frac{\mathrm{d}_6}{\alpha\left[\overline{Z}-w(1,t)-\bar{d}_0\sin(\Omega t)\right]^2}\tag{19}$$

In Equation (18), $G(x)$ is an arbitrary function satisfying the following conditions:

$$G(0)=\frac{dG}{dx}(0)=G(1)=\frac{dG}{dx}(1)=\frac{d^2G}{dx^2}(1)=0,\ \frac{d^3G}{dx^3}(1)=1\tag{20}$$

A suitable assumption for the function $G(x)$ is chosen as

$$G(x)=\frac{-1}{6}x^2+\frac{1}{2}x^3-\frac{1}{2}x^4+\frac{1}{6}x^5\tag{21}$$

Substituting Equation (18) into Equations (16) and (17), the governing equation is derived as:

$$\begin{aligned}\frac{\partial^2 \mathrm{W}}{\partial t^2}+c\frac{\partial \mathrm{W}}{\partial t}\quad&+\alpha\frac{\partial^4 \mathrm{W}}{\partial x^4}-\beta\left(\frac{\partial^2 \mathrm{W}}{\partial x^2}\right)\left(\frac{\partial \mathrm{W}}{\partial x}\right)^2+G\frac{d^2F}{dt^2}+cG\frac{dF}{dt}+\alpha F\frac{d^4G}{dx^4}\\ &-2\beta F\frac{dG}{dx}\left(\frac{\partial^2 \mathrm{W}}{\partial x^2}\right)\left(\frac{\partial \mathrm{W}}{\partial x}\right)-\beta F^2\left(\frac{dG}{dx}\right)^2\left(\frac{\partial^2 \mathrm{W}}{\partial x^2}\right)\\ &-\beta F\left(\frac{d^2G}{dx}\right)\left(\frac{\partial \mathrm{W}}{\partial x}\right)^2-2\beta F^2\left(\frac{dG}{dx}\right)\left(\frac{d^2G}{dx}\right)\left(\frac{\partial \mathrm{W}}{\partial x}\right)\\ &-\beta F^3\left(\frac{dG}{dx}\right)^2\left(\frac{d^2G}{dx}\right)=\bar{d}_0\,\Omega^2\sin(\Omega t)-c\bar{d}_0\,\Omega\cos(\Omega t)\end{aligned}\tag{22}$$

Also, the corresponding boundary condition for Equation (22) is as follows:

$$\mathrm{W}(0,t)=\mathrm{W}'(0,t)=\mathrm{W}''(1,t)=\mathrm{W}'''(1,t)=0\tag{23}$$

The prime notation is derivative with respect to the dimensionless axial coordinate.

It is seen that the non-homogeneous boundary condition has been transformed into a homogeneous one. Now, the Galerkin method is applied to Equations (22) and (23). Since the contribution of the first mode is dominant in comparison to higher modes, the first mode is adopted in the present work.

By applying the variables separation technique, the response of the new variable $U(x,t)$ is supposed to be as:

$$W(x,t) = q(t)\phi(x) \tag{24}$$

in which $q(t)$ denotes the generalized coordinate, and $\phi(x)$ stands for the eigenfunction for a clamped-free beam, given by

$$\begin{aligned}\phi(x) = \{\cos(\beta_1 x) - \cos\mathrm{h}(\beta_1 x)\} \\ -\left(\frac{\{\cos(\beta_1)+\cos\mathrm{h}(\beta_1)\}}{\{\sin(\beta_1)+\sin\mathrm{h}(\beta_1)\}}\{\sin(\beta_1 x) - \sin\mathrm{h}(\beta_1 x)\}\right)\end{aligned} \tag{25}$$

where $\beta_1 = 1.8751$.

Substituting Equation (24) into Equation (22) and multiplying the resultant equation by $\phi(x)$, then integrating with respect to x from 0 to 1, the following governing equation in the time domain is derived as:

$$\begin{aligned}\ddot{q} + c\dot{q} + \omega_0^2 q - r_2 q^3 + r_3\ddot{F} + cr_3\dot{F} + r_4 F - r_5 F q^2 - r_6 F^2 q - r_7 F q^2 \\ -r_8 F^2 q - r_9 F^3 = r_{10}\,\Omega^2 \sin(\Omega t) - cr_{10}\,\Omega\cos(\Omega t)\end{aligned} \tag{26}$$

in which

$$\begin{aligned} r_1 &= \int_0^1 \alpha\,\phi\phi''''dx = \omega_0^2, & r_2 &= \int_0^1 \beta\phi\phi''(\phi')^2 dx \\ r_3 &= \int_0^1 \phi\, G dx, & r_4 &= \int_0^1 \alpha\,\phi\, G'''' dx \\ r_5 &= \int_0^1 2\beta\phi G'\phi'\phi'' dx, & r_6 &= \int_0^1 \beta\phi G' G'\phi'' dx \\ r_7 &= \int_0^1 \beta\phi\phi'\phi' G'' dx, & r_8 &= \int_0^1 2\beta\phi\phi' G' G'' dx \\ r_9 &= \int_0^1 \beta\phi G' G' G'' dx, & r_{10} &= \int_0^1 \overline{d}_0\phi\, dx \end{aligned} \tag{27}$$

We now simplify the equation of motion by expanding the function $F(t)$ around zero point, i.e.,

$$F(t) = k_1 + k_2 q(t) \tag{28}$$

in which

$$\begin{aligned} k_1 = -\frac{h_1}{\left(\overline{Z}-g(t)\right)^2},\ k_2 = -\frac{h_2}{\left(\overline{Z}-g(t)\right)^3},\ g(t) = \overline{d}_0\sin(\Omega t),\ h_1 = \frac{d_6}{\alpha}, \\ h_2 = \frac{2d_6\phi(1)}{\alpha}\end{aligned} \tag{29}$$

Substituting Equation (28) into Equation (26), we get

$$M\ddot{q} + C\dot{q} + K_l q + K_q q^2 + K_c q^3 = r_{10}\,\Omega^2\sin(\Omega t) - cr_{10}\,\Omega\cos(\Omega t) \tag{30}$$

in which

$$\begin{aligned} M &= 1 - \frac{h_2 r_3}{\left(\overline{Z}-g(t)\right)^3} \\ C &= c - \frac{c\,h_2 r_3}{\left(\overline{Z}-g(t)\right)^3} \\ K_L &= r_1 + \frac{3h_1^2 h_2 r_9}{\left(\overline{Z}-g(t)\right)^7} - \frac{h_1^2 r_6}{\left(\overline{Z}-g(t)\right)^4} - \frac{h_1^2 r_8}{\left(\overline{Z}-g(t)\right)^4} - \frac{h_2 r_4}{\left(\overline{Z}-g(t)\right)^3} \\ K_q &= \frac{3h_1 h_2^2 r_9}{\left(\overline{Z}-g(t)\right)^8} - \frac{2h_1 h_2 r_6}{\left(\overline{Z}-g(t)\right)^5} - \frac{2h_1 h_2 r_8}{\left(\overline{Z}-g(t)\right)^5} + \frac{h_1 r_5}{\left(\overline{Z}-g(t)\right)^2} + \frac{h_1 r_7}{\left(\overline{Z}-g(t)\right)^2} \\ K_c &= -r_2 + \frac{h_2^3 r_9}{\left(\overline{Z}-g(t)\right)^9} - \frac{h_2^2 r_6}{\left(\overline{Z}-g(t)\right)^6} - \frac{h_2^2 r_8}{\left(\overline{Z}-g[t]\right)^6} + \frac{h_2 r_5}{\left(\overline{Z}-g[t]\right)^3} + \frac{h_2 r_7}{\left(\overline{Z}-g(t)\right)^3} \end{aligned} \tag{31}$$

To distinguish between parametric and non-parametric resonances, the resultant equation is multiplied by $\left(\overline{Z} - g(t)\right)^9$, resulting in

$$\begin{aligned} M\left(\overline{Z} - g(t)\right)^9\ddot{q} \quad &+C\left(\overline{Z} - g(t)\right)^9\dot{q} + K_l\left(\overline{Z} - g(t)\right)^9 q + K_q\left(\overline{Z} - g(t)\right)^9 q^2 \\ &+K_c\left(\overline{Z} - g(t)\right)^9 q^3 \\ &= r_{10}\,\Omega^2\sin(\Omega t)\left(\overline{Z} - g(t)\right)^9 - cr_{10}\,\Omega\cos(\Omega t)\left(\overline{Z} - g(t)\right)^9 \end{aligned} \tag{32}$$

Simplifying the equations by assuming that $\overline{d}_0$ is too small compared to $\overline{Z} = 1$.

$$\boldsymbol{M_1}\ddot{q} + \boldsymbol{C_1}\dot{q} + \boldsymbol{K_{L1}}q + \boldsymbol{K_{q1}}q^2 + \boldsymbol{K_{c1}}q^3 = r_{10}\,\Omega^2\sin(\Omega t) - cr_{10}\,\Omega\cos(\Omega t) \tag{33}$$

in which

$$\begin{aligned} \boldsymbol{M_1} &= 1 - h_2 r_3 + \left(6\overline{d}_0 h_2 r_3 - 9\overline{d}_0\right)\sin(\Omega t) \\ \boldsymbol{C_1} &= c - c\,h_2 r_3 + \left(6\overline{d}_0 c\,h_2 r_3 - 9\overline{d}_0 c\right)\sin(\Omega t) \\ \boldsymbol{K_{L1}} &= 3h_1^2 h_2 r_9 - h_1^2 r_6 - h_1^2 r_8 - h_2 r_4 + r_1 \\ &+\left(-6\overline{d}_0 h_1^2 h_2 r_9 + 5\overline{d}_0 h_1^2 r_6 + 5\overline{d}_0 h_1^2 r_8 + 6\overline{d}_0 h_2 r_4 - 9\overline{d}_0 r_1\right)\sin(\Omega t) \\ \boldsymbol{K_{q1}} &= 3h_1 h_2^2 r_9 - 2h_1 h_2 r_6 - 2h_1 h_2 r_8 + h_1 r_5 + h_1 r_7 \\ &+\left(-3\overline{d}_0 h_1 h_2^2 r_9 + 8\overline{d}_0 h_1 h_2 r_6 + 6\overline{d}_0 h_1 h_2 r_8 - 7\overline{d}_0 h_1 r_5 - 7\overline{d}_0 h_1 r_7\right)\sin(\Omega t) \\ \boldsymbol{K_{c1}} &= -r_2 + h_2 r_5 + h_2 r_7 - h_2^2 r_6 - h_2^2 r_8 + h_2^3 r_9 \\ &+\left(3\overline{d}_0 h_2^2 r_6 + 3\overline{d}_0 h_2^2 r_8 - 6\overline{d}_0 h_2 r_5 - 6\overline{d}_0 h_2 r_7 + 9\overline{d}_0 r_2\right)\sin(\Omega t) \end{aligned} \tag{34}$$

4. Result and Discussion

In this section, the numerical results for the system under consideration shown in Figures 1 and 2 are illustrated. Unless stated otherwise, the material and geometrical parameters of the AFM cantilever are listed in Table 1. Because the influence of the size effect has been well addressed in previous studies, the material length-scale parameter is taken as zero in the numerical simulation and only has been expressed in the mathematical formulation. The main aim of this section is to investigate the principal parametric resonance of the system [47].

Table 1. The material and geometrical parameters of the AFM cantilever.

Parameter	Value
Modulus of elasticity, E	3 GPa
Length, L	225 μm
Cross-section area, A	7.02×10^{-11} m^2
The second moment of area, I	3.57×10^{-23} m^4
Hamaker constant, H	2.96×10^{-19} J
Tip radius, R	10 nm
Initial tip–sample distance, $\hat{Z}$	60 nm
Poisson's ratio ν	0.49

As seen in Equations (33) and (34), the combination of van der Waals force and base-excitation results in time-dependent linear inertia, stiffness, damping terms, and harmonically time-varying nonlinear stiffness terms, creating parametric parameters excitations. Moreover, the AFM is affected by two external excitations having the same frequencies, which are equal to the frequency of the base excitation. It is observed that the frequency of all parametric excitations appearing in the linear, quadratic, and cubic nonlinearities is equal to that of the external excitation. In this work, the quality factor of the system is assumed to be 200, defining a low-damped vibrating system. Under this condition, if the frequency of the parametric excitation varies twice the fundamental natural frequency of the AFM, principal parametric resonance is activated for small values of the base-excitation amplitude, d_0. In other words, for the quality factor of 200, the activation level of parametric resonance is reached at small values of d_0 that can let us neglect the contribution of the time-varying term existing in the inertia term. Another point that is worth mentioning is that, while the frequency of the base excitation is close to two times the first natural frequency, the sub-harmonic resonance of the first mode can be activated by the external excitation; however, its activation level is much greater than that of the principal parametric resonance. Therefore, it can be concluded that this type of nonlinear resonance does not contribute to the system's response within the variation range of d_0, which results in parametric resonance. With this in mind, the frequency-response behavior of the AFM is captured for different values of the base excitation amplitude and illustrated in Figure 3.

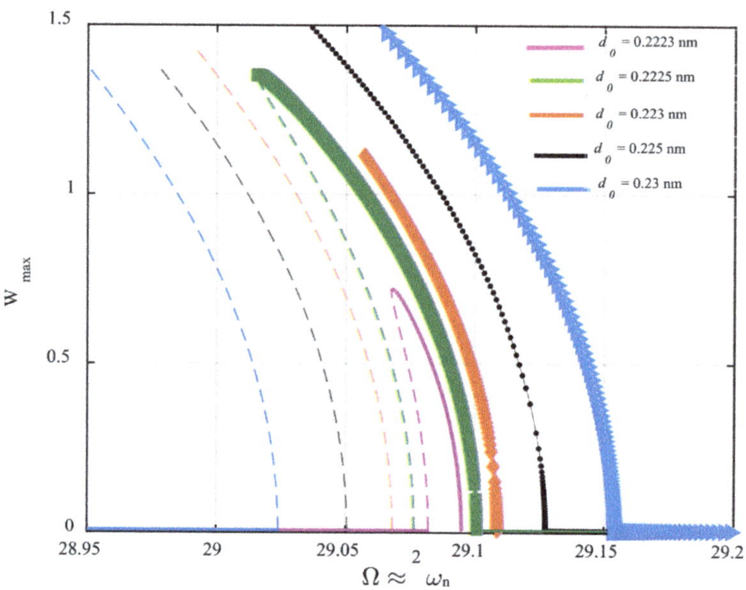

Figure 3. Frequency-response behavior of the AFM in the neighbor of its principal parametric resonance for different values of the base-excitation amplitude.

The figure shows that the frequency–response curves are composed of trivial stable, trivial unstable, nontrivial stable, and nontrivial unstable branches. The term *trivial* refers to the periodic response with zero amplitude; however, *nontrivial* phrase returns to the periodic orbit with non-zero amplitude. Moreover, the term *stable* clarifies that the system's states are absorbed by the periodic orbit after the system is disturbed; however, this is not the case for the *unstable* term, meaning that any small disturbance applied to the system's dynamics causes the system's states to get off the periodic orbit without the ability of returning back to it. The system's dynamics begin with stable zero-amplitude solutions and continue until primary Hopf bifurcation (sub-critical) occurs, and the stable branch loses its stability. Further increasing the parametric frequency, the unstable trivial solutions retrieve their stability at super-critical Hopf bifurcation points. In addition, for the smallest value of d_0 parameter close to the activation level, the stable and unstable nontrivial branches meet at the cyclic-fold bifurcation point at $\Omega \approx 29.07$; however, this is not the case for larger values of the parameter. As the parametric pump enhances, the resonance bandwidth becomes broader. The non-zero stable and unstable solutions meet at a displacement value beyond the gap distance between the microbeam and the substrate. Furthermore, it can be inferred that the quadratic and cubic nonlinearities arising from the intermolecular force induce a softening effect on the steady-state dynamics of the AFM, making the nontrivial solution branches bend to the left. It should be mentioned that W_{max} is the displacement of the cantilever tip. For better describing the frequency–response curves, it should be noted that there has been considered a set of two curves corresponding to each color. For instance, the blue plot drawn in Figure 3 is composed of two curves; the one which is marked by a star showing the stable solution, and the one that is mark-free shows the unstable solution branch corresponding to the same value of d_0.

Figure 4 demonstrates the loci of the primary Hopf bifurcation points for different values of the excitation amplitude. The vertical axis shows the ratio between the excitation amplitude to its threshold, leading to parametric resonance activation. As seen in the figure, while this ratio is close to one from below, there is still no bifurcation in the system's dynamics. However, while this ratio is close to one from above, two bifurcation points are occurring at two frequencies so close to each other that the left ones return to the sub-critical, and the right ones introduce the position of the super-critical Hopf bifurcation points. As the amplitude of the parametric excitation grows, the distance between these two points increases. The grey area bounded by the fitted blue curve displays the parametric resonance region, known as the instability tongue. It is worth noting that if the parametric excitations are not accompanied by external excitation, the parametric noise squeezing effect is potent to happen while the excitation amplitude is below the activation level. However, this phenomenon cannot occur in the present dynamics, because the parametric excitation is collaborated by external excitation with the same excitation frequency. Inspecting the numerical values on the vertical axis, one can find that the parametric resonance bandwidth is extremely sensitive to the excitation amplitude.

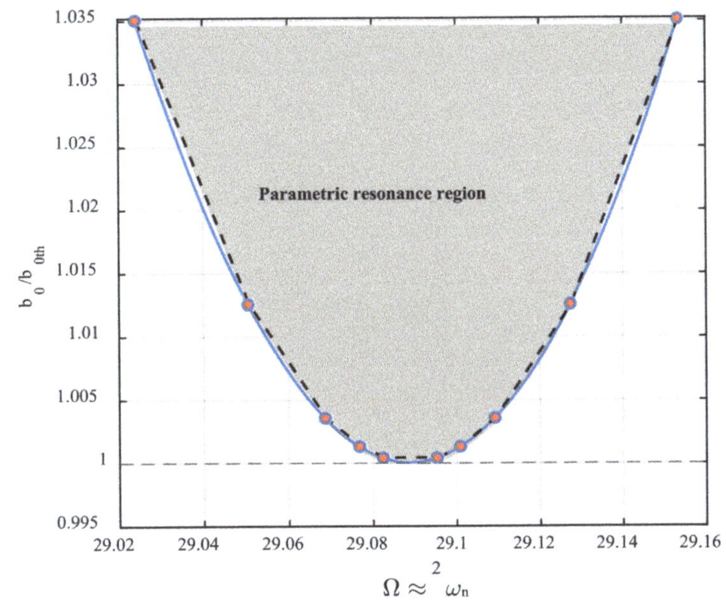

Figure 4. Instability tongue. The loci of both sub- and super-critical primary Hopf bifurcation points define the boundary of the parametric resonance region, where the threshold value for the base excitation amplitude is $d_{th} = 0.222$ nm.

Figure 5 illustrates the influence of the incompressibility (ν) on the frequency-response behavior of the AFM system by choosing two different values for Poisson's ratio, for $d_0 = 0.23$ nm. Increasing the value of Poisson's ratio causes the structure to become stiffer and more rigid, and therefore, its natural frequencies grow remarkably. With this in mind, the frequency of the parametric excitation is swept in the neighbour of twice the first natural frequency, which is evaluated for that specific Poisson's ratio. In order to have a precise evaluation of how the impact of Poisson's ratio influences the slope of the frequency–response curve, the displacement amplitude is plotted versus the difference between the excitation frequency and two times the natural frequency. It is found from the figure that, not only the resonance region becomes more expansive, but also the softening trend is intensified. At the same time, the resonator is made from a more incompressible material.

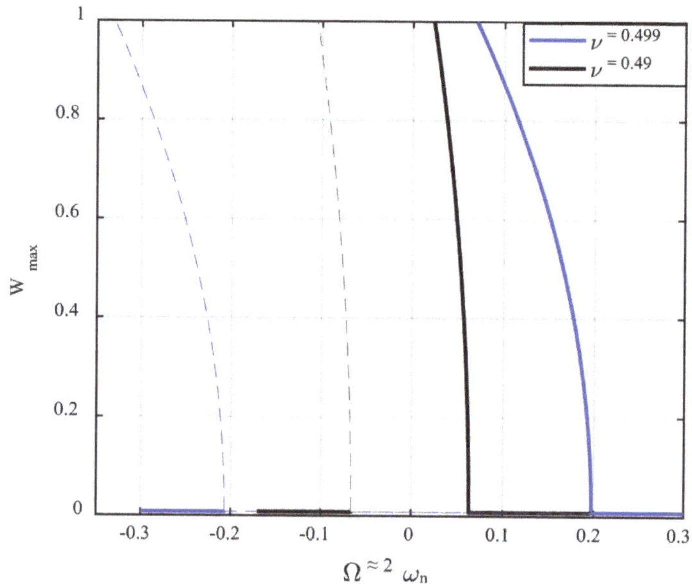

Figure 5. The impact of Poisson's ratio on the frequency-displacement behavior of the AFM, for $d_0 = 0.23$ nm. The dashed and solid lines are unstable and stable branches, respectively.

Profile Height Detection Mechanism

In this section, a sensing mechanism to detect the height of a sample's profile that is implemented based on the principal parametric resonance characteristics is proposed. First, it is shown that while the AFM is operating near its parametric resonance, the bifurcation points' position is highly affected by the gap distance between the microcantilever and its sample underneath. Figure 6 demonstrates the frequency-response behavior of the AFM for the excitation amplitude of $d_0 = 0.23$ nm. Here, the initial gap distance is set to be 60 nm. As seen in the figure, the sub-critical and super-critical Hopf bifurcation points shift to the left and to the right, respectively, while there is a 1 nm rise in the height of the sample profile (orange curve). An increment in the height of the sample profile causes the gap distance between the cantilever tip and the sample top surface to decrease. Therefore, the impact of van der Waals's force enhances, and consequently, the activation level decreases so that the resonance region becomes broader while the excitation amplitude is kept constant. This 1 nm rise in the height of the sample causes a 0.01 nondimensional frequency shift at primary Hopf bifurcation points. On the other hand, if the sample profile undergoes a 1 nm surface depression, extending the gap between the cantilever tip and the top surface of the sample, weakening the van der Waals force, the activation level of the parametric resonance increases. Hence, the resonance region shrinks while the parametric pump is kept constant, meaning the sub-critical and super-critical Hopf bifurcation points shift to the right and the left, respectively (black curve). As observed in the figure, the frequency shift caused by the rise in the height of the sample profile is quite a bit smaller than that caused by surface depression. Here, the frequency shift at bifurcation points caused by a 1 nm profile depression is obtained at about 0.0125. In Figure 6, PH stands for primary Hopf bifurcation.

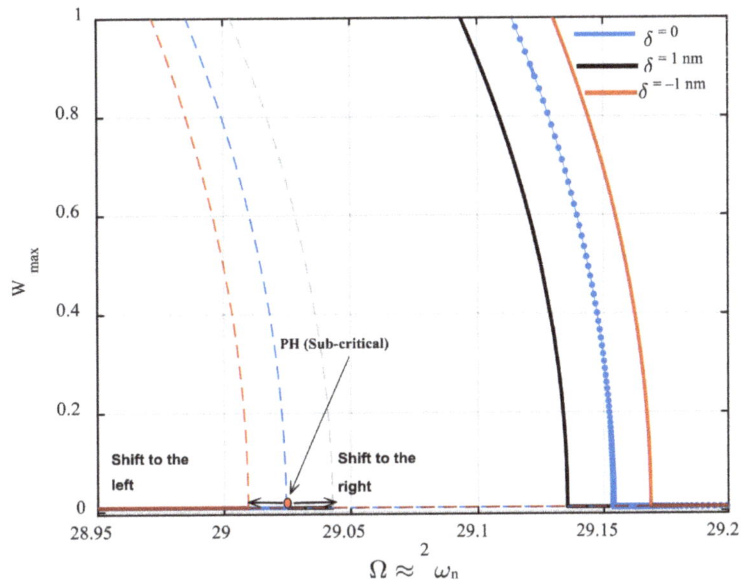

Figure 6. The influence of the surface profile variations on the primary Hopf bifurcation points near parametric resonance for the base excitation amplitude of 0.23 nm. The dashed and solid lines are unstable and stable branches, respectively.

The frequency-response behavior of the AFM device near the primary resonance of its first mode is illustrated in Figure 7. To monitor the steady-state dynamic response of the system near its primary resonance, the external excitation frequency is set to be changing in the vicinity of the first natural frequency, and the displacement amplitude of the periodic orbit is recorded for each specific value of the excitation frequency. As seen in the figure, the frequency-displacement behavior does not experience bifurcation for the excitation amplitude of $d_0 = 0.15$ nm; however, cyclic-fold bifurcation occurs beyond the gap distance for larger values of the base excitation amplitude, which is not meaningful, and we prevented presenting that result here. It is worth mentioning that for this case, the secondary parametric resonance of the first mode ($\Omega \approx \omega_n$) of oscillation is probable to happen because, as stated before, the frequency of the parametric terms is identical to that of the external stimulus. While the excitation frequency varies near the natural frequency of the first mode itself, the system has the potential to experience the combination of both primary resonance (due to external/direct excitation) and secondary parametric resonance (due to parametric excitation) of the first mode. However, the secondary parametric resonance ($\Omega \approx \omega_n$) requires a higher level of activation compared to the primary parametric resonance ($\Omega \approx 2\omega_n$). Therefore, it cannot be motivated for the proposed AFM, because the cantilever experiences tapping mode, which is caused by large-amplitude oscillation due to direct excitation. Inspecting the figure, it is seen that while there is a 1 nm increase in the height of the sample profile, the resonance amplitude grows slightly (orange curve) compared to the case in which there is no variation in the surface height (blue curve). On the other hand, the displacement amplitude at resonance drops quite a bit while the surface of the sample undergoes a 1 nm depression. Conversely, these variations in the profile configuration do not result in a change in the primary resonance frequency. In this work, the authors propose an effective detection mechanism based on amplitude shift at bifurcation points that are highly sensitive to detecting the high variations of samples' surface profiles. The AFM needs to be run near its parametric resonance zone instead of the primary resonance region.

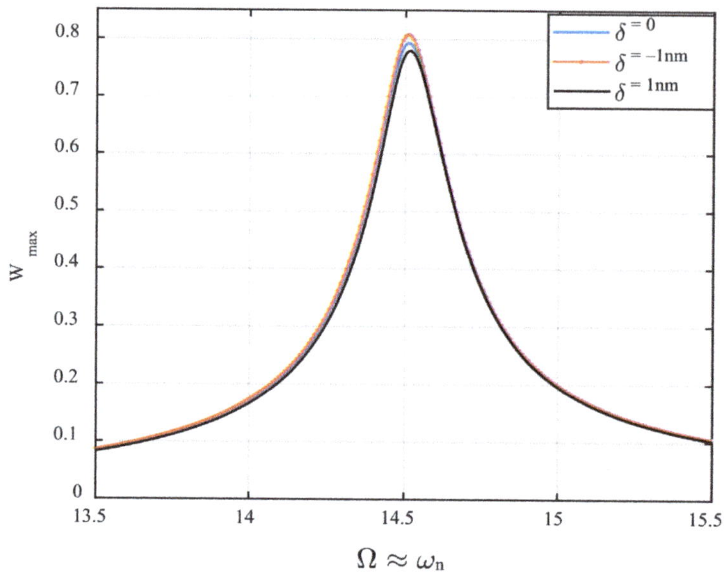

Figure 7. The influence of the surface profile variations on the resonance amplitude of the microcantilever near primary resonance, for $d_0 = 0.15$ nm.

Figure 8 depicts the frequency-displacement amplitude of the AFM near its parametric resonance for a parametric pump which is slightly above the activation level, $\frac{d_0}{d_{th}} = 1.001$. For the case in which there is neither bulge nor surface depression on the surface profile (violet curve), a cyclic-fold bifurcation exists at $\Omega = 29.068$. This dynamic behavior corresponds to the case where the cantilever tip is precisely on top of the sample area from which the initial gap distance, $\hat{Z}$, is measured. This profile status is considered the surface baseline, provided that any variation in the height of the sample profile concerning this baseline is denoted by δ. As observed in the figure, the frequency–response curve shifts to the right once the distance between the microcantilever tip and the surface profile increases by 15 pm; namely, the AFM faces a surface depression of 15 pm on the sample profile. Because the AFM is already set to be operating at point A, this right shift in the steady-state behavior of the device causes a significant drop in the displacement amplitude, jumping from point A down to point B with zero amplitude. The AFM's sensitivity can be obtained by evaluating the dimensional displacement drop to the change in the profile height. Accordingly, the sensitivity of the proposed method is as follows:

$$S_{D_\delta^w} = \frac{\Delta w}{\Delta \delta} = \frac{0.71 \times 60 \text{ nm}}{15 \text{ pm}} = 2840 \frac{\text{amplitude (nm)}}{\delta \text{ (nm)}} \tag{35}$$

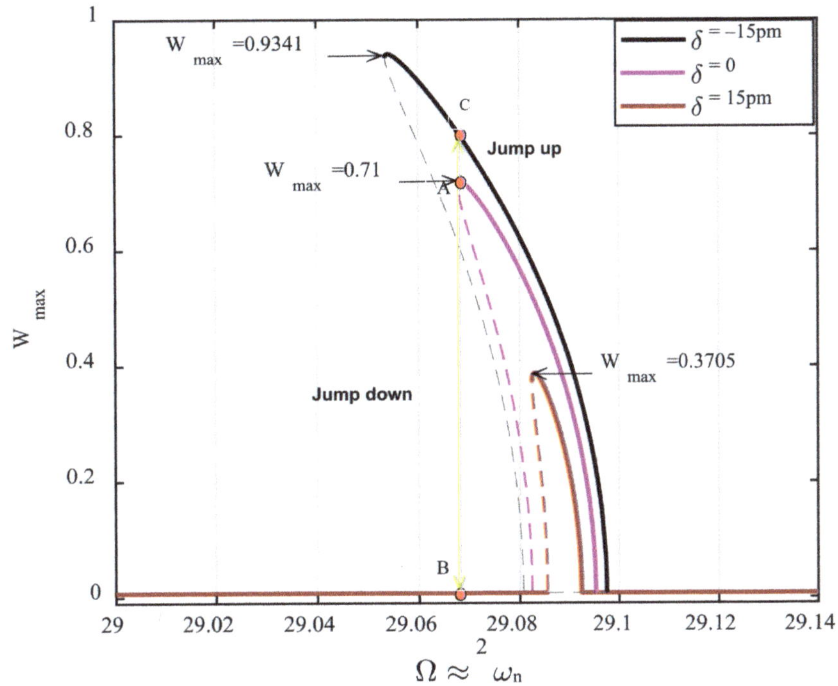

Figure 8. The influence of the surface profile variations on the amplitude of the resonator's displacement at cyclic-fold bifurcation near parametric resonance, for $\frac{d_0}{d_{th}} = 1.001$. The dashed and solid lines are unstable and stable branches, respectively.

On the other hand, the frequency–response curve shifts to the left while there is a 15 pm rise in the sample profile, so the resonator's response experiences a jump up to point C. Although this enhancement in the displacement amplitude is smaller than what is achieved for surface depression, the detection sensitivity is still considerable compared to the case where the cantilever is actuated near its primary resonance (comparing Figures 7 and 8). Similarly, for the case in which the distance between the AFM tip and the sample profile diminishes, the sensitivity of the device is obtained as follows,

$$S_{R_\delta^-}^{w} = \frac{\Delta \text{w}}{\Delta\delta} = \frac{(0.8 - 0.71) \times 60 \text{ nm}}{15 \text{ pm}} = 360 \frac{\text{amplitude (nm)}}{\delta \text{ (nm)}} \tag{36}$$

Scientifically speaking, because of the softening behavior of the AFM, the response amplitude drops from the stable nontrivial branch down to the stable trivial branch. At the same time, there is a positive variation in the height of the sample (surface depression), leading to ultrahigh sensitivity. However, this is not the case while there are negative variations in the height of the sample profile, meaning that the system's response jumps from a stable nontrivial branch up to a new generated stable nontrivial solution branch, degrading the AFM sensitivity. This can be improved by designing a tuneable AFM that can show both softening and hardening behavior near its parametric resonance regime. It is worth noting that the most challenging part in achieving ultrahigh sensitivity in AFMs returns to the employment of the base excitation amplitude. The finer the resolution of the parametric pump (parametric excitation amplitude), the higher the possibility for the implementation of the proposed detection mechanism. This means that the suggested method requires the capability of changing the base excitation amplitude with fine resolution. The proposed method can be feasibly employed once this concern is responded to.

In this paper, a micropolar model Equation (4) suitable for hyperelastic materials was utilized. However, other hyperelastic models can be utilized for hyperelastic microcantilevers; for example, the Gent model [48], generalized neo-Hookean models [49], Gent-Gent model [50], etc.

The foremost aspect of the analysis in the present work is that the system operates in parametric resonance regime. We have identified such a regime for non-contact AFM that is common in many real setups. For other kinds of AFM, it is essential to seek the presence of parametric resonance in the system first. Then, the analysis of the current study can be developed for the system.

5. Conclusions

This work examined the nonlinear resonance of a hyperelastic-based cantilever AFM in the non-contact region so that the non-contact force is modeled using the van der Waals force. A vertical base excitation is used to vibrate the AFM. A hyperelastic model combined with the modified couple stress theory is proposed to incorporate the elastic energy stored in the microcantilever caused by deformation. The obtained equation of motion is discretized via a developed Galerkin method. Furthermore, an efficient detection mechanism based on principal parametric resonance, assessing surface depression and height increase of a sample profile, is proposed. The following concluding remarks can be deduced from the presented numerical results.

- The hyperelastic microcantilever of the AFM device undergoes softening behavior near its principal parametric resonance.
- The frequency–displacement curve governing the resonator's dynamics comprises stable trivial, unstable trivial, stable nontrivial, and unstable nontrivial branches.
- The resonance of the AFM exhibits both super- and sub-critical Hopf bifurcations for the considerable value of d_0, and cyclic-fold bifurcation, for a small value of d_0.
- Increasing the incompressibility condition (higher values of the Poisson's ratio) results in stronger softening nonlinearity, and the resonance bandwidth becomes wider.
- Surface profile depression and rise in the height of a surface profile can be detected by inspecting the bifurcation points' position.
- According to the sensitivity analysis presented in Equation (35), the proposed AFM can detect surface depression in the order of a picometer, providing ultrahigh sensitivity.

Author Contributions: Conceptualization, A.A. and S.D.; methodology, A.A., S.R. and S.D.; software, A.A. and S.R.; validation, S.D., M.M., B.K., B.A. and Ö.C.; formal analysis, A.A. and S.R.; investigation, M.M., B.K., B.A. and Ö.C.; resources, A.A. and S.D.; data curation, A.A. and S.R.; writing—original draft preparation, A.A. and S.D.; writing—review and editing, A.A., S.D., M.M., S.R., B.K., B.A. and Ö.C.; visualization, A.A., S.D. and S.R.; supervision, M.M., B.K., B.A. and Ö.C.; funding acquisition, S.D. and M.M.; project administration, Ö.C. All authors have read and agreed to the published version of the manuscript.

Funding: This research received no external funding. The APC was funded by S.D. and M.M.

Institutional Review Board Statement: Not applicable.

Informed Consent Statement: Not applicable.

Data Availability Statement: There is no data for sharing and all data are available within the paper.

Conflicts of Interest: There are no conflicts of interest.

References

1. Kuo, C.-F.J.; Huy, V.Q.; Chiu, C.-H.; Chiu, S.-F. Dynamic modeling and control of an atomic force microscope probe measurement system. *J. Vib. Control* **2011**, *18*, 101–116. [CrossRef]
2. Sajjadi, M.; Chahari, M.; Pishkenari, H.N.; Vossoughi, G. Designing nonlinear observer for topography estimation in trolling mode atomic force microscopy. *J. Vib. Control* **2021**, 10775463211038140. [CrossRef]
3. Jalili, N.; Laxminarayana, K. A review of atomic force microscopy imaging systems: Application to molecular metrology and biological sciences. *Mechatronics* **2004**, *14*, 907–945. [CrossRef]

4. Alunda, B.O.; Lee, Y.J. Review: Cantilever-Based Sensors for High Speed Atomic Force Microscopy. *Sensors* **2020**, *20*, 4784. [CrossRef]
5. Sugimoto, Y.; Pou, P.; Abe, M.; Jelínek, P.; Perez, R.; Morita, S.; Custance, O. Chemical identification of individual surface atoms by atomic force microscopy. *Nature* **2007**, *446*, 64–67. [CrossRef] [PubMed]
6. Payton, O.D.; Picco, L.; Scott, T.B. High-speed atomic force microscopy for materials science. *Int. Mater. Rev.* **2016**, *61*, 473–494. [CrossRef]
7. Alessandrini, A.; Facci, P. AFM: A versatile tool in biophysics. *Meas. Sci. Technol.* **2005**, *16*, R65–R92. [CrossRef]
8. Zypman, F.R. Charge-Regulated Interactions: The Case of a Nanoparticle and a Sphere of Arbitrary Dielectric Constants. *Langmuir* **2022**, *38*, 3561–3567. [CrossRef] [PubMed]
9. Russillo, A.F.; Failla, G.; Amendola, A.; Luciano, R. On the Free Vibrations of Non-Classically Damped Locally Resonant Metamaterial Plates. *Nanomaterials* **2022**, *12*, 541. [CrossRef] [PubMed]
10. Mahmure, A.; Tornabene, F.; Dimitri, R.; Kuruoglu, N. Free Vibration of Thin-Walled Composite Shell Structures Reinforced with Uniform and Linear Carbon Nanotubes: Effect of the Elastic Foundation and Nonlinearity. *Nanomaterials* **2021**, *11*, 2090. [CrossRef] [PubMed]
11. Selim, M.M.; Musa, A. Nonlinear Vibration of a Pre-Stressed Water-Filled Single-Walled Carbon Nanotube Using Shell Model. *Nanomaterials* **2020**, *10*, 974. [CrossRef]
12. Civalek, Ö.; Akbaş, Ş.D.; Akgöz, B.; Dastjerdi, S. Forced Vibration Analysis of Composite Beams Reinforced by Carbon Nanotubes. *Nanomaterials* **2021**, *11*, 571. [CrossRef] [PubMed]
13. Yagasaki, K. Bifurcations and chaos in vibrating microcantilevers of tapping mode atomic force microscopy. *Int. J. Non-Linear Mech.* **2007**, *42*, 658–672. [CrossRef]
14. Bahrami, A.; Nayfeh, A.H. On the dynamics of tapping mode atomic force microscope probes. *Nonlinear Dyn.* **2012**, *70*, 1605–1617. [CrossRef]
15. Kahrobaiyan, M.; Ahmadian, M.; Haghighi, P. Sensitivity and resonant frequency of an AFM with sidewall and top-surface probes for both flexural and torsional modes. *Int. J. Mech. Sci.* **2010**, *52*, 1357–1365. [CrossRef]
16. Arafat, H.N.; Nayfeh, A.H.; Abdel-Rahman, E.M. Modal interactions in contact-mode atomic force microscopes. *Nonlinear Dyn.* **2008**, *54*, 151–166. [CrossRef]
17. Dastjerdi, S.; Abbasi, M. A vibration analysis of a cracked micro-cantilever in an atomic force microscope by using transfer matrix method. *Ultramicroscopy* **2018**, *196*, 33–39. [CrossRef] [PubMed]
18. Mahmoudi, M.S.; Ebrahimian, A.; Bahrami, A. Higher modes and higher harmonics in the non-contact atomic force microscopy. *Int. J. Non-Linear Mech.* **2019**, *110*, 33–43. [CrossRef]
19. Saeidi, H.; Zajkani, A.; Ghadiri, M. Nonlinear micromechanically analysis of forced vibration of the rectangular-shaped atomic force microscopes incorporating contact model and thermal influences. *Mech. Based Des. Struct. Mach.* **2020**, *50*, 609–629. [CrossRef]
20. Ahmadi, M.; Ansari, R.; Darvizeh, M. Free and forced vibrations of atomic force microscope piezoelectric cantilevers considering tip-sample nonlinear interactions. *Thin-Walled Struct.* **2019**, *145*, 106382. [CrossRef]
21. Kouchaksaraei, M.G.; Bahrami, A. High-resolution compositional mapping of surfaces in non-contact atomic force microscopy by a new multi-frequency excitation. *Ultramicroscopy* **2021**, *227*, 113317. [CrossRef] [PubMed]
22. Alibakhshi, A.; Heidari, H. Nonlinear dynamic responses of electrically actuated dielectric elastomer-based microbeam resonators. *J. Intell. Mater. Syst. Struct.* **2021**, *33*, 558–571. [CrossRef]
23. Alibakhshi, A.; Imam, A.; Haghighi, S.E. Effect of the second invariant of the Cauchy–Green deformation tensor on the local dynamics of dielectric elastomers. *Int. J. Non-Linear Mech.* **2021**, *137*, 103807. [CrossRef]
24. Alibakhshi, A.; Chen, W.; Destrade, M. Nonlinear Vibration and Stability of a Dielectric Elastomer Balloon Based on a Strain-Stiffening Model. *J. Elast.* **2022**, 1–16. [CrossRef]
25. Zeng, C.; Gao, X. Stability of an anisotropic dielectric elastomer plate. *Int. J. Non-Linear Mech.* **2020**, *124*, 103510. [CrossRef]
26. He, L.; Lou, J.; Kitipornchai, S.; Yang, J.; Du, J. Peeling mechanics of hyperelastic beams: Bending effect. *Int. J. Solids Struct.* **2019**, *167*, 184–191. [CrossRef]
27. Bacciocchi, M.; Tarantino, A.M. Finite bending of hyperelastic beams with transverse isotropy generated by longitudinal porosity. *Eur. J. Mech.-A/Solids* **2020**, *85*, 104131. [CrossRef]
28. Alibakhshi, A.; Dastjerdi, S.; Malikan, M.; Eremeyev, V.A. Nonlinear Free and Forced Vibrations of a Hyperelastic Micro/Nanobeam Considering Strain Stiffening Effect. *Nanomaterials* **2021**, *11*, 3066. [CrossRef] [PubMed]
29. Alibakhshi, A.; Dastjerdi, S.; Fantuzzi, N.; Rahmanian, S. Nonlinear free and forced vibrations of a fiber-reinforced dielectric elastomer-based microbeam. *Int. J. Non-Linear Mech.* **2022**, *144*, 104092. [CrossRef]
30. Amabili, M.; Balasubramanian, P.; Breslavsky, I.D.; Ferrari, G.; Garziera, R.; Riabova, K. Experimental and numerical study on vibrations and static deflection of a thin hyperelastic plate. *J. Sound Vib.* **2016**, *385*, 81–92. [CrossRef]
31. Du, P.; Dai, H.-H.; Wang, J.; Wang, Q. Analytical study on growth-induced bending deformations of multi-layered hyperelastic plates. *Int. J. Non-Linear Mech.* **2019**, *119*, 103370. [CrossRef]
32. Amabili, M.; Breslavsky, I.; Reddy, J. Nonlinear higher-order shell theory for incompressible biological hyperelastic materials. *Comput. Methods Appl. Mech. Eng.* **2018**, *346*, 841–861. [CrossRef]

33. Mihai, L.A.; Alamoudi, M. Likely oscillatory motions of stochastic hyperelastic spherical shells and tubes. *Int. J. Non-Linear Mech.* **2021**, *130*, 103671. [CrossRef]
34. Alibakhshi, A.; Dastjerdi, S.; Malikan, M.; Eremeyev, V.A. Nonlinear free and forced vibrations of a dielectric elastomer-based microcantilever for atomic force microscopy. *Contin. Mech. Thermodyn.* **2022**, 1–18. [CrossRef]
35. Schmidt, R.; Reddy, J.N. A Refined Small Strain and Moderate Rotation Theory of Elastic Anisotropic Shells. *J. Appl. Mech.* **1988**, *55*, 611–617. [CrossRef]
36. Şimşek, M.; Reddy, J. Bending and vibration of functionally graded microbeams using a new higher order beam theory and the modified couple stress theory. *Int. J. Eng. Sci.* **2013**, *64*, 37–53. [CrossRef]
37. Srinivasa, A.; Reddy, J. A model for a constrained, finitely deforming, elastic solid with rotation gradient dependent strain energy, and its specialization to von Kármán plates and beams. *J. Mech. Phys. Solids* **2012**, *61*, 873–885. [CrossRef]
38. Reddy, J.; Srinivasa, A. Non-linear theories of beams and plates accounting for moderate rotations and material length scales. *Int. J. Non-Linear Mech.* **2014**, *66*, 43–53. [CrossRef]
39. Alibakhshi, A.; Dastjerdi, S.; Akgöz, B.; Civalek, Ö. Parametric vibration of a dielectric elastomer microbeam resonator based on a hyperelastic cosserat continuum model. *Compos. Struct.* **2022**, *287*, 115386. [CrossRef]
40. Mindlin, R. *Influence of Couple-Stresses on Stress Concentrations*; Columbia University: New York, NY, USA, 1962.
41. Choi, J.-H.; Kim, H.; Kim, J.-Y.; Lim, K.-H.; Lee, B.-C.; Sim, G.-D. Micro-cantilever bending tests for understanding size effect in gradient elasticity. *Mater. Des.* **2022**, *214*, 110398. [CrossRef]
42. Dastjerdi, S.; Akgöz, B.; Civalek, Ö. On the effect of viscoelasticity on behavior of gyroscopes. *Int. J. Eng. Sci.* **2020**, *149*, 103236. [CrossRef]
43. Ghayesh, M.H.; Farokhi, H.; Amabili, M. Nonlinear behaviour of electrically actuated MEMS resonators. *Int. J. Eng. Sci.* **2013**, *71*, 137–155. [CrossRef]
44. Malikan, M.; Krasheninnikov, M.; Eremeyev, V.A. Torsional stability capacity of a nano-composite shell based on a nonlocal strain gradient shell model under a three-dimensional magnetic field. *Int. J. Eng. Sci.* **2020**, *148*, 103210. [CrossRef]
45. Farokhi, H.; Ghayesh, M.H.; Amabili, M. Nonlinear dynamics of a geometrically imperfect microbeam based on the modified couple stress theory. *Int. J. Eng. Sci.* **2013**, *68*, 11–23. [CrossRef]
46. Aranda-Ruiz, J.; Fernández-Sáez, J. On the use of variable-separation method for the analysis of vibration problems with time-dependent boundary conditions. *Proc. Inst. Mech. Eng. Part C J. Mech. Eng. Sci.* **2012**, *226*, 2912–2924. [CrossRef]
47. Rahmanian, S.; Hosseini-Hashemi, S.; Rezaei, M. Out-of-plane motion detection in encapsulated electrostatic MEMS gyroscopes: Principal parametric resonance. *Int. J. Mech. Sci.* **2020**, *190*, 106022. [CrossRef]
48. Gent, A.N. A New Constitutive Relation for Rubber. *Rubber Chem. Technol.* **1996**, *69*, 59–61. [CrossRef]
49. Anssari-Benam, A.; Bucchi, A. A generalised neo-Hookean strain energy function for application to the finite deformation of elastomers. *Int. J. Non-Linear Mech.* **2020**, *128*, 103626. [CrossRef]
50. Alibakhshi, A.; Heidari, H. Nonlinear dynamics of dielectric elastomer balloons based on the Gent-Gent hyperelastic model. *Eur. J. Mech.-A/Solids* **2020**, *82*, 103986. [CrossRef]

nanomaterials

MDPI

Article

Free Vibration of Thin-Walled Composite Shell Structures Reinforced with Uniform and Linear Carbon Nanotubes: Effect of the Elastic Foundation and Nonlinearity

Avey Mahmure [1], Francesco Tornabene [2,*], Rossana Dimitri [2] and Nuri Kuruoglu [3]

1 Division of Mathematics, Graduate School of Natural and Applied Sciences, Usak University, Usak 64200, Turkey; mahmureavey@gmail.com
2 Department of Innovation Engineering, University of Salento, 73100 Lecce, Italy; rossana.dimitri@unisalento.it
3 Department of Civil Engineering of Faculty of Engineering and Architecture, Istanbul Gelisim University, Istanbul 34310, Turkey; nkuruoglu@gelisim.edu.tr
* Correspondence: francesco.tornabene@unisalento.it

Abstract: In this work, we discuss the free vibration behavior of thin-walled composite shell structures reinforced with carbon nanotubes (CNTs) in a nonlinear setting and resting on a Winkler–Pasternak Foundation (WPF). The theoretical model and the differential equations associated with the problem account for different distributions of CNTs (with uniform or nonuniform linear patterns), together with the presence of an elastic foundation, and von-Karman type nonlinearities. The basic equations of the problem are solved by using the Galerkin and Grigolyuk methods, in order to determine the frequencies associated with linear and nonlinear free vibrations. The reliability of the proposed methodology is verified against further predictions from the literature. Then, we examine the model for the sensitivity of the vibration response to different input parameters, such as the mechanical properties of the soil, or the nonlinearities and distributions of the reinforcing CNT phase, as useful for design purposes and benchmark solutions for more complicated computational studies on the topic.

Keywords: CNT; elastic foundations; nonlinear free vibration; nonlinear frequency; shallow shell structures

Citation: Mahmure, A.; Tornabene, F.; Dimitri, R.; Kuruoglu, N. Free Vibration of Thin-Walled Composite Shell Structures Reinforced with Uniform and Linear Carbon Nanotubes: Effect of the Elastic Foundation and Nonlinearity. *Nanomaterials* **2021**, *11*, 2090. https://doi.org/10.3390/nano11082090

Academic Editor: Yang Tse Cheng

Received: 21 July 2021
Accepted: 15 August 2021
Published: 17 August 2021

1. Introduction

The fast development of nanotechnology in recent years has encouraged the production of nanotubes, increasing their application in many engineering areas. The CNTs produced for the first time by Iijima in 1993, are increasingly used in various industries and commonly proposed as novel material due to their great potential [1,2]. One of the most important application areas of CNTs stems from their large use as reinforcement phase in traditional composites and polymers. The mechanical, thermal and electrical properties of composites reinforced with CNTs are significantly improved compared to more classical composites, along with an increased level of strength in their structural application [3–5]. For such reasons, CNTs are used in some areas of the defense industry, especially in rocket, aerospace and aviation industries, where high-precision computations are required [6–8]. Among various problems is the linear and nonlinear vibration behavior of composite shell structures involving the presence of different distributions of CNTs. Composite shells, indeed, can include uniform or nonuniform patterns of CNTs, depending on the desired mechanical properties of the structures [9–24]. In this framework, a pioneering work on the nonlinear vibrations of composite shell structures was represented by [9], which considered a linear distribution of CNTs within the material. Following this work, some linear and nonlinear free vibration problems were proposed in [10–17] and [18–24], respectively, for unconstrained shallow shells and panels reinforced by CNTs, while proposing different numerical methods to solve the related problems.

The technological evolution of artificial materials and their manufacturing has expanded the application areas for such materials, improving the interest towards even more complicated and coupled problems, as well as the possible interactions of a structural member with its surrounding medium. Composite CNT-based shell structures resting on elastic foundations can be found in different civil and mechanical engineering applications, in nuclear power plants, etc. Among different possibilities to model an elastic foundation, the Pasternak and Winkler models are two common ways of describing its mechanical behavior, based on a different number of input parameters [25,26]. When modeling the vibrations of structural members containing CNTs, it is important to study the effect of the reinforcement phase and elastic foundations on the frequency–amplitude relationships. Up to date, most works from the literature have been devoted to the solution of linear vibration problems, by means of different numerical techniques [27–35]. More specifically, Tornabene et al. [27] examined the influence of Winkler–Pasternak foundations on the static and dynamic analysis of laminated double-curved shells and panels using the differential quadrature method. The same numerical approach was successfully proposed in [28] to study the vibration response of functionally graded carbon nanotube reinforced composite (FG-CNTRC) spherical shells on an elastic foundation. Zhang and Liew [29] applied an element-free approach to study the large deflection response of FG-CNTRC plates. Dinh and Nguyen [30] applied a fourth-order Runge–Kutta method and Galerkin method to solve the dynamic and vibration problem of FG-CNTRC truncated conical shells on elastic foundations. Shen and He [31] performed a large amplitude vibration analysis of FG-CNTRC double-curved panels on elastic foundation by applying a two-step perturbation approach, as also implemented in [32] to analyze the large amplitude vibration of FG shallow arches on a nonlinear elastic foundation. A further linear formulation was proposed by Sobhy and Zenkour [33] to study the vibrations of FG graphene platelet reinforced composite double-curved shallow shells on an elastic foundation; Sofiyev et al. [34,35] investigated the stability of CNTRC conical shells resting on an elastic foundation under hydrostatic pressure and combined loads in different settings.

Despite the considerable attention paid by the scientific literature to the linear vibration of shell structures, the nonlinear vibrations of CNT shallow shells resting on elastic foundations have not been adequately investigated. In this context, this paper aims to study the nonlinear free vibration behavior of thin-walled shell structures reinforced with CNTs and resting on an elastic Winkler- or Pasternak-type foundation, while proposing a Grigolyuk method to handle the problem. The organization of the rest of the paper is as follows: Section 2 recalls the basic theoretical aspects for both the shell-foundation interaction and nonlinear structural problem. Section 3 illustrates the analytical methodology applied to solve the problem, whose numerical investigation is presented and discussed in Section 4, while Section 5 closes the work with main comments and remarks.

2. Theoretical Formulation

2.1. Description of Shell-Foundation Interaction Model

Let us consider a composite spherical and hyperbolic paraboloidal (hypar) shallow shell reinforced with CNTs with length a, width b, thickness h and curvature radii R_1 and R_2, respectively (see Figure 1a,b). The Cartesian coordinate system (x_1, x_2, x_3) is here assumed to define the shell geometry in its length, width and thickness direction, respectively. As also shown in Figure 1, both the spherical and hypar shallow shells are immersed in an elastic WPF, here modeled as follows [25,26]:

$$K(w) = k_w w - k_p \left(\frac{\partial^2 w}{\partial x_1^2} + \frac{\partial^2 w}{\partial x_2^2} \right) \quad (1)$$

where k_w (in Pa/m) is the Winkler spring stiffness and k_P (in $Pa \cdot m$) refers to the shear layer stiffness. When $k_P = 0$, the foundation reverts to a Winkler-type elastic foundation (WF). The FG-CNTRC shell structures feature the following properties [9]

$$\begin{gathered} Y_{11}^{\overline{x}_3} = \eta_1 V_{CN}^{\overline{x}_3} Y_{11}^{CN} + V_m E^m, \ \frac{\eta_2}{Y_{22}^{\overline{x}_3}} = \frac{V_{CN}^{\overline{x}_3}}{Y_{22}^{CN}} + \frac{V_m}{E^m}, \ \frac{\eta_3}{Y_{12}^{\overline{x}_3}} = \frac{V_{CN}^{\overline{x}_3}}{Y_{12}^{CN}} + \frac{V_m}{Y^m}, \\ \nu_{12} = V_{CN}^{*} \nu_{12}^{CN} + V_m \nu^m, \ \rho_1^{\overline{x}_3} = V_{CN}^{*} \rho^{CN} + V_m \rho^m, \ \overline{x}_3 = x_3/h \end{gathered} \tag{2}$$

where the elastic properties for CNTs and matrix denoted as $Y_{ij}^{CN}(i,j=1,2)$, and Y^m,G^m, respectively; $\eta_j(j = 1,2,3)$ refers to the efficiency parameters for CNTs; $V_{CN}^{\overline{x}_3}$ and V^m stand for the volume fraction of CNTs and matrix, respectively, such that $V_{CN}^{\overline{x}_3} + V_m = 1$. The density can be defined as

$$V_{CN}^{*} = \frac{w_{CN}}{w_{CN} + (\rho^{CN}/\rho^m)(1 - w_{CN})} \tag{3}$$

whereas the volume fraction for shallow shells takes the following form (see Figure 2)

$$V_{CN}^{\overline{x}_3} = \begin{cases} UD \ \text{at} \ V_{CN}^{*} \\ VD \ \text{at} \ (1 - \overline{x}_3) V_{CN}^{*} \\ OD \ \text{at} \ (1 + \overline{x}_3) V_{CN}^{*} \\ XD \ \text{at} \ 4|\overline{x}_3| \, V_{CN}^{*} \end{cases} \tag{4}$$

The strain field on the reference surface is governed by the following kinematic relations [36]

$$\begin{gathered} \varepsilon_{11} = \frac{\partial u}{\partial x_1} - \frac{w}{R_1} + \frac{1}{2}\left(\frac{\partial w}{\partial x_1}\right)^2, \ \varepsilon_{22} = \frac{\partial v}{\partial x_2} - \frac{w}{R_2} + \frac{1}{2}\left(\frac{\partial w}{\partial x_2}\right)^2 \\ \gamma_{12} = \frac{\partial v}{\partial x_1} + \frac{\partial u}{\partial x_2} + \frac{\partial w}{\partial x_1}\frac{\partial w}{\partial x_2} \end{gathered} \tag{5}$$

and the constitutive relations accounting for the von Karman nonlinearity within a classical shell framework are defined as [21]

$$\begin{bmatrix} \tau_{11} \\ \tau_{22} \\ \tau_{12} \end{bmatrix} = \begin{bmatrix} E_{11}^{\overline{x}_3} & E_{12}^{\overline{x}_3} & 0 \\ E_{21}^{\overline{x}_3} & E_{22}^{\overline{x}_3} & 0 \\ 0 & 0 & E_{66}^{\overline{x}_3} \end{bmatrix} \begin{bmatrix} e_{11} - \overline{x}_3 \frac{\partial^2 w}{\partial x_1^2} \\ e_{22} - \overline{x}_3 \frac{\partial^2 w}{\partial x_2^2} \\ \gamma_{12} - 2\overline{x}_3 \frac{\partial^2 w}{\partial x_1 \partial x_2} \end{bmatrix} \tag{6}$$

with

$$E_{ii}^{\overline{x}_3} = \frac{Y_{ii}^{\overline{x}_3}}{1 - \nu_{ij}\nu_{ji}}, \ E_{ij}^{\overline{x}_3} = \frac{\nu_{ji} Y_{ii}^{\overline{x}_3}}{1 - \nu_{ij}\nu_{ji}} = E_{ji}^{\overline{x}_3}, \ E_{66}^{\overline{x}_3} = Y_{ij}^{\overline{x}_3}(i,j=1,2) \tag{7}$$

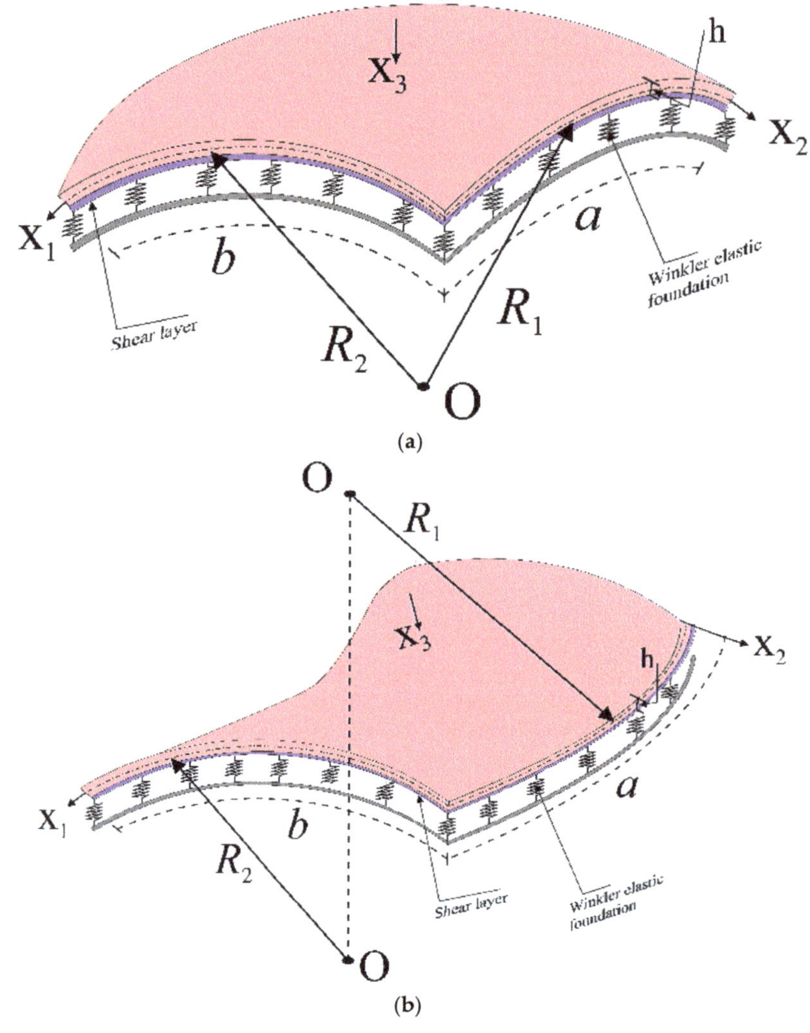

Figure 1. (**a**) Spherical and (**b**) hypar shallow shells reinforced with CNTs, resting on elastic foundations.

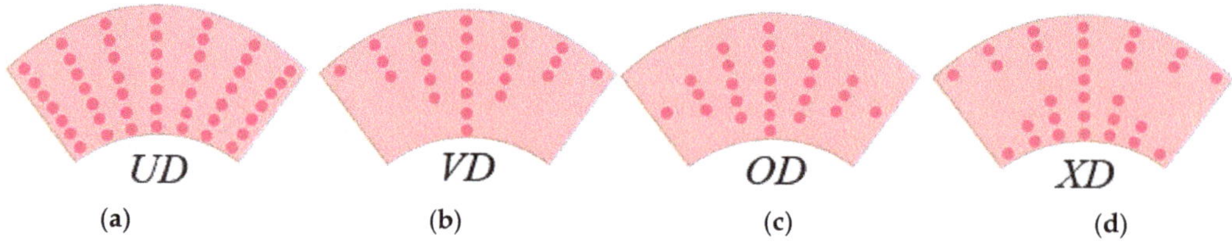

Figure 2. Cross-section of shallow shells with uniform and linearly patterned CNTs (**a**) UD, (**b**) VD, (**c**) OD and (**d**) XD.

2.2. Nonlinear Structural Model in the Presence of a PF

By using relations (1), (2), (5) and (6), the nonlinear governing equations for doubly curved shallow shells reinforced with a linear pattern of CNTs and resting on a WPF, attain the following form

$$L_{11}(F) + L_{12}(w) + L_{13}(F,w) + K(w) = 0 \tag{8}$$

$$L_{21}(F) + L_{22}(w) + L_{13}(w,w) = 0 \tag{9}$$

where F is a stress function and $L_{ij}(i = 1,2, j = 1,2,3)$ are differential operators defined as

$$\begin{aligned}
L_{11}(F) &= h\left[u_{12}\frac{\partial^4}{\partial x_1^4} + (u_{11} - 2u_{31} + u_{22})\frac{\partial^4}{\partial x_1^2\partial x_2^2} + u_{21}\frac{\partial^4}{\partial x_2^4} + \left(\frac{1}{R_2}\frac{\partial^2}{\partial x_1^2} + \frac{1}{R_1}\frac{\partial^2}{\partial x_2^2}\right)\right], \\
L_{12}(w) &= -u_{13}\frac{\partial^4}{\partial x_1^4} - (u_{14} + 2u_{32} + u_{23})\frac{\partial^4}{\partial x_1^2\partial x_2^2} - u_{24}\frac{\partial^4}{\partial x_1^4} - \rho_1\frac{\partial^2}{\partial t^2}, \\
L_{13}(F,w) &= h\left(\frac{\partial^2}{\partial x_2^2}\frac{\partial^2}{\partial x_1^2} - 2\frac{\partial^2}{\partial x_1\partial x_2}\frac{\partial^2}{\partial x_1\partial x_2} + \frac{\partial^2}{\partial x_1^2}\frac{\partial^2}{\partial x_2^2}\right),\ K(w) = k_w - k_p\left(\frac{\partial^2}{\partial x_1^2} + \frac{\partial^2}{\partial x_2^2}\right) \\
L_{21}(F) &= h\left[q_{11}\frac{\partial^4}{\partial x_2^4} + (q_{12} + q_{21} + q_{31})\frac{\partial^4}{\partial x_1^2\partial x_2^2} + q_{22}\frac{\partial^4}{\partial x_1^4}\right], \\
L_{22}(w) &= -q_{23}\frac{\partial^4}{\partial x_1^4} - (q_{24} + q_{13} - q_{32})\frac{\partial^4}{\partial x_1^2\partial x_2^2} - q_{14}\frac{\partial^4}{\partial x_2^4} + \left(\frac{1}{R_2}\frac{\partial^2}{\partial x_1^2} + \frac{1}{R_1}\frac{\partial^2}{\partial x_2^2}\right), \\
L_{23}(w,w) &= -\left(\frac{\partial^2}{\partial x_1\partial x_2}\right)^2 + \frac{\partial^2}{\partial x_1^2}\frac{\partial^2}{\partial x_2^2}
\end{aligned} \tag{10}$$

being t the time variable, and $\rho_1 = \int_{-h/2}^{h/2} \rho_1^{\overline{x}_3} d\overline{x}_3$. Moreover, u_{ij} are defined as

$$\begin{aligned}
&u_{11} = A_{11}^1 q_{11} + A_{12}^1 q_{21},\ u_{12} = A_{11}^1 q_{12} + A_{12}^1 q_{11},\ u_{13} = A_{11}^1 q_{13} + A_{12}^1 q_{23} + A_{11}^2, \\
&u_{14} = A_{11}^1 q_{14} + A_{12}^1 q_{24} + A_{12}^2,\ u_{21} = A_{21}^1 q_{11} + A_{22}^1 q_{21},\ u_{22} = A_{21}^1 q_{12} + A_{22}^1 q_{22}, \\
&u_{23} = A_{21}^1 q_{13} + A_{22}^1 q_{23} + A_{21}^1,\ u_{24} = A_{21}^1 q_{14} + A_{22}^1 q_{24} + A_{22}^1,\ u_{31} = A_{66}^1 q_{35}, \\
&u_{32} = A_{66}^1 q_{32} + 2A_{66}^2.
\end{aligned} \tag{11}$$

with

$$\begin{aligned}
&q_{11} = \frac{A_{22}^0}{\Pi},\ q_{12} = -\frac{A_{12}^0}{\Pi},\ q_{13} = \frac{A_{12}^0 A_{21}^1 - A_{11}^1 A_{22}^0}{\Pi},\ q_{14} = \frac{A_{12}^0 A_{22}^1 - A_{12}^1 A_{22}^0}{\Pi},\ q_{21} = -\frac{A_{21}^0}{\Pi},\ q_{22} = \frac{A_{11}^0}{\Pi}, \\
&q_{23} = \frac{A_{11}^1 A_{21}^0 - A_{21}^1 A_{11}^0}{\Pi},\ q_{24} = \frac{A_{12}^1 A_{21}^0 - A_{22}^1 A_{11}^0}{\Pi},\ q_{31} = \frac{1}{A_{66}^0},\ q_{32} = -\frac{2A_{66}^1}{A_{66}^0},\ \Pi = A_{11}^0 A_{22}^0 - A_{12}^0 A_{21}^0, \\
&A_{ij}^{i_1} = \int_{-h/2}^{h/2} E_{kk}^{\overline{x}_3} \overline{x}_3{}^{i_1} d\overline{x}_3, (k = 1,2,6;\ i_1 = 0,1,2),\ A_{ij}^{i_1} = \int_{-h/2}^{h/2} \nu_{ji} E_{ii}^{\overline{x}_3} \overline{x}_3{}^{i_1} d\overline{x}_3, (i,j = 1,2).
\end{aligned} \tag{12}$$

3. Solution Procedure

In what follows, we provide an analytical solution to the problem of a simply-supported doubly-curved shell. Thus, the structural deflection can be approximated as [21,36]

$$w = \overline{w}(t)\ \sin(\alpha_1 x_1)\ \sin(\alpha_2 x_2) \tag{13}$$

where $\overline{w}(t)$ is a function of time, $\alpha_1 = \frac{m\pi}{a}$, $\alpha_2 = \frac{n\pi}{b}$, in which m and n are the wave numbers in directions x_1 and x_2, respectively. By substitution of Equation (13) into Equation (9), we get the following expression for the stress function F

$$F = \overline{w}(t)[c_1\overline{w}(t)\cos(2\alpha_1 x_1) + c_2\overline{w}(t)\cos(2\alpha_2 x_2) + c_3\sin(\alpha_1 x_1)\sin(\alpha_2 x_2)] \tag{14}$$

where $c_j(j = 1,2,3)$ are defined as

$$c_1 = \frac{\alpha_2^2}{32\alpha_1^2 q_{22} h},\ c_2 = \frac{\alpha_1^2}{32\alpha_2^2 q_{11} h},\ c_3 = \frac{q_{23}\alpha_1^4 + (q_{24} + q_{13} - q_{32})\alpha_1^2\alpha_2^2 + q_{14}\alpha_2^4 + \alpha_1^2/R_2 + \alpha_2^2/R_1}{h[q_{11}\alpha_2^3 + (q_{12} + q_{21} + q_{31})\alpha_1^2\alpha_2^2 + q_{22}\alpha_1^4]} \tag{15}$$

By substituting Equations (13) and (15) into Equation (8) and by applying the Galerkin procedure in the domain $0 \leq x_1 \leq a$ and $0 \leq x_2 \leq b$, we obtain

$$L(t) \equiv \frac{d^2\widetilde{w}(t)}{dt^2} + \left(\omega_{wp}^L\right)\widetilde{w}(t) + \theta_1\widetilde{w}^2(t) + \theta_2\widetilde{w}^3(t) = 0 \tag{16}$$

where $\widetilde{w}(t) = \overline{w}(t)/h$, the quantities $\theta_1,\ \theta_2$ are defined as

$$\begin{gathered}\theta_1 = \frac{64h^2}{3ab}\frac{1}{\rho_1}\left(\frac{u_{12}\alpha_1^4 c_1 + u_{21}\alpha_2^4 c_2}{\alpha_2\alpha_1} - \frac{\alpha_1\alpha_2 c_3}{8} - \frac{c_1}{4R_2}\frac{\alpha_1}{\alpha_2} - \frac{c_2}{4R_1}\frac{\alpha_2}{\alpha_1}\right)\left[1 - (-1)^m - (-1)^n + (-1)^{m+n}\right] \\ \theta_2 = \frac{2h^3\alpha_1^2\alpha_2^2(c_1 + c_2)}{\rho_1}\end{gathered} \tag{17}$$

and ω_{wp}^L is the frequency associated to the shallow structure resting on the PF at small deflections, defined as

$$\begin{aligned}\omega_{wp}^L = \frac{1}{\sqrt{\rho_1}}\Bigg\{&\left[\frac{\alpha_1^2}{R_2} + \frac{\alpha_2^2}{R_1} - u_{12}\alpha_1^4 - (u_{11} - 2u_{31} + u_{22})\alpha_1^2\alpha_2^2 - u_{21}\alpha_2^4\right]hs_3 \\ &+u_{13}\alpha_1^4 + (u_{14} + 2u_{32} + u_{24})\alpha_1^2\alpha_2^2 + u_{24}\alpha_2^4 + k_w + k_p(\alpha_1^2 + \alpha_2^2)\Bigg\}^{1/2}\end{aligned} \tag{18}$$

The approximate solution of Equation (16) reads as follows

$$\widetilde{w}(t) = w_0 \cos(\varpi_{NL} t) \tag{19}$$

where w_0 is the dimensionless amplitude, ϖ_{NL} is the nonlinear frequency and the initial conditions are defined as

$$\widetilde{w}(t)\Big|_{t=0} = w_0 \text{ and } \frac{d\widetilde{w}(t)}{dt}\Big|_{t=0} = 0 \tag{20}$$

By combining the relations (16) and (19), we obtain an equation of the type $L(t) = 0$. Thus, by applying the Grigolyuk method [37], one obtains

$$\int_0^{\pi/2\varpi_{NL}} L(t)\cos(\varpi_{NL} t)dt = 0 \tag{21}$$

After integrating this last relation, we obtain the following nonlinear amplitude–frequency dependence

$$\varpi_{wp}^{NL} = \sqrt{\left(\omega_{wp}^L\right)^2 + \frac{8}{3\pi}\theta_1 w_0 + 0.75\theta_2 w_0^2} \tag{22}$$

and

$$\varpi_{1wp}^{NL} = \varpi_{wp}^{NL} h\sqrt{\rho^m/Y^m} \tag{23}$$

The rational nonlinear-to-linear free vibration frequency (NLFVF / LFVF), $\varpi_{wp}^{NL}/\omega_L$, becomes

$$\frac{\varpi_{wp}^{NL}}{\omega_L} = \sqrt{\left(\frac{\omega_{wp}^L}{\omega_L}\right)^2 + \frac{8}{3\pi}\frac{\theta_1 w_0}{\omega_L^2} + \frac{0.75\theta_2 {w_0}^2}{\omega_L^2}} \tag{24}$$

where the linear-free vibration frequency ω_L for the unconstrained structure is defined as (18) and $k_w = k_p = 0$ represents a special case. Based on Equations (22) and (24), we can treat different cases, namely shallow spherical shells (for $R_1 = R_2$) or shallow hypar shells (for $R_1 = -R_2$) resting on a PF, as well as shallow cylindrical panels ($R_1 \to \infty$) or plates ($R_1 \to \infty,\ R_2 \to \infty$), on a WPF.

The lowest values of ϖ_{wp}^{NL}, ϖ_{1wp}^{NL} and $\frac{\varpi_{wp}^{NL}}{\omega_L}$ for shallow spherical and hypar shells on a PF are determined by minimizing Equations (20)–(22) depending on the vibration modes (m, n), for fixed values of the dimensionless amplitude w_0.

4. Results and Discussion

The numerical investigation starts with a comparative evaluation of the dimensionless linear frequency parameters, $\omega_{1L} = \omega_L h\sqrt{\rho^m/Y^m}$, for isotropic shallow shells with respect to predictions from the literature [38] (see Table 1).

Table 1. Comparative evaluation with the literature of ω_{1L} for different shallow structural members made of an isotropic material.

Structural Members	$\frac{a}{R_1}$	$\frac{b}{R_2}$	$\omega_{1L}=\omega_L h\sqrt{\rho^m/Y^m}$	
			Alijani [38]	Present Study
Spherical shell	0.5	0.5	0.0779	0.0781
Hypar shell	0.5	−0.5	0.0597	0.0600

Thus, we use the expression (18), while keeping $k_w = k_p = 0$, $Y_{11}^m = Y_{22}^m = Y^m = 70$ Gpa, $\nu_{12} = \nu_{21} = \nu^m = 0.3177$,$\rho^m = 2.702 \times 10^3\,\mathrm{kg/m^3}$ and the geometrical ratios a/b = 1, a/h = 10. As visible from Table 1, our results match very well predictions from [38], for both spherical and hypar shell members; this proves the reliability and consistency of the proposed formulation. A further comparison with the literature [39,40] is also provided in terms of dimensionless linear frequencies for isotropic square plates resting on PFs with $h/b = 0.01$, $a/b = 1$, $k_w = 100D^m$ and $k_p = 10D^m$. For this subcase, the relation (16) is computed for $R_1 \to \infty$, $R_2 \to \infty$ and the dimensionless linear frequency parameter is determined as $\Omega^{Lwp} = \omega_{wp}^L (b/\pi)^2\sqrt{\rho^m h/D^m}$ in which $D^m = \frac{Y^m h^3}{12\left[1-(\nu^m)^2\right]}$, see [40]. Table 2 summarizes the results based on different approaches, with a consistent agreement between our formulation and findings from [39,40].

Table 2. Comparison of dimensionless frequency parameters for square plates on a PF ($h/b = 0.01$, $a/b = 1$, $k_w = 100D^m$, $k_p = 10D^m$).

Studies	Mode Number		
	$\Omega_{1,1}^{Lwp}$	$\Omega_{1,2}^{Lwp}$	$\Omega_{2,1}^{Lwp}$
Zhou et al. [39]	2.6551	5.5717	5.5717
Wang et al. [40]	2.6551	5.5717	5.5717
Present study	2.6557	5.5761	5.5761

After this preliminary validation, we continue the analysis by computing the NLFVF of shallow spherical and hypar shells reinforced with a uniform and linear distribution of CNTs, and resting on a WF and PF. The selected shell members are made of polymethyl methacrylate (PMMA), as matrix, and single-walled CNTs, with geometrical properties $r_1 = 9.26\,\mathrm{nm}$, $a_1 = 6.8 \times 10^{-1}\,\mathrm{nm}$, $h_1 = 6.7 \times 10^{-2}\,\mathrm{nm}$, as reinforcement. The mechanical properties for the CNT phase are $Y_{11}^{CN} = 5646.6$ Gpa, $Y_{22}^{CN} = 7080$ Gpa, $Y_{12}^{CN} = 1944.5$ Gpa, $\nu_{12}^{CN} = 0.175$, $\rho^{CN} = 1.4 \times 10^3\,\mathrm{kg/m^3}$; for PMMA, it is $Y^m = 2.5$ Gpa, $\nu^m = 0.34$, $\rho^m = 1.15 \times 10^3\,\mathrm{kg/m^3}$. In line with [9], we also consider different efficiency parameters of CNT/matrix depending on the selected value of V_{CN}^*, as summarized in Table 3. As also listed in Table 4, we check for the variation of NLFVFs for both the selected shallow shells, while keeping different distributions of CNTs (i.e., UD, VD, OD and XD), and by varying the stiffness constants (k_w, k_p) for the elastic foundation under the three fixed values of V_{CN}^* (0.12, 0.17, and 0.28). The frequency values are computed for $R_1 = 20h$, $a = b$, $a = 20h$, $(m, n) = (1, 1)$ and $w_0 = 1.5$, with a clear increase in results for an increased value of the stiffness parameters, k_p and/or k_w, for all the reinforcement assumptions. For fixed values of k_p, k_w, and V^*_{CN}, it also seems that OD and XD patterns of CNTs always provide the lowest and highest frequency values, respectively, independently of the selected shell geometry.

Table 3. Typical properties of CNT/matrix efficiency parameters depending on the volume fraction of CNTs.

V^*_{CN}	η_1	η_2	η_3
0.12	0.137	1.022	0.715
0.17	0.142	1.626	1.138
0.28	0.141	1.585	1.109

Table 4. Variation in NLFVF for shallow spherical and hypar shells with CNTs resting on a W-EF and PF with various foundation elastic parameters, versus V^*_{CN}.

		$\omega^{NL}_{1wp} \times 10$ **(R_2=R_1)**											
	V^*_{CN}	**0.12**				**0.17**				**0.28**			
k_p	k_w	*UD*	*VD*	*OD*	*XD*	*UD*	*VD*	*OD*	*XD*	*UD*	*VD*	*OD*	*XD*
0	0	0.445	0.388	0.362	0.540	0.569	0.506	0.476	0.680	0.615	0.517	0.493	0.782
	0.7×10^9	0.503	0.452	0.431	0.589	0.614	0.557	0.530	0.719	0.656	0.566	0.544	0.815
0	1.0×10^9	0.526	0.478	0.457	0.608	0.633	0.577	0.551	0.735	0.673	0.586	0.564	0.829
	1.3×10^9	0.548	0.502	0.482	0.627	0.651	0.597	0.572	0.750	0.690	0.605	0.584	0.843
	0.7×10^9	0.583	0.540	0.522	0.658	0.681	0.629	0.605	0.776	0.717	0.636	0.66	0.685
9×10^4	1.0×10^9	0.602	0.561	0.544	0.676	0.697	0.647	0.624	0.791	0.733	0.653	0.634	0.878
	1.3×10^9	0.622	0.581	0.565	0.693	0.714	0.665	0.642	0.805	0.748	0.670	0.652	0.891
	0.7×10^9	0.599	0.557	0.540	0.672	0.694	0.644	0.621	0.788	0.730	0.650	0.631	0.876
11×10^4	1.0×10^9	0.618	0.578	0.561	0.690	0.711	0.662	0.639	0.803	0.745	0.667	0.649	0.889
	1.3×10^9	0.637	0.598	0.581	0.706	0.727	0.679	0.657	0.817	0.760	0.684	0.666	0.901
	0.7×10^9	0.615	0.574	0.557	0.687	0.708	0.659	0.636	0.800	0.743	0.664	0.646	0.886
13×10^4	1.0×10^9	0.634	0.594	0.578	0.703	0.724	0.676	0.654	0.815	0.758	0.681	0.663	0.899
	1.3×10^9	0.652	0.614	0.598	0.720	0.740	0.693	0.671	0.829	0.773	0.698	0.680	0.912
		$\omega^{NL}_{1wp} \times 10$ **(R_2=$-R_1$)**											
0	0	0.911	0.878	0.878	0.960	1.095	1.055	1.055	1.155	1.376	1.325	1.324	1.453
	0.7×10^9	0.941	0.909	0.909	0.988	1.119	1.080	1.080	1.178	1.395	1.345	1.344	1.471
0	1.0×10^9	0.953	0.921	0.922	1.000	1.129	1.091	1.091	1.188	1.403	1.353	1.352	1.479
	1.3×10^9	0.965	0.934	0.934	1.012	1.140	1.101	1.101	1.198	1.411	1.361	1.361	1.486
	0.7×10^9	0.986	0.955	0.955	1.031	1.157	1.119	1.119	1.214	1.425	1.375	1.375	1.499
9×10^4	1.0×10^9	0.997	0.967	0.967	1.043	1.167	1.129	1.129	1.223	1.433	1.384	1.383	1.507
	1.3×10^9	1.009	0.979	0.979	1.054	1.177	1.139	1.140	1.233	1.440	1.392	1.391	1.514
	0.7×10^9	0.995	0.965	0.965	1.041	1.165	1.127	1.128	1.122	1.431	1.382	1.382	1.505
11×10^4	1.0×10^9	1.007	0.977	0.977	1.052	1.175	1.138	1.138	1.231	1.439	1.390	1.390	1.513
	1.3×10^9	1.018	0.989	0.989	1.063	1.185	1.148	1.148	1.240	1.447	1.398	1.398	1.520
	0.7×10^9	1.005	0.975	0.975	1.050	1.173	1.136	1.136	1.229	1.438	1.389	1.388	1.512
13×10^4	1.0×10^9	1.016	0.987	0.987	1.061	1.183	1.146	1.146	1.239	1.445	1.397	1.396	1.519
	1.3×10^9	1.028	0.999	0.999	1.072	1.193	1.156	1.156	1.248	1.453	1.405	1.405	1.527

The highest sensitivity of the response to the CNT dispersion within the matrix is observed for a fixed value of $V^*_{CN} = 0.28$, with a maximum percentage variation with respect to a UD of 21.6%. At the same time, the largest foundation effect on ω^{NL}_{1wp} occurs at $V^*_{CN} = 0.12$ and OD patterns with a percentage variation of 51.5%. A PF also seems to affect the response more significantly compared to a W-EF, reaching the highest sensitivity with an OD-type reinforcement and $V^*_{CN} = 0.12$, whereas the lowest sensitivity is obtained for a XD of CNTs with $V^*_{CN} = 0.28$.

As far as the sensitivity to the volume fraction is concerned, the largest influence is noticed for structures on a PF reinforced by XD CNTs with $V^*_{CN} = 0.28$ and the lowest effect is obtained with an OD pattern and $V^*_{CN} = 0.12$, respectively. In Figure 3, we plot the variation in NLFVFs for shallow spherical and hypar shells reinforced with a UD and VD of CNTs versus w_0, for a fixed value of $V^*_{CN} = 0.28$, for three different geometrical ratios $R_1/a = 2.0, 2.5, 3.0$, accounting (or not) for the presence of a surrounding WPF. The other parametric data are: $a/b = 0.5$, $a/h = 15$, $(m,n) = (1,1)$, $k_w = 4 \times 10^9 (\mathrm{N/m^3})$ and $k_p = 1.6 \times 10^4 (\mathrm{N/m})$. As visible in Figure 3, the NLFVF for hypar shells resting on a PF increases monotonically with w_0, whereas it varies non-monotonically for spherical shells on a PF with an initial decrease for $w_0 \leq 0.5$, and a further increase for $w_0 > 0.5$. By comparing results among unconstrained spherical and hypar members, it seems that NLFVFs for hypar shells always reach higher values than spherical ones for all w_0; the NLFVFs for shallow spherical shells are usually higher for $w_0 \leq 0.5$.

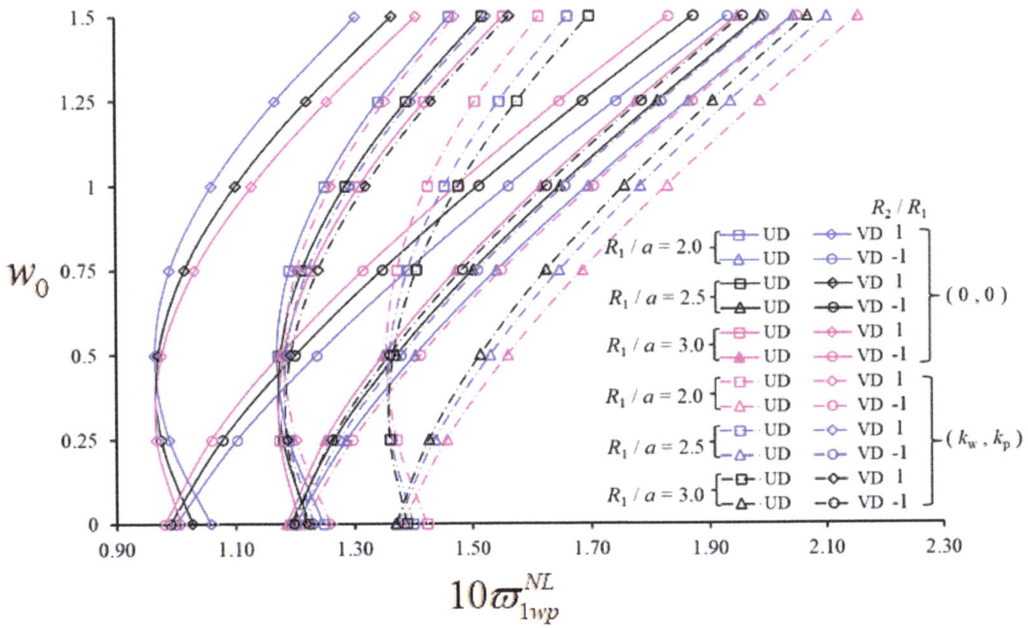

Figure 3. Variation in NLFVF for shallow spherical and hypar shells with UD- and VD-patterned CNTs on a PF versus w_0, with different geometrical ratios R_1/a.

The magnitude of NLFVFs for shell members with and without a PF can vary significantly under the same geometrical assumption R_1/a and the same value of w_0. In addition, for an increasing rational value of R_1/a, the NLFVF values of spherical shells decrease for $w_0 \leq 0.5$ and increase for $w_0 > 0.5$. The influence of CNT patterns on the NLFVF of hypar shells in the presence, or not, of a PF, decreases with a varying w_0. For spherical shells, instead, such an effect increases for $w_0 \leq 0.5$ and decreases for $w_0 > 0.5$. Such sensitivity becomes more pronounced for $w_0 > 0.5$. More specifically, the influence of CNT patterns on the NLFVF for unconstrained hypar shells decreases from −16.71% to −5.41% due to the increase in w_0. For unconstrained spherical shells, the influence of CNT patterns increases from −15.20% to −17.51% in the range of $w_0 \leq 0.5$, and decreases from −17.51% to −10.82% for $w_0 > 0.5$, under a fixed ratio $R_1/a = 2$. The influence of different CNT patterns is more pronounced for unconstrained hypar shells, with the largest difference being approximately 1.00%; for spherical shells on a PF, the largest difference becomes approximately 1.8% due to an increased ratio R_1/a.

The influence of CNT patterns on NLFVFs reduces with a maximum percentage of 4.76% and 4.61%, for spherical and hypar shells on PF, respectively. A pronounced effect of

CNT patterns is also observed for both shallow shells in the presence, or not, of a PF, which is quantified as a percentage by 4.63% and 7.19%, respectively. The effect of a PF on NLFVF for both shells is approximately 6% greater for a VD pattern compared to a UD pattern.

Variation in the NLFVF / LFVF ratio, for shallow spherical and hypar shells with UD and OD patterns versus w_0, is plotted in Figure 4, for $V_{CN}^* = 0.12; 0.17; 0.28$, while keeping $a/h = 10$ $R_1/a = 2$, $a/b = 2$, $(m,n) = (1,1)$, $k_w = 3 \times 10^9 (\text{N/m}^3)$ and $k_p = 1.5 \times 10^4 (\text{N/m})$. As clearly visible in Figure 4, the NLFVF / LFVF ratio varies nonmonotonically for shallow spherical shells and monotonically for hypar shells, with an increased value of w_0 for all selected values of V_{CN}^*.

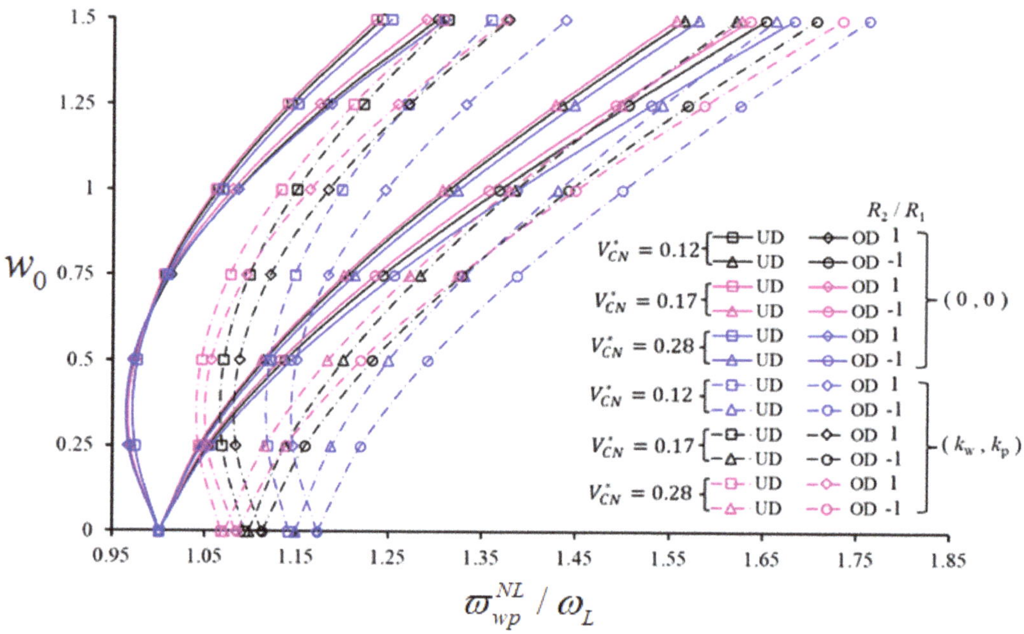

Figure 4. Variation in the NLFVF / LFVF ratio for shallow spherical and hypar shells with UD- and OD-patterned CNTs, in the presence/absence of a PF versus w_0, with different V_{CN}^*.

Based on a comparison of the NLFVF / LFVF ratio for hypar shells with UD and OD patterns, in presence or absence of a PF, a higher variation is noticed for an OD pattern. More specifically, the NLFVF / LFVF ratio for spherical shells with an OD pattern becomes higher for all w_0 in presence of a PF, and for $w_0 > 0.5$ in the absence of a surrounding elastic medium. When the NLFVF / LFVF ratio is evaluated comparatively for spherical shells with different V_{CN}^*, the largest NLFVF / LFVF ratio for unconstrained spherical shells occurs at $V_{CN}^* = 0.28$, and for spherical shells on PF at $V_{CN}^* = 0.12$. For hypar shells in presence or not of a surrounding elastic medium, the highest NLFVF / LFVF ratio is always obtained for $V_{CN}^* = 0.28$. It is also noticeable that this ratio becomes higher for hypar shells with and without the PF, as spherical and hypar shells are compared. The pattern effect on the NLFVF / LFVF ratio increases for unconstrained spherical shells (4%) and unconstrained hypar shells (5%), whereas the pattern effect on the $\varpi_{wp}^{NL} / \omega_L$ ratio in both shells on PF increases with w_0, accordingly. The most pronounced increase seems to be approximately equal to 2.5% for spherical shells, and approximately equal to 2.1% for hypar shells. In absence of a surrounding elastic medium, the influence of a CNT pattern on $\varpi_{wp}^{NL} / \omega_L$ for hypar shells becomes 1.9% more pronounced than unconstrained spherical shells on a PF.

The effect of a PF on $\varpi_{wp}^{NL}/\omega_L$ ratio decreases for an increased value of w_0, for both spherical and hypar shells, with a maximum percentage variation of 6% and 10%, respectively.

The variation in $\varpi_{wp}^{NL}/\omega_L$ with the PF is about 3.5% for spherical shells with an OD pattern of CNTs; this effect is 2.4% more pronounced for hypar shells. In addition, for a reinforcement phase with $V_{CN}^* = 0.12$, the percentage variation of $\varpi_{wp}^{NL}/\omega_L$ for both shells under a PF is approximately 6% (or 9%) greater than that one for $V_{CN}^* = 0.17$ (or $V_{CN}^* = 0.28$).

Figure 5 shows the variation in the NLFVF / LFVF ratio of spherical shells on PF, reinforced with UD- and XD-patterned CNTs, against w_0, for $V_{CN}^* = 0.17$. In this parametric study we also consider different values of a/b (i.e., $a/b = 0.5$, 1.0, 1.5), along with $R_1/a = 2$, $a/h = 15$, $(m, n) = (1, 1)$, $k_w = 3 \times 10^9$(Pa/m) and $k_p = 1.5 \times 10^5$(Pa.m). As visible in Figure 5, for an increased value of a/b, the NLFVF / LFVF ratios of spherical shells with and without a PF vary nonmonotonically, with an increase after an initial decrease up to a minimum value. The NLFVF / LFVF ratio for UD patterns is larger than XD patterns for all w_0 (in presence of PF) and for $w_0 > 1$ (in absence of an elastic ground). The same ratio, for XD patterns, is larger than UD patterns, as $w_0 \leq 1$ only in the absence of ground.

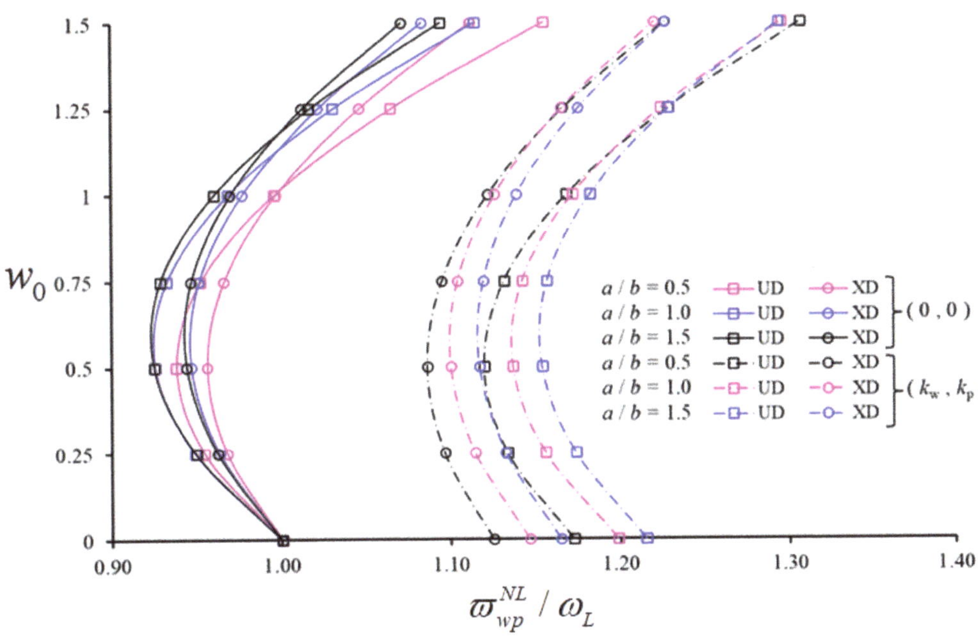

Figure 5. Variation in the NLFVF / LFVF ratio for shallow spherical shells containing UD- and XD-patterned CNTs in the presence/absence of a PF versus w_0, with different values of a/b.

For an increased value of a/b, the NLFVF / LFVF ratio of unconstrained spherical shells decreases for both patterns, while it decreases (or increases) when $w_0 > 1.25$ (or $w_0 \leq 1.25$), for spherical shells on a PF. It is also observed that XD patterns effect on the NLFVF / LFVF ratio is higher in presence of a PF; it decreases/increases depending on the value of w_0, while it continuously decreases depending on the increase in a/b. Although the effect of PF on NLFVF / LFVF ratio is more pronounced for UD patterns, it decreases depending on the increase in w_0 and increases for an increased value of a/b. The minimum and maximum influence of PF on the NLFVF / LFVF ratio for spherical shells corresponds to a percentage variation of 10.24% and 24.71%, respectively.

Finally, in Figure 6 we plot the variation of NLFVF / LFVF ratio for UD- and VD-patterned spherical and hypar shells (in the presence or absence of a PF), versus w_0 for different R_1/a ratios (i.e., $R_1/a = 2.0, 2.5, 3.0$), while keeping $V_{CN}^* = 0.28$, $a/b = 0.5$, $a/h = 15$, $(m,n) = (1,1)$, $k_w = 3 \times 10^9 (\mathrm{N/m^3})$ and $k_p = 1.5 \times 10^5 (\mathrm{N/m})$. The NLFVF / LFVF ratio of hypar shells on PF increases with w_0, while it varies nonmonotonically with R_1/a. Similarly, the NLFVF / LFVF ratio of spherical shells on PF decreases first and then increases for an increased value of w_0, while always increasing for an increased R_1/a ratio. The NLFVF / LFVF ratio of hypar shells for VD patterns with and without a PF, as well as for spherical shells resting on a PF, is greater than the same shells reinforced uniformly by CNTs. The NLFVF / LFVF ratio of unconstrained spherical shells with VD patterns is higher than those with UD patterns, at least when $w_0 \leq 0.75$; the contrary occurs for $w_0 > 0.75$. Based on a comparative evaluation of both geometries, the NLFVF / LFVF ratios of UD- and VD-reinforced spherical shells on a PF are lower than hypar shells. The influence of VD patterns on $\varpi_{wp}^{NL}/\omega_L$ for spherical shells is higher than hypar shells, with a maximum increase of 3.2% or 0.6%, respectively, depending on the increase in R_1/a. Looking at the influence of the foundation on $\varpi_{wp}^{NL}/\omega_L$ for spherical shells with a UD of CNTs, it decreases first, up to a minimum value, then increases for an increased value of w_0. A monotonic decrease is differently observed for hypar shells. Depending on the increase of R_1/a, the effect of PF is lower than 2% for spherical shells, and lower than 1.5% for hypar shells.

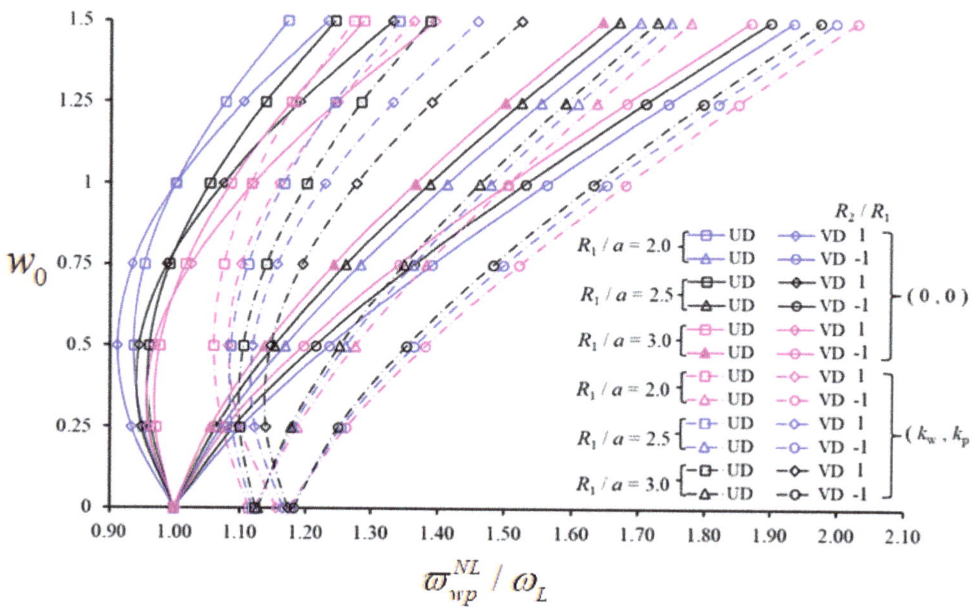

Figure 6. Variation in the NLFVF/LFVF ratio for shallow spherical and hypar shells with UD- and VD-patterned CNTs, in the presence/absence of a PF versus w_0 with different values of R_1/a.

5. Conclusions

In this work, the Donnell's nonlinear shell theory is applied to study the free vibration behavior of composite shell structures reinforced by uniform and linearly patterned CNTs resting on a PF. Once the basic relations for composite shallow shells reinforced by CNTs on WPFs are established, the partial differential equations of nonlinear motion are derived, taking into account the von Karman nonlinearity. These equations are solved here by means of the Galerkin and Grigolyuk methods in terms of linear and nonlinear free vibrations for inhomogeneous nanocomposite construction members such as plates, panels, spherical and hyperbolic paraboloidal (hypar) shallow shells. The accuracy of the results in the current study has been confirmed by means of a successful comparison with reliable predictions from the literature. After this preliminary validation, a detailed numerical analysis is performed, including the effect of nonlinearity, CNT patterns and volume fraction on the nonlinear frequency response. Based on a large systematic investigation, the analytical results could serve as valid benchmark solutions for further computational studies on the topic, as well as for design purposes. Among the most useful insights, it is found that the variation rate of NLFVFs for both shallow shells with linearly patterned CNTs decreases, while remaining constant for different elastic foundations with an increased stiffness. For both shallow shells, a single- or dual-parameter elastic foundation yields an increase in NLFVFs, where the NLFVF decreases for VD, OD and XD patterns, as foundation coefficients increase. Moreover, the influence of PF on the ω_{1wp}^{NL} for shallow spherical and hypar shells reinforced with CNTs has revealed as more pronounced than that of a WEF. The highest influence of PF on NLFVF values is observed with an OD pattern of CNTs for $V_{CN}^{*} = 0.12$, whereas the smallest effect is observed with an XD pattern of CNTs for $V_{CN}^{*} = 0.28$, respectively, when the influences of PF on NFVFs for spherical or hypar shells are compared to each other. Based on a comparative evaluation of the nonlinear vibration response for both shallow shells, the largest effect of the PF on NLFVFs is observed for a XD CNT-based reinforcement with a volume fraction, $V_{CN}^{*} = 0.28$, whereas the smallest effect occurs for an OD pattern and $V_{CN}^{*} = 0.12$. At the same time, the NLFVF of hypar shells on PF increases continuously for an increased w_0; it decreases when $w_0 \leq 0.5$ for spherical shells on PF and increases for $w_0 > 0.5$. The pattern effect on $\omega_{wp}^{NL}/\omega_L$ ratio for spherical shells (4%) and hypar shells (5%) increases with w_0 in absence of a PF. Based on the parametric study, the influence of the PF on $\omega_{wp}^{NL}/\omega_L$ ratio seems to be more pronounced for a UD pattern, but it decreases depending on the increase in w_0 and increases for an increased geometrical ratio a/b. Moreover, the rational value of $\omega_{wp}^{NL}/\omega_L$ for a UD pattern is higher than a XD pattern, for all w_0 in presence of an elastic medium, and for $w_0 > 1$ in absence of an elastic medium. The same ratio for an XD pattern is larger than the one for a UD pattern, as $w_0 \leq 1$ only in absence of an elastic ground. Finally, for a VD pattern, the rational value of $\omega_{wp}^{NL}/\omega_L$ for spherical shells is greater than the hypar shells, with a maximum increase of 3.2% in lieu of 0.6%, as found for hypar shells, depending on the increase in the geometrical ratio R_1/a.

Author Contributions: Conceptualization, A.M., F.T., R.D. and N.K.; Formal analysis, A.M., F.T., R.D. and N.K.; Investigation, A.M. and N.K.; Validation, A.M., F.T., R.D. and N.K.; Writing—Original Draft, A.M. and N.K.; Writing—Review & Editing, F.T. and R.D. All authors have read and agreed to the published version of the manuscript.

Funding: This research received no external funding.

Conflicts of Interest: The authors declare no conflict of interest.

References

1. Iijima, S.; Ichihashi, T. Single-shell carbon nanotubes of 1-nm diameter. *Nat. Cell Biol.* **1993**, *363*, 603–605. [CrossRef]
2. Xie, S.; Li, W.; Pan, Z.; Chang, B.; Sun, L. Mechanical and physical properties on carbon nanotube. *J. Phys. Chem. Solids* **2000**, *61*, 1153–1158. [CrossRef]
3. Curtin, W.A.; Sheldon, B.W. CNT-reinforced ceramics and metals. *Mater. Today* **2004**, *7*, 44–49. [CrossRef]
4. Esawi, A.M.; Farag, M.M. Carbon nanotube reinforced composites: Potential and current challenges. *Mater. Des.* **2007**, *28*, 2394–2401. [CrossRef]
5. Sofiyev, A.H.; Tornabene, F.; Dimitri, R.; Kuruoglu, N. Buckling Behavior of FG-CNT Reinforced Composite Conical Shells Subjected to a Combined Loading. *Nanomaterials* **2020**, *10*, 419. [CrossRef]
6. Silvestre, J.; Silvestre, N.; De Brito, J. Polymer nanocomposites for structural applications: Recent trends and new perspectives. *Mech. Adv. Mater. Struct.* **2016**, *23*, 1263–1277. [CrossRef]
7. Park, S.-H.; Bae, J. Polymer Composite Containing Carbon Nanotubes and their Applications. *Recent Patents Nanotechnol.* **2017**, *11*, 109–115. [CrossRef] [PubMed]
8. Liew, K.M.; Pan, Z.; Zhang, L.-W. The recent progress of functionally graded CNT reinforced composites and structures. *Sci. China Ser. G Phys. Mech. Astron.* **2019**, *63*, 1–17. [CrossRef]

9. Shen, H.-S.; Xiang, Y. Nonlinear vibration of nanotube-reinforced composite cylindrical shells in thermal environments. *Comput. Methods Appl. Mech. Eng.* **2011**, *213–216*, 196–205. [CrossRef]
10. Lei, Z.; Zhang, L.; Liew, K. Vibration analysis of CNT-reinforced functionally graded rotating cylindrical panels using the element-free kp-Ritz method. *Compos. Part B Eng.* **2015**, *77*, 291–303. [CrossRef]
11. Wang, Q.; Qin, B.; Shi, D.; Liang, Q. A semi-analytical method for vibration analysis of functionally graded carbon nanotube reinforced composite double-curved panels and shells of revolution. *Compos. Struct.* **2017**, *174*, 87–109. [CrossRef]
12. Wang, Q.; Cui, X.; Qin, B.; Liang, Q. Vibration analysis of the functionally graded carbon nanotube reinforced composite shallow shells with arbitrary boundary conditions. *Compos. Struct.* **2017**, *182*, 364–379. [CrossRef]
13. Zghal, S.; Frikha, A.; Dammak, F. Free vibration analysis of carbon nanotube-reinforced functionally graded composite shell structures. *Appl. Math. Model.* **2018**, *53*, 132–155. [CrossRef]
14. Sobhy, M. Magneto-electro-thermal bending of FG-graphene reinforced polymer double-curved shallow shells with piezoelectromagnetic faces. *Compos. Struct.* **2018**, *203*, 844–860. [CrossRef]
15. Radwan, A.F.; Sobhy, M. Transient instability analysis of viscoelastic sandwich CNTs-reinforced microplates exposed to 2D magnetic field and hygrothermal conditions. *Compos. Struct.* **2020**, *245*, 112349. [CrossRef]
16. Vu, V.T.; Tran, H.Q.; Tran, M.T. Free vibration analysis of laminated functionally graded carbon nanotube-reinforced composite doubly curved shallow shell panels using a new four-variable refined theory. *J. Compos. Sci.* **2019**, *3*, 104.
17. Sofiyev, A.H.; Mammadov, Z.; Dimitri, R.; Tornabene, F. Vibration analysis of shear deformable carbon nanotubes-based functionally graded conical shells resting on elastic foundations. *Math. Methods Appl. Sci.* **2020**. [CrossRef]
18. Babaei, H.; Kiani, Y.; Eslami, M. Geometrically nonlinear analysis of functionally graded shallow curved tubes in thermal environment. *Thin-Walled Struct.* **2018**, *132*, 48–57. [CrossRef]
19. Nguyen, P.D.; Quang, V.D.; Anh, V.T.T.; Duc, N.D. Nonlinear Vibration of Carbon Nanotube Reinforced Composite Truncated Conical Shells in Thermal Environment. *Int. J. Struct. Stab. Dyn.* **2019**, *19*. [CrossRef]
20. Khoa, N.D.; Anh, V.M.; Duc, N.D. Nonlinear dynamic response and vibration of functionally graded nanocomposite cylindrical panel reinforced by carbon nanotubes in thermal environment. *J. Sandw. Struct. Mater.* **2019**. [CrossRef]
21. Avey, M.; Yusufoglu, E. On the solution of large-amplitude vibration of carbon nanotube-based double-curved shallow shells. *Math. Methods Appl. Sci.* **2020**. [CrossRef]
22. Yang, J.; Huang, X.; Shen, H.S. Nonlinear vibration of temperature-dependent FG-CNTRC laminated plates with negative Poisson's ratio. *Thin-Wall Struct.* **2020**, *148*, 106514. [CrossRef]
23. Zghal, S.; Frikha, A.; Dammak, F. Large deflection response-based geometrical nonlinearity of nanocomposite structures reinforced with carbon nanotubes. *Appl. Math. Mech.* **2020**, *41*, 1227–1250. [CrossRef]
24. Shahmohammadi, M.A.; Abdollahi, P.; Salehipour, H. Geometrically nonlinear analysis of doubly curved imperfect shallow shells made of functionally graded carbon nanotube reinforced composite (FG_CNTRC). *Mech. Based Des. Struct. Mach.* **2020**, 1–25. [CrossRef]
25. Pasternak, P.L. On a new method of analysis of an elastic foundation by means of two foundation constants. *Gos. Izd. Lib. Po Stroit. Arkhitekture Mosc.* **1054**, *1*, 1–56. (In Russian)
26. Kerr, A.D. Elastic and Viscoelastic Foundation Models. *J. Appl. Mech.* **1964**, *31*, 491–498. [CrossRef]
27. Tornabene, F.; Fantuzzi, N.; Viola, E.; Reddy, J. Winkler–Pasternak foundation effect on the static and dynamic analyses of laminated doubly-curved and degenerate shells and panels. *Compos. Part B Eng.* **2013**, *57*, 269–296. [CrossRef]
28. Ansari, R.; Torabi, J.; Shojaei, M. Faghih Vibrational analysis of functionally graded carbon nanotube-reinforced composite spherical shells resting on elastic foundation using the variational differential quadrature method. *Euro J. Mech. A-Solids* **2016**, *60*, 166–182. [CrossRef]
29. Zhang, L.; Liew, K.M. Large deflection analysis of FG-CNT reinforced composite skew plates resting on Pasternak foundations using an element-free approach. *Compos. Struct.* **2015**, *132*, 974–983. [CrossRef]
30. Dinh, D.N.; Nguyen, P.D. The Dynamic Response and Vibration of Functionally Graded Carbon Nanotube-Reinforced Composite (FG-CNTRC) Truncated Conical Shells Resting on Elastic Foundations. *Materials* **2017**, *10*, 1194. [CrossRef] [PubMed]
31. Shen, H.-S.; He, X.-Q. Large amplitude free vibration of nanotube-reinforced composite doubly curved panels resting on elastic foundations in thermal environments. *J. Vib. Control.* **2015**, *23*, 2672–2689. [CrossRef]
32. Babaei, H.; Kiani, Y.; Eslami, M.R. Large amplitude free vibration analysis of shear deformable FGM shallow arches on nonlinear elastic foundation. *Thin-Walled Struct.* **2019**, *144*, 106237. [CrossRef]
33. Sobhy, M.; Zenkour, A.M. Vibration analysis of functionally graded graphene platelet reinforced composite double-curved shallow shells on elastic foundations. *Steel Compos. Struct.* **2019**, *133*, 195–208.
34. Sofiyev, A.H.; Pirmamedov, I.T.; Kuruoglu, N. Influence of elastic foundations and carbon nanotube reinforcement on the hydrostatic buckling pressure of truncated conical shells. *Appl. Math. Mech.* **2020**, *41*, 1011–1026. [CrossRef]
35. Sofiyev, A.; Kuruoglu, N. Buckling analysis of shear deformable composite conical shells reinforced by CNTs subjected to combined loading on the two-parameter elastic foundation. *Def. Technol.* **2021**. [CrossRef]
36. Volmir, A.S. *The Nonlinear Dynamics of Plates and Shells*; Sci Edit: Moscow, Russia, 1972.
37. Grigolyuk, E.I. On vibrations of a shallow circular cylindrical panel experiencing finite deflections. *App. Math. Mech.* **1955**, *19*, 382–386.

38. Alijani, F.; Amabili, M.; Karagiozis, K.; Bakhtiari-Nejad, F. Nonlinear vibrations of functionally graded doubly curved shallow shells. *J. Sound Vib.* **2010**, *330*, 1432–1454. [CrossRef]
39. Zhou, D.; Cheung, Y.K.; Lo, S.H.; Au, F.T.K. Three-dimensional vibration analysis of rectangular thick plates on Pasternak foundation. *Int. J. Numer. Methods Eng.* **2004**, *59*, 1313–1334. [CrossRef]
40. Wang, Q.; Shi, D.; Shi, X. A modified solution for the free vibration analysis of moderately thick orthotropic rectangular plates with general boundary conditions, internal line supports and resting on elastic foundation. *Meccanica* **2015**, *51*, 1985–2017. [CrossRef]

nanomaterials

MDPI

Article

Forced Vibration Analysis of Composite Beams Reinforced by Carbon Nanotubes

Ömer Civalek [1], Şeref D. Akbaş [2], Bekir Akgöz [3] and Shahriar Dastjerdi [3,4,*]

[1] Department of Medical Research, China Medical University Hospital, China Medical University, Taichung 40447, Taiwan; omer@mail.cmuh.org.tw
[2] Department of Civil Engineering, Bursa Technical University, 16310 Bursa, Turkey; seref.akbas@btu.edu.tr
[3] Department of Civil Engineering, Akdeniz University, 07058 Antalya, Turkey; bekirakgoz@akdeniz.edu.tr
[4] Department of Mechanical Engineering, Shahrood Branch, Islamic Azad University, 36199-43189 Shahrood, Iran
* Correspondence: shahriar.dastjerdi@iau-shahrood.ac.ir or dastjerdi_shahriar@yahoo.com

Abstract: This paper presents forced vibration analysis of a simply supported beam made of carbon nanotube-reinforced composite material subjected to a harmonic point load at the midpoint of beam. The composite beam is made of a polymeric matrix and reinforced the single-walled carbon nanotubes with their various distributions. In the beam kinematics, the first-order shear deformation beam theory was used. The governing equations of problem were derived by using the Lagrange procedure. In the solution of the problem, the Ritz method was used, and algebraic polynomials were employed with the trivial functions for the Ritz method. In the solution of the forced vibration problem, the Newmark average acceleration method was applied in the time history. In the numerical examples, the effects of carbon nanotube volume fraction, aspect ratio, and dynamic parameters on the forced vibration response of carbon nanotube-reinforced composite beams are investigated. In addition, some comparison studies were performed, with special results of published papers to validate the using formulations.

Keywords: carbon nanotube-reinforced composite; forced vibration; dynamic analysis; beam; harmonic load

Citation: Civalek, Ö.; Akbaş, Ş.D.; Akgöz, B.; Dastjerdi, S. Forced Vibration Analysis of Composite Beams Reinforced by Carbon Nanotubes. *Nanomaterials* **2021**, *11*, 571. https://doi.org/10.3390/nano11030571

Academic Editor: Raffaele Barretta

Received: 2 February 2021
Accepted: 22 February 2021
Published: 25 February 2021

1. Introduction

Composite material refers to any solid that consists of more than one component, in which they are in separate phases. The main advantages of composite materials are excellent strength-to-weight and stiffness-to-weight ratios. The fibrous composites, consisting of carbon, glass, aramid, and basalt fibers, have a wide range of applications in many modern engineering and industries, such as civil, automotive, bicycle, mechanical, defense, marine, aviation, and aerospace [1–9].

In the early 1990s, carbon nanotubes (CNTs) were studied by Sumio Iijima [10]. CNTs can be considered as one of the key building blocks of nanotechnology. Interest of the scientists and researchers in CNTs' potential engineering and its industrial applications, such as in aerospace, composites, electronics, computers, energy, medicine, sensors, and air and water purifications, have grown rapidly due to the unique material properties of CNTs [11–18].

Due to extraordinary electrical, thermal, and mechanical properties, besides providing good interfacial bonds, CNTs can be ideally suited for the next generation of composite materials and have recently been used instead of the traditional fibers for the reinforcing element in the reinforced composites. Some studies on the bending, buckling, and vibration responses of carbon nanotube-reinforced composite (CNTRC) beams are mentioned below.

Free vibration analysis of functionally graded single-walled CNTs reinforced aluminum alloy beam was performed by Selmi and Bisharat [19]. They obtained the natural frequencies by both the Rayleigh–Ritz method and ANSYS, a software package that implements the finite element method. Pure bending and local buckling behaviors of a nanocomposite beam reinforced by a single-walled carbon nanotube were investigated by Vodenitcharova and Zhang [20]. Ke et al. [21] surveyed the nonlinear free vibration response of functionally graded carbon nanotube-reinforced composite (FG-CNTRC) beams on the basis of first-order shear deformation beam theory with von Kármán geometric nonlinearity assumption. The Ritz method with a direct iterative method was used to obtain solutions for different boundary conditions. Yas and Heshmati [22] investigated the vibrational characteristics of FG-CNTRC beams reinforced by randomly oriented CNTs subjected to a moving load, based on Bernoulli–Euler and Timoshenko beam theories by finite element method. Yas and Samadi [23] analyzed the free vibration and buckling problems of embedded nanocomposite beams resting on an elastic foundation. SWCNTs were used as reinforcement elements. Four different CNTs distributions along the height were taken into consideration. The governing equations were derived by implementing Hamilton's principle, and the resulting equations were solved by generalized differential quadrature method. Deepak et al. [24] developed a spectral finite element formulation for uniform and tapered rotating CNTRC polymer beams. The obtained results were comparatively presented with the results of carbon fiber-reinforced laminated composite rotating beams. Ke et al. [25] examined the dynamic stability analysis of FG-CNTRC beams subjected to axial loading. The related governing equations were derived based on Timoshenko beam theory and they were solved by differential quadrature method.

Heshmati and Yas [26] perused the free vibration characteristics of FG-CNTRC beams by using an equivalent fiber based on the Eshelby–Mori–Tanaka approach via the finite element method. The influences of agglomeration and distribution of carbon nanotubes were investigated in detail. Wattanasakulpong and Ungbhakorn [27] performed the static bending, free vibration, and buckling analyses of embedded CNTRC beams lying on the Pasternak elastic foundation. Analytic solutions were achieved by Navier's solution technique for simply supported CNTRC beams. Linear and nonlinear vibrations of CNTRC beams were comprehensively studied on the basis of first and third-order shear deformation beam theories [28–32]. Mayandi and Jeyaraj [33] investigated the static and dynamic behaviors of FG-CNTRC beams subjected to various non-uniform thermal loads by employing the finite element method. Fattahi and Safaei [34] performed stability analysis of CNTRC beams based on three different beam theories. The obtained equations were solved by generalized differential quadrature method. Babu Arumugam et al. [35] surveyed the free and forced vibration responses of CNTRC beams with constant and variable cross-sections by using the finite element method. A detailed parametric work was made to show the influences of slenderness ratio, percentage of CNT constituent, ply orientation, and boundary conditions. Mohseni and Shakouri [36] studied the free vibration and buckling responses of tapered FG-CNTRC beams surrounding an elastic medium based on Timoshenko beam theory. Dynamic analysis of the pre-twisted FG-CNTRC beams subjected to thermal loading was carried out by Shenas et al. [37]. A higher-order shear deformation beam theory was employed to derive the constitutive equations, and the Chebyshev–Ritz method was applied to solve the resulting equations for various end conditions. Khosravi et al. [38] performed the thermal stability analysis of rotating CNTRC beams. Timoshenko beam theory was used in the derivation of the governing differential equations. Generalized differential quadrature method was employed to obtain some numerical results. Additionally, buckling and vibration responses of micro-and nanocomposite beams were investigated [39–41]. On the other hand, there have recently been many studies on the mechanical behaviors of CNTRC plates and shells [42–56].

As mentioned above, such structures may be subjected to various loads. Due to this fact, it is very important to understand the dynamical behavior of these structures subjected to the harmonic loads. Additionally, the aforementioned review reveals that researchers have so far examined the bending, buckling, and free vibration responses of CNTRC beams. In particular, to the best of the authors' knowledge, forced vibration of CNTRC beams due to harmonic loads has not been investigated in detail. The main purpose of the present study is to fill this gap.

In this paper, forced vibrational behavior of CNTRC beam is examined. It was considered that CNTRC beams are made of a polymeric matrix reinforced by the single-walled carbon nanotubes and is subjected to a harmonic point load at the middle. Three different distributions of CNTs are considered in the analysis. The governing equations of problem have been derived by using the Lagrange procedure based on Timoshenko beam theory. The Ritz method, in conjunction with algebraic polynomials with the trivial functions, was utilized to solve the resulting equation. Additionally, the Newmark average acceleration method was used in the time history for the solution of the forced vibration problem. A detailed parametric study was carried out to peruse the influences of CNTs volume fraction, slenderness ratio, and dynamic parameters on the forced vibration response of CNTRC beam.

2. Theory and Formulation

Consider a simply supported beam made of CNTRC material under a dynamic load, as shown in Figure 1. The composite beam was subjected to a dynamic point load, Q(t), at the midpoint. The geometry of the beam was indicated as the length, *L*; the height, *h*; and width, *b*. Additionally, three different patterns of CNTs, such as uniformly distributed (UD), Λ- and X- type distributions, were considered throughout the thickness of the composite beam.

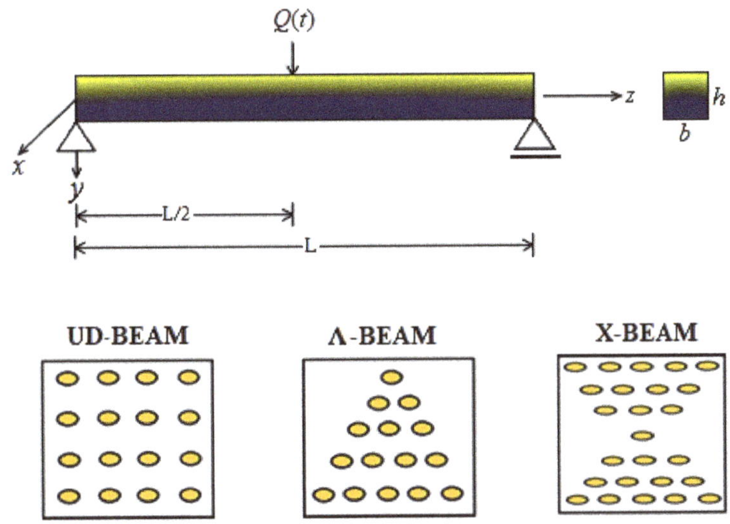

Figure 1. A schematic illustration of a simply supported carbon nanotube-reinforced composite (CNTRC) beam under a harmonic load and three different patterns of CNTs.

The dynamic point load ($Q(t)$) was assumed to be sinusoidal harmonic in time domains, such as in the following:

$$(t) = Q_0 \sin(\overline{\omega} t),\ 0 \leq t \ll \infty \tag{1}$$

where Q_0 and $\overline{\omega}$ denote the amplitude and frequency of the dynamic load, respectively.

The effective material properties for CNTRC beams are given below [27,57]:

$$E_{11} = \eta_1 V_{CNT} E_{11}^{CNT} + V_p E^p \tag{2}$$

$$\frac{\eta_2}{E_{22}} = \frac{V_{CNT}}{E_{22}^{CNT}} + \frac{V_p}{E^p} \tag{3}$$

$$\frac{\eta_3}{G_{12}} = \frac{V_{CNT}}{G_{12}^{CNT}} + \frac{V_p}{G^p} \tag{4}$$

$$V_{CNT} + V_p = 1 \tag{5}$$

$$v = V_{CNT} v^{CNT} + V_p v^p \tag{6}$$

$$\rho = V_{CNT} \rho^{CNT} + V_p \rho^p \tag{7}$$

where E, G, v, and ρ are the material properties that represent the Young's modulus of elasticity, shear modulus, Poisson's ratio, and density, respectively. The superscripts of CNT and p respectively symbolize the related material properties of the carbon nanotube and polymer matrix. η_1, η_2, η_3 can be indicated the efficiency parameters of CNT. Additionally, V_{CNT} and V_p define the volume fractions for CNT and polymer matrix, respectively.

The axial strain (ε_z) and shear strain (γ_{zy}) are given according to the first-order shear deformation beam theory, as follows:

$$\varepsilon_z = \frac{\partial u_0}{\partial z} - Y\frac{\partial \varnothing}{\partial z} \tag{8a}$$

$$\gamma_{zy} = \frac{\partial u_0}{\partial y} + \frac{\partial v_0}{\partial z} \tag{8b}$$

where u_0, v_0, and $\varnothing$ are axial and vertical displacements, and rotation, respectively. The constitute relation is given below:

$$\sigma_z = E(Y)\left[\frac{\partial u_0}{\partial z} - Y\frac{\partial \varnothing}{\partial z}\right] \tag{9a}$$

$$\sigma_{zy} = G(Y)K_s\left[\frac{\partial v_0}{\partial z} - \varnothing\right] \tag{9b}$$

where E is the Young's modulus, G is the shear modulus, σ_z is the normal stress, σ_{zy} is the shear stress, and K_s is the shear correction factor.

The strain energy (U_{i}), the kinetic energy (K), the dissipation function, and potential energy of the external loads (U_{e}) are presented as follows:

$$U_i = \frac{1}{2}\int_0^L \left[A_0\left(\frac{\partial u_0}{\partial z}\right)^2 - 2A_1\frac{\partial u_0}{\partial z}\frac{\partial \varnothing}{\partial z} + A_2\left(\frac{\partial \varnothing}{\partial z}\right)^2\right]dZ + \frac{1}{2}\int_0^L K_s B_0\left[\left(\frac{\partial v_0}{\partial z}\right)^2 - 2\frac{\partial v_0}{\partial z}\varnothing + \varnothing^2\right]dZ \tag{10a}$$

$$K = \frac{1}{2}\int_0^L \left(I_0\left(\frac{\partial u_0}{\partial t}\right)^2 - 2I_1\left(\frac{\partial u_0}{\partial t}\right)\left(\frac{\partial \varnothing}{\partial t}\right) + I_2\left(\frac{\partial \varnothing}{\partial t}\right)^2 + I_0\left(\frac{\partial v_0}{\partial t}\right)^2\right)dZ \tag{10b}$$

$$U_e = -Q(t)\, v(z_p, t) \tag{10c}$$

where

$$(A_0,\ A_1, A_2) = \int_A E(Y)\left(1, Y, Y^2\right)dA, \tag{11a}$$

$$B_0 = \int_A G(Y)dA \tag{11b}$$

$$(I_0, I_1, I_2) = \int_A \rho(Y)\left(1, Y, Y^2\right)dA \tag{11c}$$

The Lagrangian functional of the problem is presented as:

$$I = K - (U_i + U_e) \tag{12}$$

In the solution of the problem in the Ritz method, the approximate solution is given as a series of *i* terms, as in the following:

$$u_0(z,t) = \sum_{i=1}^{\infty} a_i\,(t)\alpha_i(z) \tag{13a}$$

$$v_0(z,t) = \sum_{i=1}^{\infty} b_i\,(t)\beta_i(z) \tag{13b}$$

$$\varnothing(z,t) = \sum_{i=1}^{\infty} c_i\,(t)\gamma_i(z) \tag{13c}$$

where a_i, b_i, and c_i are the unknown coefficients, and $\alpha_i(z,t)$, $\beta_i(z,t)$, and $\gamma_i(z,t)$ are the coordinate functions depending on the end conditions over the interval [*0*, *L*]. The coordinate functions for the simply supported beam are given as algebraic polynomials:

$$\alpha_i(z) = z^i \tag{14a}$$

$$\beta_i(z) = (L - z)z^{\mathrm{i}} \tag{14b}$$

$$\gamma_i(z) = z^{(i-1)} \tag{14c}$$

where *i* indicates the number of polynomials involved in the admissible functions. After substituting Equation (7) into Equation (4), and then using the Lagrange's equation, the following equation can be derived:

$$\frac{\partial I}{\partial q_i} - \frac{\partial}{\partial t}\frac{\partial I}{\partial \dot{q}_i} = 0 \tag{15}$$

where q_i is the unknown coefficients which are a_i, b_i, and c_i. After implementing the Lagrange procedure, the motion equation of the problem was obtained as follows:

$$[K]\{q(t)\} + [M]\{\ddot{q}(t)\} = \{F(t)\} \tag{16}$$

where $[K]$, $[M]$, and $\{F(t)\}$ are the stiffness matrix, mass matrix, and load vector, respectively. The details of these expressions are given as follows:

$$[K] = \begin{bmatrix} K_{11} & K_{12} & K_{13} \\ K_{21} & K_{22} & K_{23} \\ K_{31} & K_{32} & K_{33} \end{bmatrix} \tag{17}$$

where

$$K_{ij}^{11} = \sum_{i=1}^{n}\sum_{j=1}^{n}\int_0^L A_0 \frac{\partial \alpha_i}{\partial z}\frac{\partial \alpha_j}{\partial z} dz \tag{18a}$$

$$K_{ij}^{12} = 0, \tag{18b}$$

$$K_{ij}^{13} = -\sum_{i=1}^{n}\sum_{j=1}^{n}\int_0^L A_1 \frac{\partial \alpha_i}{\partial z}\frac{\partial \gamma_j}{\partial z} dz \tag{18c}$$

$$K_{ij}^{21} = 0, \tag{18d}$$

$$K_{ij}^{22} = \sum_{i=1}^{n}\sum_{j=1}^{n}\int_0^L K_s B_0 \frac{\partial \beta_i}{\partial z}\frac{\partial \beta_j}{\partial z} dz \tag{18e}$$

$$K_{ij}^{23} = -\sum_{i=1}^{n}\sum_{j=1}^{n}\int_0^L K_s B_0 \frac{\partial \beta_i}{\partial z}\gamma_j dz \tag{18f}$$

$$K_{ij}^{31} = -\sum_{i=1}^{n}\sum_{j=1}^{n}\int_0^L A_1 \frac{\partial \gamma_i}{\partial z}\frac{\partial \alpha_j}{\partial z} dz \tag{18g}$$

$$K_{ij}^{32} = -\sum_{i=1}^{n}\sum_{j=1}^{n}\int_0^L K_s B_0 \gamma_i \frac{\partial \beta_j}{\partial z} dz \tag{18h}$$

$$K_{ij}^{33} = \sum_{i=1}^{n}\sum_{j=1}^{n}\int_0^L A_2 \frac{\partial \gamma_i}{\partial z}\frac{\partial \gamma_j}{\partial z} + \sum_{i=1}^{n}\sum_{j=1}^{n}\int_0^L K_s B_0 \gamma_i \gamma_j dz \tag{18i}$$

$$[M] = \begin{bmatrix} M_{11} & M_{12} & M_{13} \\ M_{21} & M_{22} & M_{23} \\ M_{31} & M_{32} & M_{33} \end{bmatrix} \tag{19}$$

where

$$M_{11} = \sum_{i=1}^{n}\sum_{j=1}^{n}\int_0^L I_0 \alpha_i \alpha_j dz \tag{20a}$$

$$M_{12} = 0 \tag{20b}$$

$$M_{13} = -\sum_{i=1}^{n}\sum_{j=1}^{n}\int_0^L I_1 \alpha_i \gamma_j dz \tag{20c}$$

$$M_{21} = 0 \tag{20d}$$

$$M_{22} = \sum_{i=1}^{n}\sum_{j=1}^{n}\int_0^L I_0 \beta_i \beta_j dz \tag{20e}$$

$$M_{23} = 0 \tag{20f}$$

$$M_{31} = -\sum_{i=1}^{n}\sum_{j=1}^{n}\int_0^L I_1 \gamma_i \alpha_j dz \tag{20g}$$

$$M_{32} = 0 \tag{20h}$$

$$M_{33} = \sum_{i=1}^{n}\sum_{j=1}^{n}\int_0^L I_2 \gamma_i \gamma_j dz \tag{20i}$$

$$\{F(t)\} = Q\beta_j \tag{21}$$

The constitutive equation of motions was solved by implementing implicit Newmark average acceleration method with $\alpha = 0.5$ and $\beta = 0.25$ in the time domain. By this procedure, the dynamic problem will be transferred to a system of static problems in each step, as in the following:

$$[\overline{K}(t, X)]\{d_n\}_{j+1} = \{\overline{F}(t)\} \tag{22}$$

in which

$$[\overline{K}(t, X)] = [K] + \frac{[M]}{\beta \Delta t^2} + \frac{[C]\alpha}{\beta \Delta t} \tag{23a}$$

$$\{\overline{F}(t)\} = \{F(t)\}_{j+1} + B_1\{d_n\}_j + B_2\{\dot{d}_n\}_j + B_3\{\ddot{d}_n\}_j \tag{23b}$$

where

$$B_1 = \frac{[M]}{\beta \Delta t^2},\ B_2 = \frac{[M]}{\beta \Delta t},\ B_3 = [M]\left(\frac{1}{2\beta} - 1\right) \tag{24}$$

After evaluating $\{d_n\}_{j+1}$ at a time $t_{j+1} = t_j + \Delta t$, the acceleration and velocity vectors can be expressed as:

$$\left\{\ddot{d}_n\right\}_{j+1} = \frac{1}{\beta\Delta t^2}\left(\{d_n\}_{j+1} - \{d_n\}_j\right) - \frac{[M]}{\beta\Delta t}\left\{\dot{d}_n\right\}_j - \left(\frac{\alpha}{2\beta} - 1\right)\left\{\ddot{d}_n\right\}_j \tag{25a}$$

$$\left\{\dot{d}_n\right\}_{j+1} = \left\{\dot{d}_n\right\}_j + \Delta t\,(1-\alpha)\left\{\ddot{d}_n\right\}_j + \Delta t\,\alpha\left\{\ddot{d}_n\right\}_{j+1} \tag{25b}$$

3. Numerical Results

In the numerical study, the effects of carbon nanotube volume fraction, aspect ratio, and dynamic parameters on the forced vibration response of CNTRC beams are presented and discussed. The five-point Gauss rule was employed to calculate the integration. In the numerical results, the number of terms is taken as 10. Volume fractions of CNTs as functions of thickness direction for different patterns of CNTs [27] are presented in Table 1. In this table, V^*_{CNT} is the given volume fraction of CNTs.

Table 1. Volume fractions of CNTs as a function of thickness direction for different distributions of CNTs [27].

Patterns of CNTs	V_{CNT}
UD	V^*_{CNT}
FG-V	$V^*_{CNT}\left(1+2\frac{z}{h}\right)$
FG-Λ	$V^*_{CNT}\left(1-2\frac{z}{h}\right)$
FG-O	$2V^*_{CNT}\left(1-2\frac{\lvert z\rvert}{h}\right)$
FG-X	$4V^*_{CNT}\frac{\lvert z\rvert}{h}$

In this study, it is notable that CNTs are parallel to the longitudinal direction of the composite beam. Additionally, the efficiency parameters of CNTs for three different values of V^*_{CNT} were considered as [23]:

$$\eta_1 = 1.2833,\ \eta_2 = \eta_3 = 1.0556 \text{ for } V^*_{CNT} = 0.12 \tag{26a}$$

$$\eta_1 = 1.3414,\ \eta_2 = \eta_3 = 1.7101 \text{ for } V^*_{CNT} = 0.17 \tag{26b}$$

$$\eta_1 = 1.3238,\ \eta_2 = \eta_3 = 1.7380 \text{ for } V^*_{CNT} = 0.28 \tag{26c}$$

In the numerical results, the following dimensionless displacement was used:

$$\overline{v} = \frac{E^p b h^3}{P L^3} v \tag{27}$$

In the present analysis, the material properties for reinforcement and matrix constituents were [23,27]: $E_{11}^{CNT} = 600\ GPa$, $E_{22}^{CNT} = 10\ GPa$, $G_{12}^{CNT} = 17.2\ GPa$, $v^{CNT} = 0.19$, $\rho^{CNT} = 1400\ kg/m^3$, $E^p = 2.5\ GPa$, $v^p = 0.30$, and $\rho^p = 1190\ kg/m^3$.

In order to validate the present formulations and analyses, some comparative results are listed in Tables 2 and 3. Firstly, a comparison of non-dimensional fundamental frequencies ($\lambda = \frac{L^2}{h}/\sqrt{\rho_a/E_a}$) of simply supported functionally graded CNT/Aluminum (Al)-alloy composite beams with ANSYS results [19] is presented in Table 2. Here, k is the power-law index, and E_a and ρ_a represent the elastic modulus and density of pure Al-alloy material, respectively. It can be observed (according to Table 2) that the present results agree well with the ANSYS results [19]. The dimensionless fundamental frequencies ($\omega_b = \omega L/\sqrt{I_0/A_0}$) of simply supported CNTRC beams were calculated with different volume fractions of CNTs for L/h = 15 and V_{CNT} = 0.12 compared with those of Wattanasakulpong and Ungbhakorn [27], corresponding to the first-order shear deformation theory. To obtain the vibration frequency in this study, the eigenvalue process is implemented in Equation (16). It is seen from Table 3 that the present results are in good agreement with that of the results of Wattanasakulpong and Ungbhakorn [27].

Table 2. Comparison of non-dimensional fundamental frequencies of simply supported functionally graded CNT/Al-alloy composite beams with ANSYS results.

k	0	0.4	1	2	5	10	Al-Alloy
ANSYS [19]	3.4668	3.2718	3.1496	3.0795	3.0084	2.9546	2.8500
Present	3.663	3.459	3.342	3.271	3.193	3.134	2.971

Table 3. Comparative results for dimensionless fundamental frequencies of a simply supported CNTRC beam for L/h = 15, V_{CNT} = 0.12.

Volume Fractions of CNTs	Present	Wattanasakulpong and Ungbhakorn [27]
UD-Beam	0.9905	0.9976
Λ-Beam	0.8562	0.8592
X-Beam	1.1373	1.1485

In order to investigate the effects and compare different reinforcement patterns on dynamic responses, time responses of the simply supported CNTRC beams are presented in Figures 2–4 for volume fractions of CNTs of V_{CNT} = 0.12, V_{CNT} = 0.17, and V_{CNT} = 0.28, respectively. In these figures, the dimensionless midpoint displacements ($\overline{v}_m$) of the beam are obtained within time history for aspect ratio L/h = 7 and the external load frequency $\overline{\omega} = 10\ rd/s$. In addition, the dynamical dimensionless displacements of the midpoint ($\overline{v}_m$) and the frequency of the dynamic load ($\overline{w}$) relations are presented for different reinforcement patterns for L/h = 10 and t = 1 s in Figures 5–7 for volume fractions of CNTs V_{CNT} = 0.12, V_{CNT} = 0.17, and V_{CNT} = 0.28.

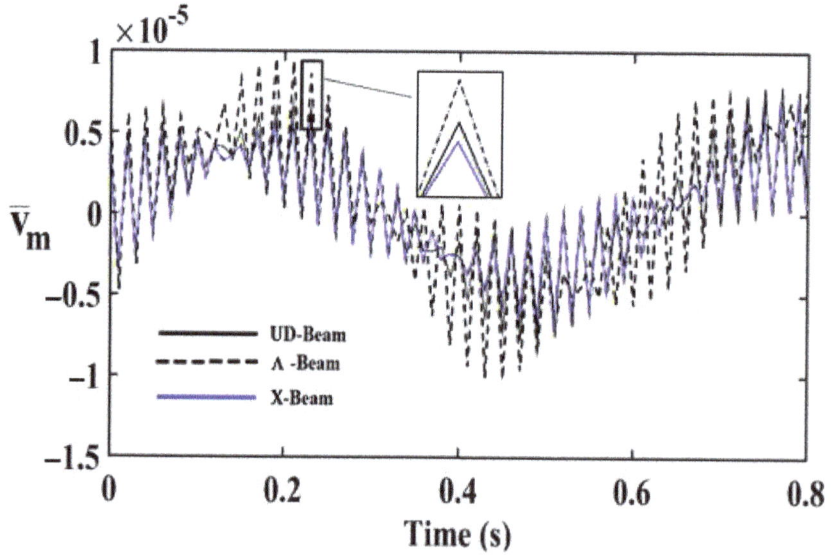

Figure 2. Time responses of the CNTRC beam with different reinforcement patterns for V_{CNT} = 0.12.

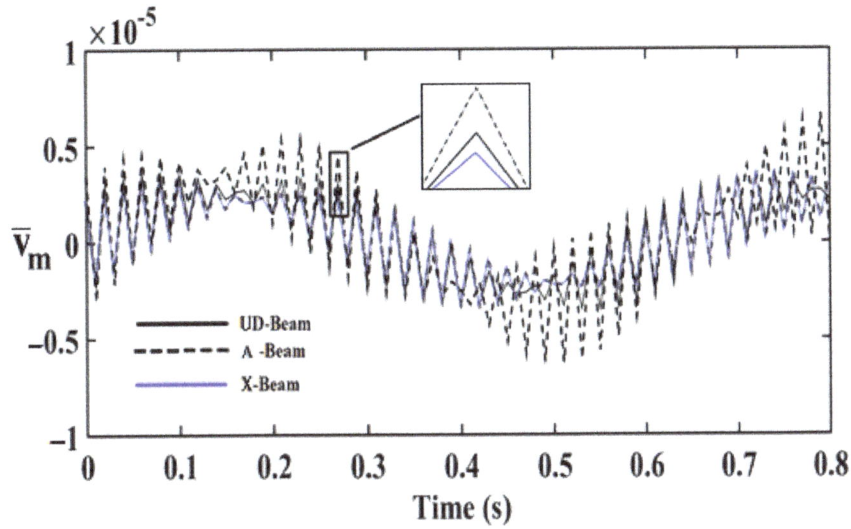

Figure 3. Time responses of the CNTRC beam with different reinforcement patterns for V_{CNT} = 0.17.

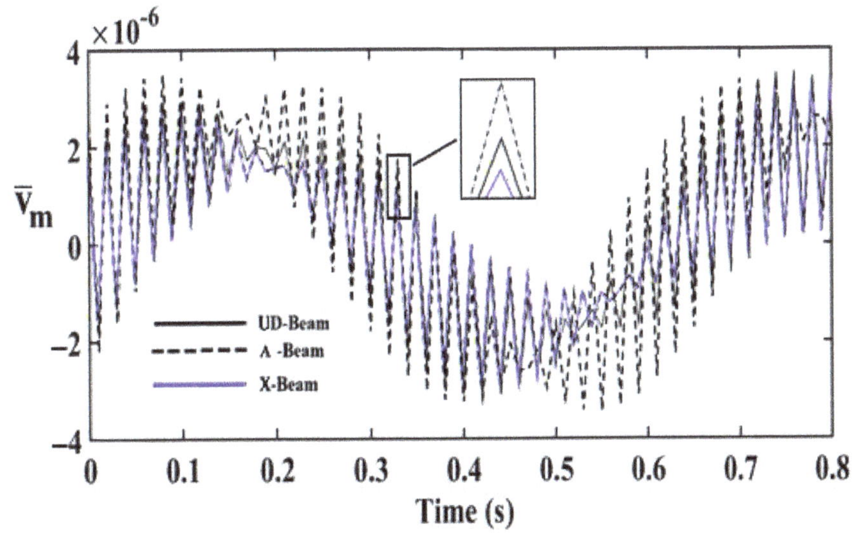

Figure 4. Time −responses of the CNTRC beam with different reinforcement pattern for V_{CNT} = 0.28.

It is shown, according to Figures 2–10, that the dynamical displacements of the Λ-Beam are the biggest of all. In Λ-Beam, the reinforcements spread only at the bottom surface, not upper surface. However, the reinforcements spread at both surfaces on the UD-Beam and X-Beam. It was known that the upper and lower surfaces of the beam have high stresses and strains. Therefore, the Λ-Beam model had the lowest rigidity in all models. As a result, more displacements occured in the Λ-Beam model. This situation could be observed in Table 3. The vibration frequency of the Λ-Beam was lower than the frequency of the other models. Additionally, the dynamical displacements of the UD-Beam were bigger than those of the X-beam. This is because of the reinforcements spread at both surfaces in the UD- and X-Beams, the X-beam has the biggest specific strength in all patterns. Therefore, dynamic response of the X-Beam is lower than all.

Influence of volume fractions of CNTs on the resonance frequencies of the reinforced composite beam with different distribution patterns are revealed in Table 4. It was found that an increase in volume fractions of CNTs gives rise to an increment in resonance frequencies. Additionally, the highest resonance frequencies occur in the X-Beam, while the lowest ones occur in Λ-Beam.

Table 4. The resonance frequencies of reinforced composite beams for various volume fractions of CNTs (L = 2 m, b = h = 0.1 m).

V_{CNT}	$\bar{w}$ (rd/s)		
	UD-Beam	Λ-Beam	X-Beam
0.12	592.98	445.39	692.32
0.17	723.80	538.75	850.10
0.28	881.36	664.42	1028.80

In Figures 5–7, the resonance phenomenon can be observed in the vertical asymptote regions. In the Λ-Beam, the resonance frequency is the lowest for all reinforcement distribution models, because the rigidity of the Λ-Beam is lowest for all. Increasing the volume fractions of CNTs (V_{CNT}) yields increased resonance frequency and decreased displacements. It can be interpreted that, by increasing the volume fractions of CNTs, the beam gets more strength. Therefore, the resonance frequencies increase and dynamically displacements decrease naturally.

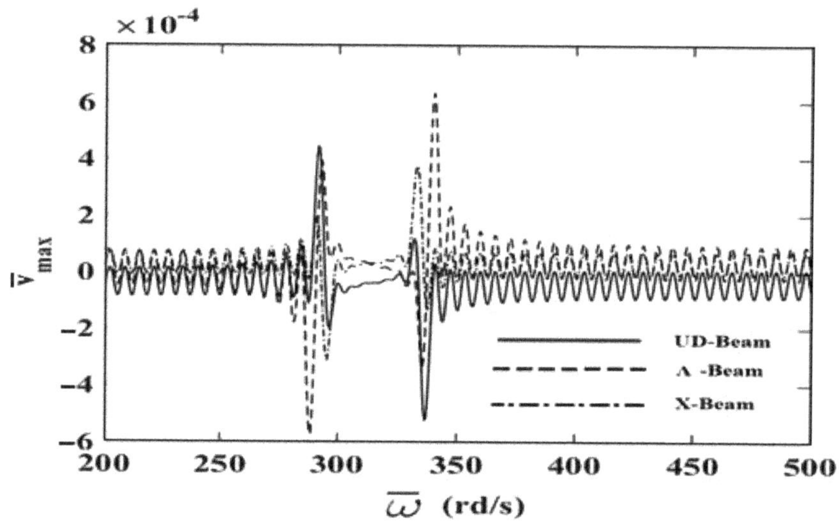

Figure 5. The relationship between the displacements and the frequency of the dynamic load ($\bar{w}$) for V_{CNT} = 0.12.

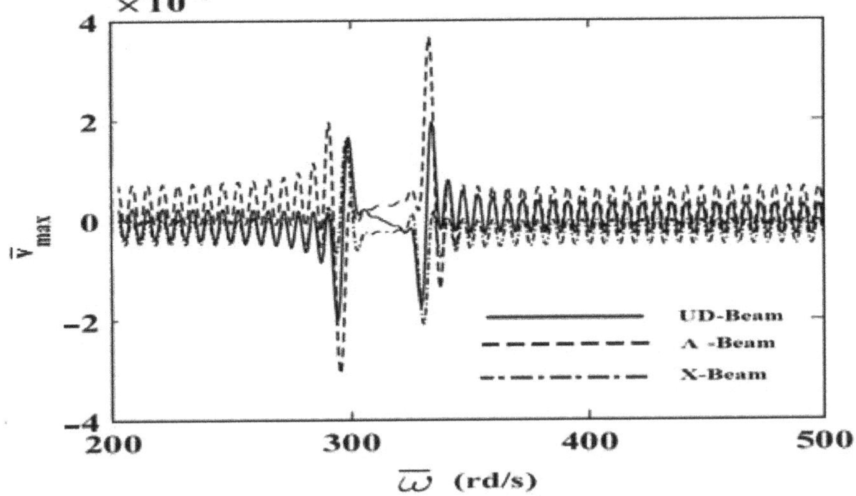

Figure 6. The relationship between the displacements and the frequency of the dynamic load ($\overline{w}$) for V_{CNT} = 0.17.

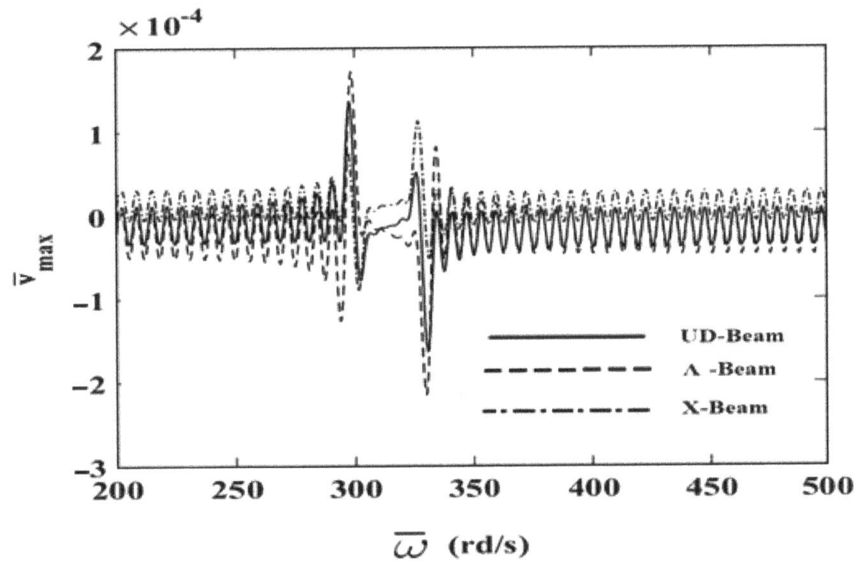

Figure 7. The relationship between the displacements and the frequency of the dynamic load ($\overline{w}$) for V_{CNT} = 0.28.

Figures 8–10 display the frequency of the dynamic load ($\overline{w}$)-dimensionless vertical displacements relationship for L/h = 10 and t = 1 s for different values of V_{CNT} for the UD-beam, Λ-Beam, and X-Beam. It was observed that the increase in values of V_{CNT} cause a decrease in the amplitudes of displacements. In the X-beam, the resonance frequencies obtained were the lowest values, in contrast with other values of reinforcement patterns. Another result is that the difference among values of V_{CNT} is the highest in the Λ-Beam. It can be concluded that the effects of volume fractions of CNTs were more effective in

Λ-Beams. It shows that the distribution of the reinforcement plays an important role on dynamic responses of CNTRC beams.

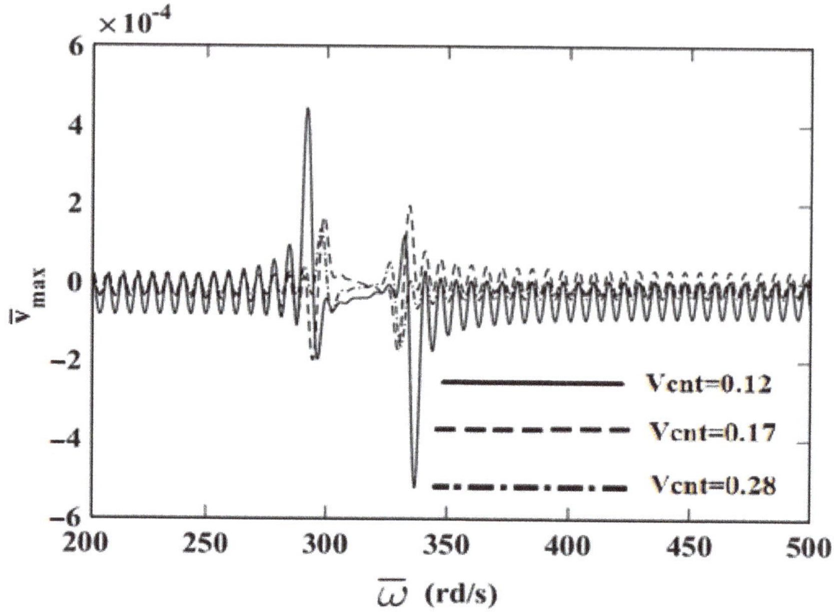

Figure 8. The relationship between the displacements and the frequency of the dynamic load ($\overline{w}$) in the simply supported uniformly distributed (UD)-beam for different values of V_{CNT}.

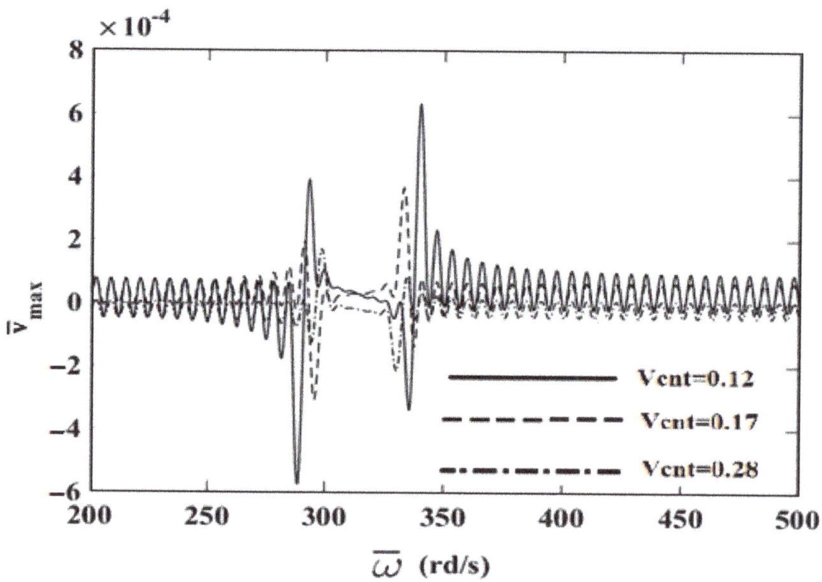

Figure 9. The relationship between the displacements and the frequency of the dynamic load ($\overline{w}$) in the simply supported Λ-Beam for different values of V_{CNT}.

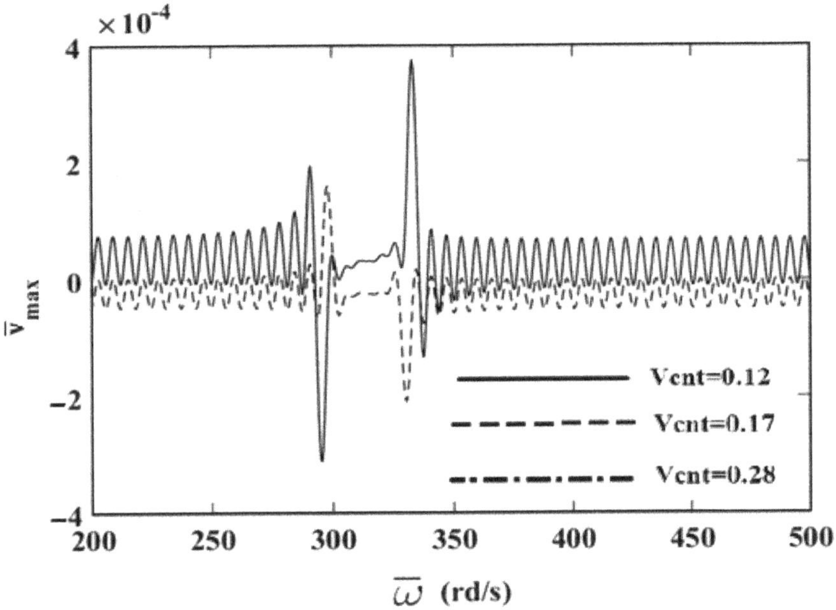

Figure 10. The relationship between of the displacements and the frequency of the dynamic load ($\overline{w}$) in the simply supported X-Beam for different values of V_{CNT}.

4. Conclusions

The forced vibration response of a simply supported CNTRC beam subjected to a harmonic point load was investigated. It was considered that the composite beam was composed of a polymeric matrix (Poly methyl methacrylate) and a reinforcing material (single-walled carbon nanotubes). Timoshenko beam theory was employed in order to take into consideration the effects of shear deformation. The Ritz and Newmark average acceleration methods were used to obtain the numerical results. The effects of volume fraction and distribution patterns of CNTs, aspect ratio, and dynamic parameters on the forced vibration behavior of CNTRC beam were investigated in detail. It was observed that the greatest dynamical displacements occurred in the Λ-Beam, dependent on having the smallest rigidity. Additionally, it was found that the lowest resonance frequencies were obtained in the X-Beam. In addition, it was revealed that an increase in the values of V_{CNT} gave rise to a decrement in the amplitudes of displacements. Moreover, it was emphasized that the distribution pattern of the reinforcement plays an important role on dynamic responses of CNTRC beams.

Author Contributions: Conceptualization, Ö.C. and Ş.D.A.; methodology, Ş.D.A.; software, Ş.D.A.; validation, Ş.D.A. and B.A.; formal analysis, Ş.D.A.; investigation, Ş.D.A.; resources, B.A. and S.D.; data curation, S.D.; writing—original draft preparation, Ş.D.A. and B.A.; writing—review and editing, Ö.C. and S.D.; visualization, B.A.; supervision, Ö.C.; project administration, Ö.C. and S.D.; funding acquisition, Ö.C. and S.D. All authors have read and agreed to the published version of the manuscript.

Funding: This research received no external funding.

Institutional Review Board Statement: Not applicable.

Informed Consent Statement: Not applicable.

Data Availability Statement: Data sharing not applicable.

Conflicts of Interest: The authors declare no conflict of interest.

References

1. Akbas, S.D. Large deflection analysis of a fiber reinforced composite beam. *Steel Compos. Struct.* **2018**, *27*, 567–576. [CrossRef]
2. Xie, Q.; Sinaei, H.; Shariati, M.; Khorami, M.; Mohamad, E.T.; Bui, D.T. An experimental study on the effect of CFRP on behavior of reinforce concrete beam column connections. *Steel Compos. Struct.* **2019**, *30*, 433–441. [CrossRef]
3. Akbas, S.D. Nonlinear behavior of fiber reinforced cracked composite beams. *Steel Compos. Struct.* **2019**, *30*, 327–336. [CrossRef]
4. Luo, Z.Y.; Sinaei, H.; Ibrahim, Z.; Shariati, M.; Jumaat, Z.; Wakil, K.; Pham, B.T.; Mohamad, E.T.; Khorami, M. Computational and experimental analysis of beam to column joints reinforced with CFRP plates. *Steel Compos. Struct.* **2019**, *30*, 271–280. [CrossRef]
5. Shariat, M.; Shariati, M.; Madadi, A.; Wakil, K. Computational Lagrangian Multiplier Method by using for optimization and sensitivity analysis of rectangular reinforced concrete beams. *Steel Compos. Struct.* **2018**, *29*, 243–256. [CrossRef]
6. Souza, P.R.; Nunes, C.S.; Freitas, A.R.; Belinato, J.R.; Pilau, E.J.; Fajardo, A.R.; da Silva, E.A.; Schreiner, W.H.; Muniz, E.C. Sub- and supercritical D-limonene technology as a green process to recover glass fibres from glass fibre-reinforced polyester composites. *J. Clean. Prod.* **2020**, *254*, 119984. [CrossRef]
7. Huang, W.J.; Yan, W.; He, W.T.; Wang, K.; Long, L.J.; He, M.; Qin, S.H.; Yu, J. Synergistic flame-retardant effect of DOPO-based derivative and organo-montmorillonite on glass-fiber-reinforced polyamide 6 T. *Polym. Advan. Technol.* **2020**, *31*, 2083–2093. [CrossRef]
8. Amiri, A.; Krosbakken, T.; Schoen, W.; Theisen, D.; Ulven, C.A. Design and manufacturing of a hybrid flax/carbon fiber composite bicycle frame. *Proc. Ins. Mech. Eng. P-J. Sport. Eng. Technol.* **2018**, *232*, 28–38. [CrossRef]
9. De Cicco, D.; Asaee, Z.; Taheri, F. Use of Nanoparticles for Enhancing the Interlaminar Properties of Fiber-Reinforced Composites and Adhesively Bonded Joints—A Review. *Nanomaterials* **2017**, *7*, 360. [CrossRef] [PubMed]
10. Iijima, S. Helical Microtubules of Graphitic Carbon. *Nature* **1991**, *354*, 56–58. [CrossRef]
11. Gao, L.; Zhou, X.F.; Ding, Y.L. Effective thermal and electrical conductivity of carbon nanotube composites. *Chem. Phys. Lett.* **2007**, *434*, 297–300. [CrossRef]
12. Li-Chung, P.J.; Rajagopal, A.K. Green's function theory of electrical and thermal transport in single-wall carbon nanotubes. *Phys. Rev. B* **2002**, *65*, 113408. [CrossRef]
13. Moisala, A.; Li, Q.; Kinloch, I.A.; Windle, A.H. Thermal and electrical conductivity of single- and multi-walled carbon nanotube-epoxy composites. *Compos. Sci. Technol.* **2006**, *66*, 1285–1288. [CrossRef]
14. Yakobson, B.I.; Avouris, P. Mechanical properties of carbon nanotubes. *Appl. Phys.* **2001**, *80*, 287–327. [CrossRef]
15. Salvetat, J.P.; Bonard, J.M.; Thomson, N.H.; Kulik, A.J.; Forro, L.; Benoit, W.; Zuppiroli, L. Mechanical properties of carbon nanotubes. *Appl. Phys. A-Mater.* **1999**, *69*, 255–260. [CrossRef]
16. Kuo, C.Y. Water purification of removal aqueous copper (II) by as-grown and modified multi-walled carbon nanotubes. *Desalination* **2009**, *249*, 781–785. [CrossRef]
17. Star, A.; Hwang, S.I. Picking Flowers with Carbon Nanotube Sensors. *ACS. Cent. Sci.* **2020**, *6*, 461–463. [CrossRef]
18. Santos, A.; Amorim, L.; Nunes, J.P.; Rocha, L.A.; Silva, A.F.; Viana, J.C. Aligned carbon nanotube-based sensors for strain sensing applications. *Sens. Actuat. A-Phys.* **2019**, *289*, 157–164. [CrossRef]
19. Selmi, A.; Bisharat, A. Free vibration of functionally graded SWNT reinforced aluminum alloy beam. *J. Vibroeng.* **2018**, *20*, 2151–2164. [CrossRef]
20. Vodenitcharova, T.; Zhang, L.C. Bending and local buckling of a nanocomposite beam reinforced by a single-walled carbon nanotube. *Int. J. Solids Struct.* **2006**, *43*, 3006–3024. [CrossRef]
21. Ke, L.L.; Yang, J.; Kitipornchai, S. Nonlinear free vibration of functionally graded carbon nanotube-reinforced composite beams. *Compos. Struct.* **2010**, *92*, 676–683. [CrossRef]
22. Yas, M.H.; Heshmati, M. Dynamic analysis of functionally graded nanocomposite beams reinforced by randomly oriented carbon nanotube under the action of moving load. *Appl. Math. Model.* **2012**, *36*, 1371–1394. [CrossRef]
23. Yas, M.H.; Samadi, N. Free vibrations and buckling analysis of carbon nanotube-reinforced composite Timoshenko beams on elastic foundation. *Int. J. Pres. Ves. Pip.* **2012**, *98*, 119–128. [CrossRef]
24. Deepak, B.P.; Ganguli, R.; Gopalakrishnan, S. Dynamics of rotating composite beams: A comparative study between CNT reinforced polymer composite beams and laminated composite beams using spectral finite elements. *Int. J. Mech. Sci.* **2012**, *64*, 110–126. [CrossRef]
25. Ke, L.L.; Yang, J.; Kitipornchai, S. Dynamic Stability of Functionally Graded Carbon Nanotube-Reinforced Composite Beams. *Mech. Adv. Mater. Struc.* **2013**, *20*, 28–37. [CrossRef]
26. Heshmati, M.; Yas, M.H. Free vibration analysis of functionally graded CNT-reinforced nanocomposite beam using Eshelby-Mori-Tanaka approach. *J. Mech. Sci. Technol.* **2013**, *27*, 3403–3408. [CrossRef]
27. Wattanasakulpong, N.; Ungbhakorn, V. Analytical solutions for bending, buckling and vibration responses of carbon nanotube-reinforced composite beams resting on elastic foundation. *Comp. Mater. Sci.* **2013**, *71*, 201–208. [CrossRef]
28. Lin, F.; Xiang, Y. Numerical Analysis on Nonlinear Free Vibration of Carbon Nanotube Reinforced Composite Beams. *Int. J. Struct. Stab. Dyn.* **2014**, *14*, 1350056. [CrossRef]
29. Lin, F.; Xiang, Y. Vibration of carbon nanotube reinforced composite beams based on the first and third order beam theories. *Appl. Math. Model.* **2014**, *38*, 3741–3754. [CrossRef]
30. Ansari, R.; Shojaei, M.F.; Mohammadi, V.; Gholami, R.; Sadeghi, F. Nonlinear forced vibration analysis of functionally graded carbon nanotube-reinforced composite Timoshenko beams. *Compos. Struct.* **2014**, *113*, 316–327. [CrossRef]

31. Heshmati, M.; Yas, M.H.; Daneshmand, F. A comprehensive study on the vibrational behavior of CNT-reinforced composite beams. *Compos. Struct.* **2015**, *125*, 434–448. [CrossRef]
32. Heidari, M.; Arvin, H. Nonlinear free vibration analysis of functionally graded rotating composite Timoshenko beams reinforced by carbon nanotubes. *J. Vib. Control.* **2019**, *25*, 2063–2078. [CrossRef]
33. Mayandi, K.; Jeyaraj, P. Bending, buckling and free vibration characteristics of FG-CNT-reinforced polymer composite beam under non-uniform thermal load. *Proc. Ins. Mech. Eng. L-J. Mat. Des. Appl.* **2015**, *229*, 13–28. [CrossRef]
34. Fattahi, A.M.; Safaei, B. Buckling analysis of CNT-reinforced beams with arbitrary boundary conditions. *Microsyst. Technol.* **2017**, *23*, 5079–5091. [CrossRef]
35. Babu Arumugam, A.; Rajamohan, V.; Bandaru, N.; Sudhagar, P.E.; Kumbhar, S.G. Vibration Analysis of a Carbon Nanotube Reinforced Uniform and Tapered Composite Beams. *Arch. Acoust.* **2019**, *44*, 309–320. [CrossRef]
36. Mohseni, A.; Shakouri, M. Vibration and stability analysis of functionally graded CNT-reinforced composite beams with variable thickness on elastic foundation. *Proc. Ins. Mech. Eng. L-J. Mat. Des. Appl.* **2019**, *233*, 2478–2489. [CrossRef]
37. Shenas, A.G.; Malekzadeh, P.; Ziaee, S. Vibration analysis of pre-twisted functionally graded carbon nanotube reinforced composite beams in thermal environment. *Compos. Struct.* **2017**, *162*, 325–340. [CrossRef]
38. Khosravi, S.; Arvin, H.; Kiani, Y. Interactive thermal and inertial buckling of rotating temperature-dependent FG-CNT reinforced composite beams. *Compos. Part B-Eng.* **2019**, *175*, 107178. [CrossRef]
39. Civalek, Ö.; Dastjerdi, S.; Akbaş, Ş.D.; Akgöz, B. Vibration analysis of carbon nanotube-reinforced composite microbeams. *Math. Meth. Appl. Sci* **2021**. [CrossRef]
40. Jalaei, M.; Civalek, O. On dynamic instability of magnetically embedded viscoelastic porous FG nanobeam. *Int. J. Eng. Sci.* **2019**, *143*, 14–32. [CrossRef]
41. Akgöz, B.; Civalek, O. Buckling analysis of functionally graded microbeams based on the strain gradient theory. *Acta Mech.* **2013**, *224*, 2185–2201. [CrossRef]
42. Van Do, V.N.; Jeon, J.T.; Lee, C.H. Dynamic analysis of carbon nanotube reinforced composite plates by using Bezier extraction based isogeometric finite element combined with higher-order shear deformation theory. *Mech. Mater.* **2020**, *142*, 103307. [CrossRef]
43. Boulal, A.; Bensattalah, T.; Karas, A.; Zidour, M.; Heireche, H.; Bedia, E.A.A. Buckling of carbon nanotube reinforced composite plates supported by Kerr foundation using Hamilton's energy principle. *Struct. Eng. Mech.* **2020**, *73*, 209–223. [CrossRef]
44. Bouazza, M.; Zenkour, A.M. Vibration of carbon nanotube-reinforced plates via refined nth-higher-order theory. *Arch. Appl. Mech.* **2020**, *90*, 1755–1769. [CrossRef]
45. Tornabene, F.; Fantuzzi, N.; Bacciocchi, M. Linear static response of nanocomposite plates and shells reinforced by agglomerated carbon nanotubes. *Compos. Part B-Eng.* **2017**, *115*, 449–476. [CrossRef]
46. Thang, P.T.; Nguyen, T.T.; Lee, J. A new approach for nonlinear buckling analysis of imperfect functionally graded carbon nanotube-reinforced composite plates. *Compos. Part B-Eng.* **2017**, *127*, 166–174. [CrossRef]
47. Fantuzzi, N.; Tornabene, F.; Bacciocchi, M.; Dimitri, R. Free vibration analysis of arbitrarily shaped Functionally Graded Carbon Nanotube-reinforced plates. *Compos. Part B-Eng.* **2017**, *115*, 384–408. [CrossRef]
48. Civalek, O. Free vibration of carbon nanotubes reinforced (CNTR) and functionally graded shells and plates based on FSDT via discrete singular convolution method. *Compos. Part B-Eng.* **2017**, *111*, 45–59. [CrossRef]
49. Ansari, R.; Torabi, J.; Shojaei, M.F. Buckling and vibration analysis of embedded functionally graded carbon nanotube-reinforced composite annular sector plates under thermal loading. *Compos. Part B-Eng.* **2017**, *109*, 197–213. [CrossRef]
50. Vinyas, M. A higher-order free vibration analysis of carbon nanotube-reinforced magneto-electro-elastic plates using finite element methods. *Compos. Part B-Eng.* **2019**, *158*, 286–301. [CrossRef]
51. Civalek, Ö.; Dastjerdi, S.; Akgöz, B. Buckling and free vibrations of CNT-reinforced cross-ply laminated composite plates. *Mech. Based Des. Struc. Mach.* **2020**, 1–18. [CrossRef]
52. Gholami, R.; Ansari, R. Geometrically nonlinear resonance of higher-order shear deformable functionally graded carbon-nanotube-reinforced composite annular sector plates excited by harmonic transverse loading. *Eur. Phys. J. Plus* **2018**, *133*, 56. [CrossRef]
53. Gholami, R.; Ansari, R. Nonlinear bending of third-order shear deformable carbon nanotube/fiber/polymer multiscale laminated composite rectangular plates with different edge supports. *Eur. Phys. J. Plus* **2018**, *133*, 282. [CrossRef]
54. Mohammadi, M.; Arefi, M.; Dimitri, R.; Tornabene, F. Higher-Order Thermo-Elastic Analysis of FG-CNTRC Cylindrical Vessels Surrounded by a Pasternak Foundation. *Nanomaterials* **2019**, *9*, 79. [CrossRef]
55. Sofiyev, A.H.; Tornabene, F.; Dimitri, R.; Kuruoglu, N. Buckling Behavior of FG-CNT Reinforced Composite Conical Shells Subjected to a Combined Loading. *Nanomaterials* **2020**, *10*, 419. [CrossRef]
56. Farajpour, M.R.; Karimi, M.; Shahidi, A.R.; Farajpour, A. Smart reinforced nano/microscale plates for mass detection at ultrasmall levels: A nonlocal continuum-based approach. *Eur. Phys. J. Plus* **2019**, *134*, 568. [CrossRef]
57. Shen, H.S. Nonlinear bending of functionally graded carbon nanotube-reinforced composite plates in thermal environments. *Compos. Struct.* **2009**, *91*, 9–19. [CrossRef]

MDPI
St. Alban-Anlage 66
4052 Basel
Switzerland
Tel. +41 61 683 77 34
Fax +41 61 302 89 18
www.mdpi.com

Nanomaterials Editorial Office
E-mail: nanomaterials@mdpi.com
www.mdpi.com/journal/nanomaterials

MDPI

www.ingramcontent.com/pod-product-compliance
Lightning Source LLC
Chambersburg PA
CBHW061610210926
51850CB00009B/2278
* 9 7 8 3 0 3 6 5 5 7 7 8 6 *